THE ECONOMICS OF POLLUTION HAVENS

NEW HORIZONS IN ENVIRONMENTAL ECONOMICS

Series Editors: Wallace E. Oates, *Professor of Economics, University of Maryland, USA* and Henk Folmer, *Professor of General Economics, Wageningen University and Professor of Environmental Economics, Tilburg University, The Netherlands*

This important series is designed to make a significant contribution to the development of the principles and practices of environmental economics. It includes both theoretical and empirical work. International in scope, it addresses issues of current and future concern in both East and West and in developed and developing countries.

The main purpose of the series is to create a forum for the publication of high quality work and to show how economic analysis can make a contribution to understanding and resolving the environmental problems confronting the world in the twenty-first century.

Recent titles in the series include:

Econometrics Informing Natural Resources Management
Selected Empirical Analyses
Phoebe Koundouri

The Theory of Environmental Agreements and Taxes
CO_2 Policy Performance in Comparative Perspective
Martin Enevoldsen

Modelling the Costs of Environmental Policy
A Dynamic Applied General Equilibrium Assessment
Rob B. Dellink

Environment, Information and Consumer Behaviour
Edited by Signe Krarup and Clifford S. Russell

The International Yearbook of Environmental and Resource Economics 2005/2006
A Survey of Current Issues
Edited by Henk Folmer and Tom Tietenberg

The Greening of Markets
Product Competition, Pollution and Policy Making in a Duopoly
Michael Kuhn

Managing Wetlands for Private and Social Good
Theory, Policy and Cases from Australia
Stuart M. Whitten and Jeff Bennett

Amenities and Rural Development
Theory, Methods and Public Policy
Edited by Gary Paul Green, Steven C. Deller and David W. Marcouiller

The Evolution of Markets for Water
Theory and Practice in Australia
Edited by Jeff Bennett

Integrated Assessment and Management of Public Resources
Edited by Joseph C. Cooper, Federico Perali and Marcella Veronesi

Climate Change and the Economics of the World's Fisheries
Examples of Small Pelagic Stocks
Edited by Rognvaldur Hannesson, Manuel Barange and Samuel F. Herrick Jr.

The Theory and Practice of Environmental and Resource Economics
Edited by Thomas Aronsson, Roger Axelsson and Runar Brännlund

The International Yearbook of Environmental and Resource Economics 2006/2007
A Survey of Current Issues
Edited by Tom Tietenberg and Henk Folmer

The Economics of Pollution Havens

Edited by

Don Fullerton

University of Texas at Austin, USA

NEW HORIZONS IN ENVIRONMENTAL ECONOMICS

Edward Elgar

Cheltenham, UK • Northampton, MA, USA

Published by
Edward Elgar Publishing Limited
Glensanda House
Montpellier Parade
Cheltenham
Glos GL50 1UA
UK

Edward Elgar Publishing, Inc.
136 West Street
Suite 202
Northampton
Massachusetts 01060
USA

A catalogue record for this book
is available from the British Library

Library of Congress Cataloguing in Publication Data

The economics of pollution havens / edited by Don Fullerton.
p. cm. – (New horizons in environmental economics)
Includes bibliographical references
1. Pollution—Economic aspects. 2. Environmental policy—Economic aspects. 3. International trade—Environmental aspects. 4. Industries—Environmental aspects. 5. Environmental economics. I. Fullerton, Don. II. Series.

HC79.P55E265 2006
363.73'2—dc22

2005052832

ISBN-13: 978 1 84542 535 7
ISBN-10: 1 84542 535 9

Printed and bound in Great Britain by MPG Books Ltd, Bodmin, Cornwall.

Contents

Contributors

Edward B. Barbier, Department of Economics and Finance, University of Wyoming, USA

Randy A. Becker, Center for Economic Studies, U.S. Bureau of the Census, USA

Corrado Di Maria, Department of Economics and CentER, Tilburg University, The Netherlands

Josh Ederington, Department of Economics, University of Kentucky, USA

Chris Elbers, Vrije Universiteit, Amsterdam, The Netherlands

Raymond J.G.M. Florax, Department of Agricultural Economics, Purdue University, and Vrije Universiteit, Amsterdam, The Netherlands

Per G. Fredriksson, Department of Economics, Southern Methodist University, USA

Alberto Gallegos, Departamento de Economía, Instituto Tecnológico de Monterrey, México

Rolf Golombek, The Frisch Centre, Norway

Michael Hoel, Department of Economics, University of Oslo, Norway

Patrick T. Hultberg, Department of Economics and Business, Kalamazoo College, USA

Beata Smarzynska Javorcik, The World Bank, USA

Matthew E. Kahn, Fletcher School, Tufts University, USA

Arik Levinson, Department of Economics, Georgetown University, USA

John A. List, Department of Agricultural and Resource Economics, University of Maryland, College Park, USA

Muthukumara Mani, Policy and Economics Team, Environment Department, The World Bank, USA

Carol McAusland, Department of Economics, University of California, Santa Barbara, USA

Warren McHone, Department of Economics, University of Central Florida, USA

Daniel L. Millimet, Department of Economics, Southern Methodist University, USA

Jenny Minier, Department of Economics, University of Kentucky, USA

Abay Mulatu, Department of Economic History, London School of Economics, UK

Pierre M. Regibeau, Department of Economics, University of Essex, UK

Hilary Sigman, Department of Economics, Rutgers University, USA

Sjak A. Smulders, Department of Economics and CentER, Tilburg University, The Netherlands

M. Scott Taylor, Department of Economics, University of Calgary, Canada

Shang-Jin Wei, Research Department, International Monetary Fund, USA

Cees Withagen, Vrije Universiteit, Amsterdam, The Netherlands

Xiaodong Wu, Department of Economics, University of North Carolina, Chapel Hill, USA

Yutaka Yoshino, Department of Economics, University of Virginia, USA

Preface

A 'pollution haven' may arise if environmental regulations differ between countries, if capital is mobile, and if trade rules allow firms to relocate and still sell their products to the same customers. The three appearances of the word 'if' in that sentence suggest a whole host of theoretical and empirical questions under the general heading of 'pollution havens,' and many such questions have been addressed in the economics literature in the past ten years since Brian Copeland and M. Scott Taylor introduced the first complete model of pollution havens (Copeland and Taylor, 1994). To categorize those questions, they later use 'pollution haven effect' to refer to the relatively unambiguous effect of environmental regulations on the incentives to relocate, and 'pollution haven hypothesis' to refer to the more ambiguous effects of trade liberalization on those incentives (Copeland and Taylor, 2004). This distinction is important, because the existence of a 'pollution haven effect' of environmental policy is a necessary condition for the validity of the 'pollution haven hypothesis' that trade liberalization will drive firms away from countries with strict pollution laws.

The 'pollution haven hypothesis' is highly controversial, both to trade economists and to environmental economists. It takes as given an unequal worldwide distribution of income, and it assumes that environmental protection is a normal good. These assumptions are then used to generate predictions about trade patterns and pollution emissions in each country. At its simplest, the hypothesis states that liberalization of international trade in goods will lead polluting producers to move away from countries with tight environmental protection and towards countries with lax protection. If so, then the environment will worsen in countries with weak regulation (those with relatively low income per capita) and improve in countries with tight regulation (those with relatively high income per capita). Total worldwide pollution is then predicted to rise.

In the ten years following the statement of this hypothesis, further research has faced both theoretical and empirical challenges. One empirical difficulty is the assumption that differences in regulations are the key determinant of relative production costs and location decisions. The empirical literature must deal with other motivations that are likely to loom large in actual economies. In addition, the hypothesis was originally construed to apply to the problem of pollution caused by industrial production, and hence it has unclear implications for the use of renewable resources or for the amount of pollution from consumption activities such as driving. The empirical literature has used data on trade flows, and data on foreign direct investment. Some of it provides evidence on the size of any 'pollution haven effect' – the sensitivity of firm location to environmental regulations – and some of it has tried to test the pollution haven hypothesis directly by looking at changes in trade rules.

In contrast, the theoretical literature has tried to relax various assumptions in the original model of Copeland and Taylor, in order to extend it to answer other questions. Their original model focuses on environmental policy differences between rich and poor countries, and it therefore assumes that these differences determine location decisions of firms in different industries. It assumes that neither rich nor poor countries control trade by the use of tariffs or by strategic use of environmental policy. The model assumes no trans-boundary pollution, and it employs the concept of a social planner in each country who corrects all pollution externalities within each country. Thus, this literature asks about the implications of other effects on firm location decisions, of simultaneous tariff and environmental policies, of strategic use of environmental policy, of trans-boundary pollution, and of political economy considerations such as lobbies, corruption, or just simple majority rule in a nation of heterogeneous voters.

Given this flurry of research activity, and great policy interest, it is time to take stock of the economic literature on pollution havens. Ten years after Copeland and Taylor's original article, *The B.E. Journals in Economic Analysis & Policy* (*BE-JEAP*) have provided on the internet a collection of 16 new research papers. Since these journals do not appear in print, we are very pleased that Edward Elgar Publishers agreed to publish a collection of these papers. All 16 papers first appeared in 2004 at http://www.bepress.com/bejeap/, and now all are collected in this book.

In the first chapter, Scott Taylor provides a particularly useful introduction to the topic and a summary of just how each of the other 15 chapters fits into the literature. He 'unbundles' the pollution haven hypothesis into a series of five links. Some of the articles in this volume relate to the first link, in which country characteristics determine environmental rules; some relate to the second link in which those rules affect production costs; some relate to the next link in which production costs affect trade or FDI flows; some relate to the effect of those flows on pollution, prices, and incomes; and, finally, some relate to the ways those last considerations feed back into the determination of environmental rules.

Rather than try to summarize all 16 chapters, this Preface will just use a few examples to demonstrate the importance of the collection. One theoretical study points out how the pollution haven effect can be reversed: if more stringent environmental protection in one country induces firms to do more research in abatement technology, and if other countries can freely use the new technology, then the outcome may be *less* pollution abroad (Chapter 8). Another points out that the effect of trade integration on environmental policy is conditional on the degree of political stability (Chapter 5). For another example, one of the empirical studies uses annual industry-level data to find that the USA has shifted toward cleaner production *and* toward cleaner imports, and it finds no evidence that pollution-intensive industries have been disproportionately affected by tariff changes (Chapter 9).

The collection of studies in this book will be appreciated by all researchers in 'trade and the environment,' and by many others who work only in one field or the other – those in international trade and those in environmental economics. It will also be useful to multidisciplinary audiences in political science, public policy, and environmental studies.

References

Copeland, B.R. and Taylor, M.S. (1994), 'North–South trade and the environment', *Quarterly Journal of Economics*, **109**(3), August, 755–87.

Copeland, B.R. and Taylor, M.S. (2004), 'Trade, growth, and the environment', *Journal of Economic Literature*, **42**(1), March, 7–71.

Acknowledgements

The editor and publishers wish to thank The Berkeley Electronic Press for kindly giving their permission to print, in their entirety, Volume **3**(2) of *Contributions to Economic Analysis & Policy*, 2004 and Volume **4**(2) of *Advances in Economic Analysis & Policy*, 2004.

[1]

Unbundling the Pollution Haven Hypothesis*

M. Scott Taylor

Abstract

The "Pollution Haven Hypothesis" (PHH) is one of the most hotly debated predictions in all of international economics. This paper explains the theory behind the PHH by dividing the hypothesis into a series of logical steps linking assumptions on exogenous country characteristics to predictions on trade flows and pollution levels. I then discuss recent theoretical and empirical contributions investigating the PHH to show how each contribution either questions the logical inevitability, or the empirical significance of one or more steps in the pollution haven chain of logic. Suggestions for future research are also provided.

KEYWORDS: trade, pollution haven hypothesis, environment

*I am grateful to Don Fullerton and two anonymous referees for helpful comments on an earlier draft. All remaining errors are my own.

One of the most important debates in trade policy concerns the environmental impact of trade liberalization. Should trade rules be altered to ensure international trade has a benign environmental impact? Should countries be allowed to countervail goods originating in countries with less strict controls on pollution? Should trade agreements and multilateral environmental agreements be linked, and if so how? Despite a flurry of both theoretical and empirical research over the last ten years, the debate surrounding these questions is far from over.

The debate started in the early 1990s when the North American Free Trade Agreement placed firms located in rich and tightly regulated countries (the U.S. and Canada) and firms located in a poor and laxly regulated country (Mexico) in direct competition for the North American market. Critics of the accord pointed to the obvious divergence in environmental control costs across countries, together with the already poor environmental record in the Maquiladora region to argue that NAFTA was sure to be an environmental disaster for Mexico, and a jobs disaster for the U.S. and Canada. Since then environmental issues have continued to play a major role in trade debates, while job issues have waned in importance. For example, the Doha Declaration lists several trade and environment issues for negotiation. Ongoing discussions are to clarify the relationship between WTO rules and trade obligations set out in Multilateral Environmental Agreements (MEAs); to establish procedures for regular information exchange between MEA secretariats and the relevant WTO Committees; and to reduce or eliminate tariff and non-tariff barriers on environmental goods and services. These negotiating objectives are additional to existing WTO discussions on the role fishery subsidies play in natural resource depletion, the benefits of accelerated liberalization in forest product trade, the usefulness of eco-labeling, and the role, if any, for the precautionary principle to play in determining allowable imports. Clearly the trade and environment debate is far from over.

It is then quite timely and very useful that *The B.E. Journals in Economic Analysis & Policy* has published a volume of studies examining the trade and environment link with special emphasis on the pollution haven hypothesis. The "Pollution Haven Hypothesis" (PHH) is one of the most contentious and hotly debated predictions in all of international economics. It is at the center of the trade and environment debate, since it makes a direct link between differences across countries in their environmental regulation and trade flows. Simply put, the hypothesis predicts liberalized trade in goods will lead to the relocation of pollution intensive production from high income and stringent environmental regulation countries, to low income and lax environmental regulation countries. In the NAFTA debate, it was used to support the prediction of both an environmental disaster in Mexico (because Mexico would specialize in pollution intensive goods) and a jobs disaster in partner countries (because jobs in pollution intensive industries would be destroyed by imports from Mexico). If the PHH is true, it can be used to argue for trade interventions such as green countervail on imported goods, or export bans on tropical timber. If false, the environmental impact of liberalized trade is likely to be far more benign. Providing evidence for or against the hypothesis should be a high priority for researchers in both international and environmental economics.

The purpose of this introduction is to set out the theory behind the Pollution Haven Hypothesis and to show how each of the contributions in this volume sheds light on one or another link in the logic of the pollution haven chain. By doing so, I hope to identify research questions requiring further scrutiny, while acknowledging the valuable contributions made in this volume. To organize my thoughts, I first restate the Pollution Haven Hypothesis as envisaged by Copeland and Taylor (1994) "North-South Trade and the Environment". I do so because CT (1994) was the first paper to link country income levels and the stringency of environmental regulation to predictions on pollution levels and trade patterns. It was in turn inspired by the ongoing NAFTA debate, and the seminal contribution of Gene Grossman and Alan Krueger (1993, 1995) on NAFTAs environmental consequences.

The CT (1994) analysis provides me with a springboard to organize my review. In what follows, I first state the assumptions underlying the Pollution Haven Hypothesis and then recount its predictions regarding trade flows and pollution levels. Following this, I unbundle the hypothesis into a series of logical steps from country characteristics to trade flows and interpret each contribution in the volume as shedding light on one or another of these links. Finally, I present a short conclusion with suggestions for future research.

1. The Pollution Haven Hypothesis

The original Copeland and Taylor analysis is an exercise in applied theory. It develops a two-country static general equilibrium model of international trade with a continuum of goods differentiated by their pollution intensity. The model has only one primary factor of production, and for the most part the authors assume countries only differ in their endowment of this factor (suggestively labelled human capital). Differences across countries in population density, assimilative capacity, and country size are considered in the penultimate section of the paper, but the focus is on how differences in human capital across countries affects their income, regulation, and resulting trade flows and pollution levels.

The production of any good in the economy creates pollution as a joint product. Pollution abatement is possible, but requires real resources. As the paper demonstrates, active abatement plus joint production leads to a final goods specification where pollution appears as if it was an input to production. Assumptions are adopted on abatement to allow for an unambiguous ranking across the continuum of industries according to their pollution intensity. The two countries are labelled North and South, with North having much greater human capital than South. Large differences in human capital across regions ensure that each country specializes in a set of either relatively clean or dirty goods in trade.[1] Pollution is assumed to be local in its effects, and the environment is a normal

[1] If countries differ only slightly in human capital levels, then factor prices will be equalized by trade and the commodity composition of trade becomes indeterminate. This complicates the analysis but does not alter the basic conclusions.

good. To make the policy process simple and transparent, a social planner in each country trades off increases in pollution, which generate higher real income, against the sum of marginal damage across the affected population.

The model is constructed to reflect three realities: one, the distribution of income is very unequal worldwide; two, industries differ greatly in their pollution intensity of production; and three, environmental quality is a normal good. It then provides predictions on trade patterns and pollution levels from these assumptions. When countries differ only in human capital levels, it generates the Pollution Haven Hypothesis: a movement from autarky to free international trade in goods leads to the relocation of dirty good production from the high-income, tight-environmental-regulation country to the low-income, lax-environmental-regulation country. It also provides two corollaries: one, pollution rises in the lax regulation country and falls in the tight regulation country; and two, overall world pollution rises with trade.

The intuition for these results is fairly clear. Trade alters the composition of output in both North and South because of differences in the stringency of their pollution regulation. Dirty goods are relatively expensive in the tight regulation North in autarky, and relatively cheap in the lax regulation South. Regulation levels are functions of the endogenous North-South income gap. Given the relative cost structure in autarky, a movement to free trade shifts dirty good production to the South and clean good production to the North. Pollution falls in the North because the composition of its industries becomes cleaner with trade. Pollution rises in the South because the composition of its industries becomes dirtier with trade. World pollution rises with trade because the world's dirtiest industries locate in the country with the lowest environmental standards.[2]

While the statement of the Pollution Haven Hypothesis is relatively simple, moving from a statement of the hypothesis to testing in the real world has proven difficult. The theory envisages an exogenous movement from autarky to completely free trade, but nowhere in the world is trade truly free, nor are trade liberalizations exogenous events. Any empirical testing of the hypothesis has to deal with data coming from marginal movements towards free trade and address the endogeneity of both trade and pollution policy. The theory was formulated in the context of industrial pollution, but testing has sometimes employed data on renewable resource use, consumption-generated pollution or transboundary pollution.[3] Finally, the theory generates the PHH as a prediction by eliminating all other motives for trade, leaving regulatory differences alone to determine trade patterns. While the original CT analysis allowed for differences in other country at-

[2]This is an oversimplification of the logic. Trade creates scale, composition and technique effects in both countries. Because of the model's construction, scale and technique effects largely cancel out, leaving the composition effect to determine whether pollution rises or falls with trade. See CT (1994) for further details.

[3]Copeland and Taylor (1995a) extend the analysis to transboundary pollution. Adding transboundary pollution does not alter the trade pattern or pollution-generation predictions of the pollution haven hypothesis. Copeland and Taylor (1995b) examine consumption-generated pollution but find an important reversal: tight regulation now leads to a comparative advantage (rather than disadvantage) in dirty goods.

tributes (population density, size, assimilative capacity) and showed how these additional factors also determined comparative advantage, the authors were satisfied in stating "In reality, of course, trade is influenced by many conflicting factors...To make inferences about the actual pattern of trade, one would have to weigh the influences derived from environmental policy against other determinants of trade" (p. 757). Unfortunately, the authors provided no guidance as to how empirical work should go about weighing these various influences.

It is not surprising then that subsequent empirical work on the PHH has sometimes confused two quite different concepts. The first concept is a "pollution haven effect". A pollution haven effect occurs if a tightening of environmental regulation deters exports (or stimulates imports) of dirty goods. The pollution haven effect relates changes in environmental policy to resulting changes in trade flows. Most conventional economic models exhibit a pollution haven effect. Typically, a tightening of regulation raises production costs and lowers exports (or raises imports). Empirical evidence for a pollution haven effect can come from many sources, but often authors establish a link between across-industry variation in the cost of meeting environmental regulations and across-industry variation in trade flows. Many of the studies in this volume provide evidence on the pollution haven effect.

The second concept is the "Pollution Haven Hypothesis".[4] The PHH is a prediction concerning trade patterns and the stringency of environmental regulation. It predicts that when trade barriers are reduced, pollution-intensive industries will shift from countries with stringent environmental regulation to countries with lax environmental regulation. Empirical evidence for the Pollution Haven Hypothesis can also come from many sources, but often researchers investigate whether countries that differ greatly in the stringency of their environmental regulation have a trade pattern matching the PHH prediction. Two studies in this volume investigate trade patterns for just this purpose.

The PHH and the pollution haven effect, although different, are related. Simply put, the trade pattern prediction of the PHH can only be true if we have a strong pollution haven effect. The existence of a pollution haven effect is necessary, but not sufficient, for the PHH to hold. Empirical researchers have used this necessity relationship to provide alternative tests of the PHH. For example, empirical work that finds a "small" impact of regulation on trade flows also provides evidence against the PHH. This is a reasonable conclusion since "small" measured pollution haven effects must surely be outweighed by other more conventional determinants of comparative advantage. Therefore, estimating the strength of pollution haven effects is of great value: failure to find them implies failure of the Pollution Haven Hypothesis; and finding only small effects is suggestive evidence against the hypothesis.

---

[4]In the body of CT (1994) differences in regulation are the only source of comparative advantage. The model exhibits pollution haven effects and generates the Pollution Haven Hypothesis. When the model is extended to include other determinants of comparative advantage (density, etc.) the model still exhibits a pollution haven effect, but the PHH may no longer hold. See CT (1994) section VIII and chapter 6 of Copeland and Taylor (2003).

It may also be useful at this point to distinguish the Pollution Haven Hypothesis from another commonly employed term: "environmental dumping." Environmental dumping refers to a situation where pollution regulation is less stringent than it would be in the absence of strategic interaction. Alternatively, it is a situation where equilibrium pollution levels in the non-cooperative Nash equilibrium exceed those in the cooperative equilibrium. Barrett (1994) and Rauscher (1991) are the earliest contributions to the environmental dumping literature. While environmental dumping and the strategic use of environmental policy may lead to a pattern of trade (or capital movements) satisfying the Pollution Haven Hypothesis, they are distinct phenomena. The definition of environmental dumping was constructed to focus attention on the difference between the level of policy instruments in a strategic setting and their first best levels. As such, environmental dumping often carries with it a negative welfare implication. The Pollution Haven Hypothesis is a positive prediction. Its focus is on the difference between policy instruments across countries and how this affects trade flows. It is silent on whether pollution haven driven trade is welfare enhancing or not. It is for this reason that the definition of the Pollution Haven Hypothesis was given in purely positive terms. It does not claim that the low income country loses from trade nor does it claim that world pollution is excessive in the trading equilibrium. The original CT (1994) analysis for example exhibits the PHH, has no environmental dumping, and trade is welfare improving.

2. Unbundling the Hypothesis

In the last ten years a flurry of work has addressed "pollution haven" questions using both empirical and theoretical methods. To a certain extent this literature questioned assumptions made in the original CT analysis or pushed into entirely new areas. The CT analysis makes a host of assumptions: differences in the regulation of pollution across countries is the sole determinant of location decisions, trade is completely free of tariffs, countries do not use environmental policy strategically, and all externalities are fully internalized leading to efficient outcomes in both autarky and trade. Not surprisingly, much of the new research examines how sensitive industry location decisions are to pollution policies, as this is a key ingredient in any pollution haven model. Others have asked how the location decision rests on the trade regime. Still others moved towards a more serious consideration of strategic issues by asking whether and when countries adopt weak environmental regulations to attract industry and jobs? Other contributions adopted a less sanguine view of government's ability to internalize all externalities, and examined the impact of pollution haven driven trade on welfare when pollution is local, regional, or global.

To clarify and classify my subsequent discussion, I present in Figure 1 a schematic representation of the logical skeleton supporting the PHH. The hypothesis takes as given country characteristics such as access to various production technologies, opportunities for abatement, and country-specific endowments of productive factors. These country

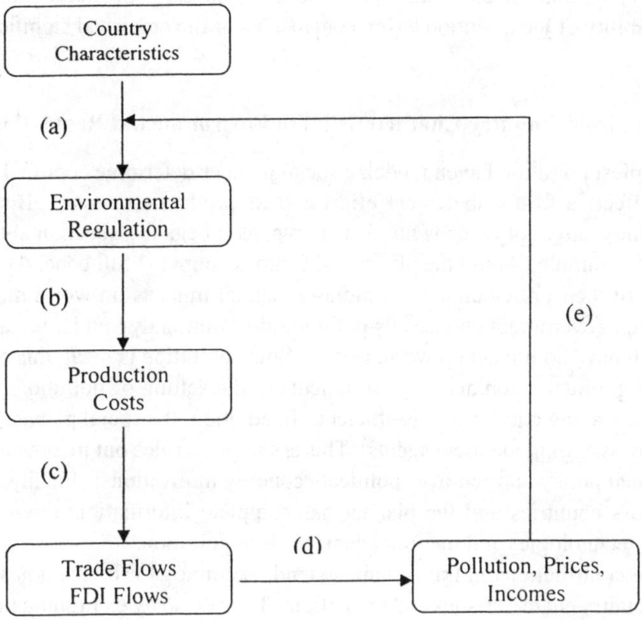

Figure 1: Unbundling the Pollution Haven Hypothesis

characteristics plus world prices determine national income, which in turn maps into the stringency of environmental regulation at step (a). In the canonical case the mapping is determined by a social planner who implements an efficient solution desired by the set of identical agents in the economy. The efficient solution is then implemented via a tax on pollution (or any other efficient instrument). This tax in turn affects the costs of goods production in the economy. Since some production technologies are more polluting than others, the impact of environmental regulation will vary across industries, and at channel (b) regulation will affect an economy's relative price structure. Relative prices define a nation's comparative advantage and determine trade flows at (c). Since the movement to trade alters production patterns, trade will also affect pollution, incomes and perhaps world prices at (d). These income and price changes feed back and modify the original mapping from country characteristics to regulation. They set in motion another round of adjustments, with the fixed point of this system being a general equilibrium where trade, pollution and regulatory stringency are all determined simultaneously.

Each of the contributions in this volume can be seen as unbundling or putting under scrutiny a piece of this logical chain. By doing so, the authors investigate either the logical inevitability of the pollution haven conclusions or the empirical significance of its predictions.

2.1 The Link from Country Characteristics to Environmental Regulation

In the simplest pollution haven model, a social planner determines optimal regulation that in turn affects a firm's abatement effort and its production costs. But even this simplest and most direct of cases is not that simple, as it begs the question about what is "optimal". For example, should the planner take into account the full general equilibrium consequences of their policy choice, including potential impacts on world markets? In CT (1994), each government chooses its pollution tax "optimally" under the assumption this choice will have no impact on world prices. Since pollution is local, this assumption rules out strategic interaction across governments in the setting of pollution policy, and ensures that the trading equilibrium is efficient. In addition, the social planner's problem is simplified by assuming identical agents. This assumption rules out income distribution affecting optimal policy and removes political economy motivations. Finally, agents are immobile across countries and the planner has complete information about abatement opportunities, technologies and marginal damage from pollution.

Five of the contributions in this volume extend, examine or criticize this simple link (a) between country characteristics and regulation. They do so by examining the relationship between regulatory stringency and factor mobility (Elbers and Withagen), information asymmetries (Wu), strategic adoption decisions (Regibeau and Gallegos), political economy motives (Fredriksson and Mani) and heterogeneity across agents (Becker).

Strategic interaction across governments is key to Elbers and Withagens' contribution "Environmental Policy, Population Dynamics and Agglomeration". These authors investigate the possibility of environmental dumping in a world where governments choose their pollution policy fully cognizant of its impact on world prices, industry location and labor migration. A commonly held view in the environment and trade literature is that factor mobility can only reinforce the tendency for polluting industries to shift towards lax regulation countries. The logic behind this view is simple. If factors are immobile, then the relocation of production is limited by the availability of productive factors in the low regulation country. If factors can chase low regulations, then this constraint is relaxed and the shift of industry reinforced. It is, however, unclear whether this is the whole story concerning factor mobility. Factors moving into the low regulation region may bring new technology or raise the host country's national income, and these changes may in turn create a tightening of standards. As Elbers and Withagen show, factor mobility may in some cases stop governments from enacting lax regulation, since skilled workers will flee a polluted country.

To generate this result, EW employ a standard two-sector model with a monopolistically competitive manufacturing sector producing under increasing returns and a perfectly competitive agricultural sector producing under constant returns. The production of manufactures pollutes, cannot abate, and manufacturing employs only skilled workers. By making these assumptions the authors tie the production of manufactures, pollution and skilled labor tightly together, so that the number of active firms in a jurisdiction fully determines its pollution levels. Workers sort across countries in accordance with both real income and environmental concerns. The authors investigate the difference between the non-cooperative Nash equilibrium in pollution taxes and the cooperative solution. They define this difference to be the extent of environmental dumping.

Surprisingly, EW find that factor mobility can reduce the tendency to concentrate polluting activities in one country. While lax pollution regulation raises the return to skilled labor and works towards the agglomeration of manufacturing in one country, the concentration of industry also raises local pollution levels. Since workers also care about environmental quality, additional local pollution lowers the welfare of skilled workers and works against agglomeration. Whereas agglomeration of manufacturing in one location is a potential equilibrium (under some parameter values), the likelihood of agglomeration is lessened by the introduction of environmental concerns.

Elbers and Withagen's contribution is important in demonstrating how factor mobility is a two-way street: factors chasing high monetary rewards because of lax regulation can also flee excessive pollution created by lax regulation. Although the basic pollution haven model assumes factor immobility, Elbers and Withagen contribution shows that altering the degree of factor mobility does not necessarily imply a worsening of industrial flight and a magnified pollution haven effect.

Another contribution examining the mapping from country characteristics to environmental policy is Regibeau and Gallegos "Managed Trade, Liberalization and Local Pollution". In contrast to other contributions in this volume, RG this paper generates its pollution haven results via a purely strategic interaction. The paper contains two main results: a) countries may want to retain some flexibility in their tariff policy for environmental reasons; and b), we may find that countries with tight WTO commitments have relatively dirty production, while those with a great deal of discretion in trade policy have clean production.

RG develop a two-firm, two-country oligopoly model where only the Home country is policy active. Pollution is local and produced via production. The Home firm can choose between an existing dirty production technology or a costly new clean production technology. For most of the paper, the only policy instrument available to the Home government is a tariff on imports. The foreign government is inactive throughout.

The model is set up so that in the absence of a tariff on imports, the Home firm would never adopt the new technology. The new technology is costly and does not lower production costs. But since the Home government dislikes pollution, it would like the firm to adopt the new technology. As RG show, this implies that the government's optimal tariff on foreign imports is higher if the Home firm adopts the clean technology. Since the Home firm benefits from tariff protection, this generates an incentive for it to adopt. If the Home government committed to zero tariffs and free trade, then the Home firm would never adopt the clean technology.

This is an interesting and thought-provoking paper. It is related to the literature on what I would call "tariff substitution" - that is, a literature where tariffs are employed to solve problems that other, more efficient instruments would typically be used to solve.[5] For example, the tariff in this case substitutes for both a direct tax on emissions and a subsidy to adoption. To be fair, the authors do consider other instruments and in the end argue that even if other instruments are available, tariffs still have a role in spurring adoption. Consequently, a commitment to free trade can be counterproductive. While the authors make a good case for their results, some remaining questions need to be answered. For example, only one country is policy active in their set up, and hence the Home country does not suffer from the Foreign country imposing trade restrictions on its own. When Foreign is inactive, it is easy to see how Home's commitment to free trade could be counterproductive. But in a setting with two policy active countries, a negotiated trade agreement between Home and Foreign may bring benefits to both countries by enlarging markets. Larger markets may spur rather than hinder adoption, and it will, in many models, also spur R&D activity. Therefore, trade liberalization in general may foster rather than deter the adoption of clean technologies.

Despite these quibbles, I think this paper is very useful. The authors demonstrate how technology adoption and strategic interaction produce a correlation between import volumes and the costs of environmental controls that are directly opposite to that expected. For example, a commitment to zero tariffs leads to dirty production, no environmental control costs, and large import volumes. Managed trade with positive tariffs leads to clean production, positive environmental control costs, and smaller import volumes. Therefore, import volumes are negatively correlated with control costs, which is opposite to the typical pollution haven prediction. This should tell us to beware simple correlations between, say, pollution abatement costs and trade flows, because tariffs, environmental protection and imports are jointly determined. Failing to account for this connection can lead us astray: any number of reasons may lead to an across-industry correlation between control costs and import volumes that has nothing to do with the competitiveness consequences of tighter regulations.

The PHH assumption of perfect information is relaxed in Wu's contribution "Pollution Havens and the Regulation of Multinationals with Asymmetric Information". This paper develops a common agency model where a continuum of multinational firms make production and location decisions. The two locations are the residence of the multinational called Home and the potential additional location denoted Foreign. The principals are the Home and Foreign governments that assign to each firm an output level and a corresponding transfer. The home and foreign country differ in two respects. Home residents exhibit a higher disutility of pollution than Foreign, and Home's government has perfect information regarding the pollution intensity of each firm, whereas the Foreign

[5]See Dixit (1985) for a discussion of the appropriate and inappropriate use of tariffs to substitute for other unavailable instruments.

government only knows the distribution of pollution intensity across firms.

The paper's main contribution is to show that the outcome of the noncooperative game between governments may produce an outcome opposite to that expected from the simple pollution haven hypothesis. The author first demonstrates that efficiency requires dirty output to be produced primarily in the country with the lower disutility for pollution. This efficient and cooperative outcome with information sharing across countries mimics a pollution haven division of output across countries. When we move to the noncooperative setting with no information sharing, the equilibrium division of dirty output across countries becomes more equal and hence less in line with the pollution haven prediction.

The logic behind Wu's results comes from the information extraction problem. Since the Foreign government doesn't know a firm's true type (its pollution intensity), any mechanism designed to extract this information will generate some informational rents for firms. However, the Home government knows all firms' pollution intensities. Home uses its superior information to improve it strategic interaction with Foreign, thereby transferring some of these informational rents to its own purse. Strikingly, in some cases, output will be higher in Home than in Foreign, which is of course directly opposite to the pollution haven prediction.

This is an interesting paper and one that identifies another area of research currently under-explored. The assumption that the Home government has superior information regarding the production and pollution process of its firms seems very natural, as does the policy game modeled across countries. It demonstrates that movements away from the perfect information set-up can produce motivations far different from those given by the pure pollution haven effect. Moreover, these offsetting forces can be large enough to reverse standard predictions.

Two remaining papers in this volume investigate the link between country characteristics and regulation using empirical methods. One of the key simplifications in the pollution haven model is that agents are identical. The motivation for this assumption is clear. By assuming away income distributional issues within countries, a researcher can focus on the country as the unit of analysis and develop a theory where income differences across countries play a key role. Alternatively, a researcher can introduce heterogeneity across agents. Countries can then be categorized by the fraction of their populous exhibiting a strong preference for protecting the environment. This division is a common one in research on the political economy determinants of regulation. However, relatively little empirical work has unbundled this mapping. As such, the contribution "Pollution Abatement Expenditure by U.S. Manufacturing Plants: Do Community Characteristics Matter?" by Randy Becker and the contribution "Trade Integration and Political Turbulence: Environmental Policy Consequences" by Fredriksson and Mani (FM) are very welcome indeed.

FM combine both theory and empirical work to investigate the interactions among government honesty, political stability, and trade liberalization. The authors develop a theory where a domestic lobby offers bribes to a regulator in exchange for less onerous regulation. The regulator values both bribes and overall social welfare and must weigh the value of additional bribes against the loss in social welfare from lax regulation. FM show that when such a country undergoes a (exogenously imposed) trade liberalization, the political equilibrium is disturbed and regulation adjusts. Their main theoretical contribution is to show that when a trade liberalization shrinks output in the polluting industry, this lessens the dirty industry's incentive to offer bribes for more lenient regulation, and in equilibrium environmental standards rise.

In their empirical implementation, the authors construct a large, cross-country dataset including 26 OECD and 92 non-OECD countries. The authors explore the predictive power of their theory by explaining the cross-country variation in environmental stringency using proxies for political stability, openness, and government honesty. Overall, the authors find results in line with their theory. Environmental stringency is higher in open economies, and is higher still in economies that are both politically stable and open, as their theory predicts. These results demonstrate a potentially important role for political stability. In typical pollution haven models, political stability plays no role whatsoever, whereas it is center stage here. In fact, the authors find that income per capita has no direct role in raising environmental standards. This is true, for example, in all of the 2SLS results reported in the paper, where they have instrumented for the endogeneity of political stability and their two measures of government honesty. The authors do not interpret this empirical finding as a contradiction to the assumed link between stringency and income made in pollution haven models, because this is too hasty a conclusion. In all of the first stage regressions of political stability and honesty on instruments, income per capita is found to be the most significant variable. Therefore, finding an important role for political stability in setting regulation may instead reflect the tight connection that exists between country income levels, political stability and government honesty.

The contribution by Becker examines the link between characteristics of local U.S. county populations and the abatement behavior of firms in the same county. To do so, Becker employs data on plant-level air pollution abatement expenditures across the U.S. from 1979-1988. Plant-level pollution abatement costs are regressed on a set of plant characteristics, local community characteristics, time dummies, and proxies for the stringency of local, state and federal regulation. The dependent variable is an establishment-level measure of costs from the PACE surveys, while the characteristics of the establishment (size, productivity, etc.) are obtained from the Census of Manufactures.

The results are quite striking. After controlling for federal, state and county-specific regulation, Becker finds four county characteristics positively and consistently linked to the intensity of plant-level abatement. These are: income per capita in the community; the extent of home ownership; the percentage who voted Democrat; and whether the county is within a metropolitan statistical area. In addition, Becker finds one factor negatively linked to abatement: the percentage of employed civilian workers in manufacturing. The findings suggest that those with Green preferences or those with much to lose if environmental quality were to deteriorate, are successful in making their views known at the plant level. How their views are made known and why firms act on them is left unanswered, but the natural inference is that community attributes are affecting the enforcement of existing regulations. Of the characteristics found to be important, income per capita is shown to have the strongest effect on abatement, although the various political economy variables also matter.

Together these contributions paint a much richer picture of the mapping from country characteristics to pollution policy than the one envisaged by the simple Pollution Haven Hypothesis. The theoretical contributions tell us this mapping may be importantly affected by the location decision of factors, the adoption decision of firms, and the lobbying behavior of politically active industry groups. Whether these additional determinants of policy turn out to be important empirically is not yet determined, but each contribution has pointed to additional and potentially important forces. The empirical contributions also suggest a richer picture of the policy process. Although income per capita is surely a determinant of policy, political economy factors are also important. The remaining empirical challenge is to determine which motive is paramount, or to give us a rough indication of their relative importance. Since many political economy measures are highly correlated with income per capita at the country level, this will prove difficult using country level data. Until this issue is resolved empirically, we can only rely on our judgement. It may be appropriate to take income per capita as the primary determinant of pollution policy in an aggregate analysis focussing on explaining cross-country variation in outcomes; alternatively, any study seeking to explain across-industry or across-county variation in outcomes would be well advised to account for elements of political economy.

2.2 The Link from Environmental Regulation to Production Costs

One of the least contentious links in the pollution haven chain of logic is the link between more stringent regulation and higher production costs at channel (b) in Figure 1. Economists are typically skeptical of free lunches that offer tighter regulation, improved environmental quality, and greater international competitiveness, although in some cases tighter regulation can indeed raise productivity levels and lower production costs - breaking the chain of logic supporting the PHH. The conditions for doing so rely on regulation playing an additional role in the economy. Two papers in this volume present theoretical arguments throwing into question the seemingly obvious link between stringent regulation and higher production costs. The first is McAusland's paper "Environmental Regulation as Export Promotion: Product Standards for Dirty Intermediate Goods" and the second is Golombek and Hoel's paper "Unilateral Emission Reductions and Cross-Country Technology Spillovers." In both papers, tighter regulation leads to productivity gains because regulation plays two roles in the model: on the one hand it raises costs and deters polluting activity; on the other hand it partially corrects an existing distortion that is reducing productivity. To break the direct regulation-to-cost link in the pollution haven chain, both authors adopt assumptions to ensure the second role outweighs the first.[6] In McAusland's

[6]It is also possible to generate the counter-intuitive result that tighter regulation lowers the relative price of dirty goods by judicious choice of assumptions in a multi-sector general equilibrium model.

paper, an intermediate input is produced with an increasing returns to scale technology. It is this pre-existing distortion that allows for her paradoxical results. In Golombek and Hoel's paper, R&D spills over across countries. Firms are not compensated for the knowledge spillovers, and we again have an existing distortion.

McAusland's paper sets out a partial equilibrium model of trade in both final and intermediate goods. The intermediate good is produced under increasing returns, and it is sold to Home and Foreign producers of the final good. The intermediate good can be produced using various emission intensities, but the dirtier the intermediate good the greater is its productivity in final goods production. Governments in the two countries set product standards limiting the emission intensity of the intermediate good used by its domestic final good producing firms. The main result of the paper is that a tightening of standards in the high standard Home country can raise the rents earned by Home firms, increase the use of intermediate goods by final goods producers, and raise the output of the final good as well. A further surprise is also offered: McAusland shows that pollution emissions may rise in the high standard country as it raises its standard.

These results follow from the interaction between market size and increasing returns. When the high standard Home country represents a large market for the intermediate good, a tightening of regulation in Home ensures that firms in the low standard Foreign country face a dilemma. On the one hand, foreign firms could comply with the less-stringent Foreign product standards and use a dirty and highly productive intermediate input. The problem with this option is that this intermediate will be relatively expensive to produce because Foreign demand is small. Alternatively, foreign firms could use the same intermediate input as the Home firms. This input is clean but less productive in final good production. The only benefit to using the clean intermediate is that it will be relatively cheap, since Home demand is large, and production is subject to increasing returns. McAusland shows that as long as the high standard country has a sufficiently large market share, then the least cost solution for Foreign firms is to use the clean input.

To generate a comparative advantage for firms in the Home country, McAusland assumes Foreign productivity falls quite quickly with increasingly clean intermediate inputs. As a consequence, a tightening of Home's standards raises Foreign production costs relatively more than Home's. Even though foreign firms "voluntarily" use the clean intermediate, they suffer disproportionately. With this assumption in place, tighter standards at Home shift world demand towards its firms and enhance its exports. If the shift is large enough, overall pollution rises in the high standard country even though its regulation tightens.

This paper is a very clever piece of applied theory. It generates two results that are counter to conventional wisdom in a setting that is not obviously strained. The results provide a theoretical argument for what some have called the "California effect." The "California effect" is the assertion that by virtue of its large market size, California has been able to export its higher product standards elsewhere, and this may have produced competitive benefits for (local) firms who are best able to meet these standards.

In "Unilateral Emission Reductions and Cross-Country Technology Spillovers" GH employ another mechanism to provide surprising results from tightened regulation. Two active countries choose both abatement and R&D expenditures to minimize the sum of environmental damage and conventional costs. Pollution is global. Abatement expenditures lower emissions directly, while R&D expenditures make abatement less costly. The authors' main result is that a shock leading to an increase in Home country abatement may lead to a complementary increase in Foreign country abatement. In a typical public goods framework with increasing marginal damages, an increase in abatement by Home should be met with a decrease in abatement by Foreign. Working against this free riding motive is the impact of R&D spillovers, which increase Foreign's incentive to abate. Given two potentially offsetting effects, it is now a horse race: whether abatement rises or falls depends on model-specific features. The main contribution of GH is to specify when the typical free-rider motive can be overwhelmed.[7]

This paper makes several contributions by examining a potentially important link between regulation and abatement costs. The Pollution Haven Hypothesis is predicated on an asymmetry in regulation and hence environmental control costs across countries. If technological knowledge concerning abatement spills across countries, as GH assume, then differences in stringency will lead to smaller differences in control costs than standard theory predicts. It also demonstrates how additional channels of adjustment - in this case enhanced R&D and technological spillovers - may undo the seemingly obvious and direct free riding effect. The free riding motive is a partial effect arising from the fall in Home emissions, lowering marginal damage. A similar logic is at work in several contributions in this volume, where the seemingly obvious and direct pollution haven effect can be overwhelmed by other motivations for trade.

At a broad level, both the McAusland and GH paper owe their surprising results to increasing returns. This is obvious in the McAusland paper, but is also true in the GH paper. Knowledge spillovers can only occur if knowledge is nonrival. Nonrivaly implies increasing returns. While it is often true that adding increasing returns to a model will produce different and sometimes paradoxical results, increasing returns may be an important feature of many polluting industries. It is well known that the distribution of pollution emissions across U.S. industries is highly concentrated: a handful of very dirty industries produce the lion's share of emissions. I would conjecture that these same heavy industries are also very concentrated, R&D intensive, and employ increasing returns to scale technologies. If this conjecture is true, then the results of these authors suggest we re-examine some of the basic pollution haven linkages within an increasing returns setting.

2.3 The Link from Production Costs to Trade or Investment Flows

This volume contains several papers presenting evidence on the importance of regulatory costs in determining either foreign direct investment or trade flows. While the costs

[7]Heal (1994) and Copeland and Taylor (2005) also demonstrate how the free rider effect can be overwhelmed by other forces.

of meeting regulation are surely important, it has long been recognized that any empirical test of the PHH must also account for other determinants of comparative advantage. The existing empirical literature linking production costs to trade flows is both large and heterogenous in terms of its results. A key early finding was that measures of environmental control costs had either no effect or often a paradoxical positive effect on a country's net exports.[8] In many cases, the small and sometimes perversely signed coefficient estimates were taken as evidence of the irrelevance of regulation in determining costs. These early studies failed to find a "pollution haven" effect, and then they employed the necessity of a pollution haven effect to argue against acceptance of the PHH.

In the late 1990's, a series of empirical studies started to overturn these results by taking explicit account of the endogeneity of regulation.[9] Since then the literature has employed a series of new estimators (such as fixed effects, instrumental variable estimators or matching estimators) and now consistently finds a negative relationship between many measures of industry health such as plant births, production and exports and measures of environmental control costs. Therefore, we now have ample evidence for a "pollution haven effect", but still little direct evidence for or against the PHH. Antweiler et al. (2001) presented evidence on the PHH, but this evidence comes from an inference linking trade liberalization to changes in pollution concentrations and not from a direct observation on trade flows. In contrast, the contribution by Ederington, Levinson and Minier (ELM) "Trade Liberalization and Pollution Havens" directly addresses the PHH by employing trade data.

ELM examine U.S. manufacturing data over the 1972-1994 period to investigate the extent to which large tariff reductions undertaken over this period have impacted production and trade flows in a manner consistent with the PHH. The paper argues that finding a negative impact of regulation on trade flows does not, in and of itself, provide support for the PHH. Instead, ELM argue that evidence for the PHH has to come from an observation that trade liberalization led the US to produce cleaner goods while replacing its dirty good production with imports from other countries. That is, the U.S. must have a demonstrated comparative advantage in clean goods for the hypothesis to hold. This is in effect the argument of Antweiler et al. (2001), although the data and methods used by ELM are very different. ELM's major contribution is to use trade data directly to investigate the PHH.

ELM examine the pollution intensity of U.S. imports, exports and production over the 1972-1994 period. To do so, they employ pollution intensity coefficients developed by the World Bank to calculate the implied pollution emissions created by U.S. production, imports and exports in manufacturing industries at the 4-digit level. The results are striking. ELM first document how U.S. manufacturing production has become cleaner over time. While overall manufactures grew by 57% over the period, the emissions of

[8]For reviews of the empirical literature, see Dean (1992), Levinson (1996), and Copeland and Taylor (2004).

[9]Becker and Henderson (2000) is perhaps the most well known of these studies.

various pollutants grew by much less. At first blush this appears to support the PHH, since the authors find US production is moving towards cleaner industries as the PHH would predict. When the authors examine import and export flows, however, they find evidence against the hypothesis. While import volumes surged over the period, growing on the order of 344%, the composition of these imports shifted towards cleaner industries as well. The pollution content of imports grew by amounts typically less than 1/3 of the change in the value of imports. If the PHH held, U.S. imports should be getting dirtier and not cleaner over time as U.S. trade was liberalized. When ELM examine U.S. exports, they find that while exports became cleaner over time, the implied pollution emissions of U.S. exports fell relatively little over this period. While U.S. exports grew by 268% over the period, the pollution content of exports also grew by over 200% for most pollutants. In total, ELM find the underlying trend is towards cleaner industries over time in U.S. production, imports and exports. This trend is greatest in U.S. imports and least in U.S. exports. ELM suggest that U.S. comparative advantage lies not in clean industries, but instead in dirty ones, and hence other determinants of comparative advantage have proven to be more important determinants of trade flows than the relatively stringent U.S. regulation.

These inferences and evidence come from simple graphs and tables documenting the pollution content of production and trade flows. To go further, the authors employ a regression framework explaining U.S. gross imports in manufacturing as a function of industry attributes, pollution abatement costs, and tariffs. The authors argue that a finding in favor of the PHH requires that imports rise most in industries with above average pollution control costs when the U.S. undergoes a trade liberalization. Recall this is the composition effect that played such a large role in the results of CT (1994). The results from the ELM regression analysis supports their earlier inferences: they find no evidence that trade liberalization has led to large impacts on dirty industries - in fact, their results are just the opposite.

A more aggregate approach to testing the PHH is given by Kahn and Yoshino's (KY) contribution "Testing for Pollution Havens Inside and Outside of Regional Trading Blocs". The authors start with a very clever way to test the PHH. Since the PHH is a statement relating trade flows to country characteristics, the authors employ variation across countries in their membership to regional trade agreements to proxy for a measure of openness to trade. In theory, countries in trading agreements have lower tariffs, all else equal, and hence this variation in their exposure to world markets coupled with knowledge of country characteristics may allow us to estimate the strength of pollution haven motives for trade.

While using this source of variation is ingenious, the authors face several obstacles in their implementation. The most important obstacle may be that membership in a trade agreement is not exogenous and instead is linked with country characteristics that also determine environmental policy. While the authors find that the pollution intensity of trade flows does vary across countries according to their trade agreement membership, these differences are not uniform across country type and escape easy explanation.

The next two studies test the PHH by examining the strength of pollution haven effects, and in particular how their strength varies across industries. In "Environmental Regulation and International Trade: Empirical Results for Germany, the Netherlands and the US, 1977-1992", the authors Mulatu, Florax and Withagen (MFW) take a slightly different approach to testing the PHH. They start by noting that the standard linear regression model of net exports against industry characteristics and pollution costs does not allow the impact of regulatory costs to vary across industries differentiated by either their pollution or natural resource intensity. Recall that regulation must in fact have a differential effect at channel (b) in Figure 1 in order for it to affect relative prices, comparative advantage, and trade flows. The magnitude of the pollution haven effect should vary across industries.

The authors argue that the impact of tighter regulation on trade flows may be greater in polluting industries that are footloose, labor intensive, and located in countries that are scarce in low skill labor. Alternatively, the impact of regulation may be less severe in industries tied to their current location because of their natural resource intensity. To test this thesis, MFW employ a new dataset on 2-digit industry trade flows from the US, Germany and the Netherlands. The authors amend the typical linear regression model of net exports on factor shares to allow the impact of regulation to vary across both industry group and reliance on natural resources. Their empirical results are somewhat mixed. On the plus side, the authors find ample evidence that the impact of regulation varies across industries. Less satisfactory is their finding that the estimated coefficients on control costs often have paradoxical signs. Overall their results suggest empirical work should take serious account of how regulatory effects differ across industries, although finding consistent and meaningful differences appears to be a difficult task.

Cross-industry differences in regulatory impact also play a large role in Javorcik and Wei's (JW) contribution "Pollution Havens and Foreign Direct Investment: Dirty Secret or Popular Myth". The authors examine the determinants of foreign direct investment into 25 transition economies in Eastern Europe and the former Soviet Union, and similar to MFW look for a differential effect of regulation across industries that vary in their pollution intensity. The study is novel in a number of ways. The authors obtain firm-level data from 143 firms who invested or had planned investments as of 1995. This rich data set allows JW to control for firm-specific attributes that may have a bearing on their propensity to invest. The empirical specification is a probit model with the dependent variable taking on a value of one if the firm had planned or undertaken investments in the region. The controls include firm-specific attributes such as size, experience in the region and R&D intensity, and host country controls such as openness, democracy, and tax rates. To investigate how sensitive FDI is to host country regulations, the authors construct two pollution haven variables. One is an indicator variable proxying for the stringency of host country regulation, and the other is an interaction of this first stringency variable with a variable proxying for the pollution intensity of the industry.

The authors find that many of the firm and country-specific variables enter significantly in all (or almost all) specifications, but the sign and significance level of the two pollution haven variables varies across specifications. In many specifications, the indicator variable for the host country's overall environmental stringency enters negatively and significantly, suggesting that FDI is deterred by tight environmental standards. In contrast, in most specifications, the interaction of this term with the industry's pollution intensity is neither significant nor uniformly signed. However, one result is robust: foreign direct investment appears to be attracted to relatively clean industries in their sample. The authors argue that they find little evidence in favor of the pollution haven hypothesis.

While I agree with the authors' conclusions, my interpretation of their results is slightly different. Tests on the sign and significance of the regulation co-variates are tests for a pollution haven effect and not a test of the PHH *per se*. The tests are akin to those conducted using trade data (as with ELM and MFW discussed above), where researchers are looking for a responsiveness of trade flows to variation in regulation. Here the variation is in FDI flows, but the logic is the same. The authors find significant deterrent effects from overall stringency, and this is evidence for a pollution haven effect. In theory it could be argued that the authors' failure to find a consistent and meaningful differential impact of regulation across industries is a rejection of the PHH, although this seems a bit extreme. For many reasons we may not be able to pick up the differential impact of regulation across industries. It may therefore have proven useful to include other industry-specific controls in their specification, to ensure that the failure to find a strong cross-industry impact is not due to other confounding factors.

Putting these three studies together, it is clear that we still have some way to go before we fully understand the determinants of trade in dirty goods. The studies by JW and MFW find empirical support for a pollution haven effect, although neither is completely successful in demonstrating that the impact of regulation varies across industries. While this is partial evidence against the PHH, it may of course reflect our inability to identify these cross-industry effects using current methods and data. At present, I would argue that failure to find a strong across-industry effect is not sufficient reason to reject the PHH. The evidence presented by both KY and ELM are directly relevant to testing the PHH, but both leave some questions unanswered. ELM is based solely on US trade flows, and underlying concerns are the endogeneity of the US trade liberalization under study and the possibility that other processes (such as growth or industrial restructuring) are responsible for their results. The work of KY is broad in cross-country coverage, but its conclusions are tentative and sometimes defy easy explanation. Here too an issue regarding endogeneity is created by the country's choice to join a free trade agreement.

While these studies represent real progress, endogeneity issues remain a concern. Testing the pollution haven hypothesis is an exercise in policy evaluation. We are trying to compare what we observe in the world regarding trade flows or pollution levels under one set of trade and pollution policies, with the unobserved counterfactual created when these policies are different. The key empirical problem is that we do not observe the counterfactual. Instead, we observe trade or investment flows under a certain state of the world X (with free trade or lower tariffs), but then wish to make statements about what would have occurred in another state Y (autarky, or higher tariffs, etc.). This observation and many others are made in the valuable contribution by List, Millimet and McHone (LMM) entitled "The Unintended Disincentive of the Clean Air Act".

The paper sets out a problem in policy evaluation. In the U.S., the Clean Air Act requires existing plants undergoing modifications comply with the New Source Review Requirements (NSR). These requirements offer very asymmetric and strange incentives for plant owners. If a modification to the plant is undertaken, not only must the new modification meet tighter environmental standards but so too must all of the remaining plant that was not modified. This all-or-nothing flavor to the regulation provides a disincentive for plants to undergo any modification or to delay closure, but measuring the size of the disincentive is not easy.

The empirical problem is to determine how the rate of plant modification was altered by NSR - that is, we would like to know what would have happened to plant modifications had NSR not been in place. To solve this problem, the authors employ a matching estimator drawn from the statistics literature. The estimator in essence constructs the counterfactual by matching each plant in a region where meeting NSR is very expensive with a set of plants in regions where meeting NSR is far less expensive. This across-region variation is then used to identify the impact of NSR on the rate of plant modifications. The regions are non-attainment counties and attainment counties in New York State.

The match across plants is made on a wide set of observable plant characteristics, but it leaves open the question of whether plants matched on observables are truly equivalent. To account for the possible impact of unobservables that may be affecting the rate of modification, the authors use another source of variation in their data: the variation across dirty and clean industries in plant modifications and closures. The idea here is simply that if the matching estimator finds a treatment effect by matching dirty plants in region (or time period) X and Y, some of this treatment effect may arise from region - or time - specific unobservables driving modifications in region (or time period) X. The authors surmount this problem by redoing the matching exercise using only plants from clean industries. Since the rate of modification in clean plants should be unaffected by NSR, any treatment effect we find across region (or time period) X and Y for clean industries can now be attributed to the impact of unobservables. If we difference out this unobservable effect, we obtain a true estimate of the treatment effect.

The paper very convincingly estimates an impact of tighter regulation on plant modification rates. As such it provides further confirmation that regulation is not irrelevant to economic activity, whether it be plant modifications, plant closures, or FDI. I take this as very strong evidence of a pollution haven effect. As well, the estimated impact of regulation is quite large, which suggests we cannot reject the PHH out of hand. Using their matching estimator, the authors find on average that three plants per non-attainment county forego modification because of NSR. Since an average county has one modification per year, this is a sizeable deterrent effect. This suggests that regulatory differences can create large impacts on behavior, although we should be careful in extending this within-country finding to an across-country setting too quickly.

2.4 Feedback Effects from Pollution, Prices and Incomes

Most general equilibrium models of pollution and trade contain an immediate and obvious feedback from trade flows to policy making. This can occur through many channels. In CT (1994) for example, the Samuelson rule makes the optimal provision of "pollution" a function of relative prices and incomes, and both of these change in the movement from autarky to trade. In other cases, trade may bring new information about trading partners or enlarge the set of issues that countries negotiate, and this may affect equilibrium pollution levels. Finally, international trade affects the incentive to accumulate capital, knowledge, or new technology, and these decisions will in turn affect policy choices even if in the distant future. This volume contains three papers where feedback effects play a major role.

In Di Maria and Smulders' (DS) contribution "Trade Pessimists vs. Technology Optimists: Induced Technical Change and Pollution Havens", the authors focus on the impact trade has on innovation and pollution levels. DS build a North-South model with trade in the two intermediate goods. The intermediates are produced from a set of differentiated capital goods (machines) and a primary factor. All innovation occurs in the North, while a fraction of innovations diffuse to the South. The authors focus on the balanced growth path throughout.

These authors show that international trade has two effects on pollution levels. The usual pollution haven effect is created when the Southern country specializes in production of the dirty intermediate, but a second effect is created when trade alters the allocation of R&D across industries and thereby alters abatement costs. The authors refer to this second effect as the technology effect.

When the authors combine these two potentially offsetting forces, they find pollution may not rise in the South even when it specializes in dirty goods. The authors are quite successful in linking this result to just two features of their framework: the initial determinants of comparative advantage and the substitutability of Northern and Southern products. Surprisingly, and in contrast to most models with feedback effects, the potential long run impact of trade on pollution is not just to dampen the short run response but to change it qualitatively.

This paper is important because it reminds us that by ignoring the impact of trade on long run accumulation processes - like innovation - we can be missing important possibilities. But it also warns us against hailing new technologies as a fix for all environmental problems, as progress in their model is not always environmentally friendly. Overall, the paper suggests we take a closer look at how the PHH is affected by the dynamic processes driving trade, innovation, and pollution levels.

Hilary Sigman's contribution "Does Trade Promote Environmental Coordination: Pollution in International Rivers" also examines a feedback effect from trade to environmental policy, but this time the goal is to estimate its magnitude. Sigman's analysis uses data on national and international river quality to ask whether greater trade contacts lead to environmental cooperation. The possibility of greater coordination and cooperation because of trading ties is a potential benefit that is rarely incorporated into theoretical analyses. While theorists could think of many reasons why "multi-market" contact may be beneficial, the empirical challenge in this paper is to identify a possible positive impact on rivers crossing country borders. Sigman's empirical work is supportive of a positive link between greater trade and improved river quality. For example, she finds that greater trade intensity between countries leads to lower levels of biological oxygen demand (BOD) measures on their international river.

This paper is novel in arguing for a link between contact in the trade sphere and cooperation over environmental issues. While the empirical results are supportive of such a link, it is difficult to pin down the causal mechanism. Is it because numerous bilateral contacts reduce the transaction costs in other negotiations over river quality? Or is it because countries have other bilateral frictions that can be traded off in these negotiations?

Finally, in "Cross-Country Policy Harmonization with Rent-Seeking", Patrik Hultberg and Edward Barbier (HB) examine an extreme form of feedback from trade flows to policy making. The authors focus on the growing sentiment in the policy community that trade and environment policy should be linked and perhaps harmonized. Harmonization of policies is often put forward as a defense against a pollution-haven-driven pattern of trade. The authors' contribution is to show that harmonization may in some cases be Pareto improving. They develop a two-country, two-firm reciprocal markets model to investigate the impact of harmonization in emission taxes. The countries differ in their disutility of pollution and in the extent to which firms can lobby the government to relax regulation. Harmonization upwards to the high-standard country necessarily benefits the high-standard country as it puts foreign firms at a competitive disadvantage in goods markets. The low-standard country can also gain (when corruption exists) because harmonization acts as a credible commitment not to bow to the pressure of lobbies.

This is an interesting paper on an important policy issue directly related to the PHH. Calls for harmonization are almost always based on the fear of industrial flight and on an acceptance of the PHH. Since many of the studies in this volume have shown that evidence for the PHH is difficult to find, calls for harmonization across very heterogenous countries cannot be based on evidence for the PHH.

3. Summary and Conclusion

The debate over trade liberalization and the environment will be going on for some time. As the world's population grows, and trade expands, future trade and environment conflicts will arise. This is, in some sense, natural. As economists, we recognize that trade-offs are involved in any decision, and that the decision to enter into a trading arrangement will require adjustment and change in many industries. Some of these changes will bring environmental consequences, and some of these consequences will be negative. No amount of research on the Pollution Haven Hypothesis will remove this conclusion, but research - such as that produced in this volume - can identify the trade-offs involved and help us estimate the magnitude of costs and benefits.

In total, the contributions in this volume contain three important messages. The first message is that environmental regulation matters to production and trade flows. Several studies produced empirical evidence of a "pollution haven effect" on both trade and investment flows. Since the existence of a pollution haven effect is a necessary condition for the PHH, this evidence rules out a quick rejection of the PHH. Some researchers have gone further, to argue that the PHH also requires the impact of regulation to differ across industries. Going one step further, they conclude that failure to find important or consistent differences across industries should be taken as evidence against the PHH. In theory this is correct: the PHH is built on the supposition that regulation affects industries differently. This is necessary for regulation to affect comparative advantage and trade flows (recall channel (b) in Figure 1). However, failure to find important differences across industries could come from many sources, and it is premature to reject the PHH on these grounds. This is especially true given the magnitude of regulatory effects found in several of the contributions. Regulatory effects are not only present, they are quite large. Therefore, work remains to be done in testing the PHH.

The second message is related to the first. Regulatory effects are important, but trade in dirty goods is also affected by a host of other factors: capital abundance, industry factor intensities, the quality of governance as measured by corruption, and other political economy determinants. Since many factors jointly determine trade flows, this finding demonstrates how important it is to distinguish between the finding of a "pollution haven effect" and a finding in support of the PHH. Given the evidence presented in this volume for pollution haven effects, future work will now need to focus on weighing the relative strength of different motives for trade in dirty goods. Antweiler et al. (2001) provides one method for doing so, but in this volume, ELM provide another more direct approach.

This type of weighing exercise is important, because even if we find that regulation has a large impact on plant-level investment (such as in LMM), we cannot conclude that trade between a tightly regulated country like the U.S. and a poorly regulated country like Mexico will generate a pollution haven pattern of trade. While it is reasonable to assume that skilled labor and capital are fairly mobile across US counties and US states, so that large differences in regulations across U.S. states may translate into large shifts in the location of production, we should not extend this conclusion to shifts of production across countries differentiated far more starkly in endowments, technologies, and resources. While studies based on U.S. data provide us with some of the most convincing evidence for a regulatory impact on economic activity - i.e., a pollution haven effect - convincing evidence for or against the PHH must employ international data.

Unfortunately, weighing all of the many relevant factors is not easy. In fact, no study in the literature provides a compelling many-country test of the PHH, and I know of no theory paper that details what such a test would look like. Future theoretical work in this area should provide such a test. But even when theorists do provide us with the necessary guidance, empirical work will remain hobbled by data constraints.

While I cannot judge exactly what data a formal multicountry test may require (since it does not yet exist!), I suspect that a fully convincing test of the hypothesis requires data on both home and partner country factor endowments and regulations (to weigh various motives for trade), trade flows, aggregate information on rest-of-world policy, and factor endowment variables (to ensure third country effects are accounted for), together with a source of exogenous variation in trade policy. The challenge for theorists is to show how these many factors interact to determine the pollution intensity of trade flows; the challenge for empiricists is to develop clever methods that eliminate the need for at least some of the data listed above. It is a cruel feature of the world that those countries where we may find the strongest evidence for the PHH typically have data of the poorest quality.

The third message I take from this volume is that the relationship between trade, technology and the environment is not well understood but potentially very important. Four contributions in this volume delivered surprising results from "technology effects." In two contributions, trade liberalization created changes in the set of technologies employed while increasing openness to international markets. In both cases, the induced technology effects produced results opposite to those expected in a pollution haven model (recall RG and DS). In two other contributions, increasing returns technologies interacted with other market features to produce another set of surprising results again at odds with conventional wisdom. Perhaps we have placed too much emphasis on the relative price component of trade liberalization, and too little on how openness to world markets affects knowledge accumulation and technology choice. At present we have very little empirical evidence linking openness to markets, pollution levels, and technology choice. This is surprising, because it is widely believed that technology transfer to poor developing countries will help them limit their pollution regardless of the stringency of their pollution policy or their income levels. If the diffusion of clean technologies is accelerating as a result of globalization, this indirect impact of trade may well become the most important for environments in the developing world.

References

Antweiler, A., B. Copeland and M. S. Taylor (2001), "Is Free Trade Good for the Environment," *American Economic Review,* Vol. 91, No. 4, September, pp. 877-908.

Barrett, S. (1994), "Strategic Environmental Policy and International Trade", *Journal of Public Economics*, Vol. 54, No. 3, July, pp. 325-338.

Becker, R., and J.V. Henderson. (2000), "Effects of Air Quality Regulations on Polluting Industries," *Journal of Political Economy,* Vol. 108, No. 2, April, pp. 379-421.

Becker, R. (2004), "Pollution Abatement Expenditures by U.S. Manufacturing Plants: Do Community Characteristics Matter?", *Contributions to Economic Analysis and Policy*, Vol. 3, No. 2, Article 6.

Copeland, B.R., and Taylor M.S. (1994), "North-South Trade and the Environment", *Quarterly Journal of Economics*, Vol. 109, No. 3, August, pp. 755-787.

Copeland, B.R. and Taylor M.S. (1995a), "Trade and Transboundary Pollution", *American Economic Review*, Vol. 85, No. 4 September, pp. 716-737.

Copeland, B.R. and Taylor M. S. (1995b), "Trade and the Environment: A Partial Synthesis", *American Journal of Agricultural Economics*, Vol. 77, No. 3, August, pp. 765-771.

Copeland, B.R., and Taylor M.S. (2003), *Trade and the Environment: Theory and Evidence*, Princeton University Press, Princeton, NJ.

Copeland, B.R., and Taylor M.S. (2004), "Trade, Growth and the Environment", *Journal of Economic Literature*, Vol. 42, No. 1, March, pp. 7-71.

Copeland, B.R., and Taylor M.S. (2005), "Free Trade and Global Warming: a Trade Theory View of the Kyoto Protocol", *Journal of Environmental Economics and Management,* Vol. 49, No. 2, March, pp. 205-234.

Dean, J.M. (1992). "Trade and Environment: A Survey of the Literature," in *International Trade and the Environment*, ed. P. Low, World Bank Discussion Papers, No. 159, World Bank, Washington, D.C.

Di Maria, C., and S.A. Smulders (2004), "Trade Pessimists vs Technology Optimists: Induced Technical Change and Pollution Havens," *Advances in Economic Analysis and Policy*, Vol. 4, No. 2, Article 7.

Dixit, A. (1985), "Tax Policy in Open Economies", chapter 6 in the *Handbook of Public Economics*, vol. I, ed. A.J. Auerbach and M. Feldstein, Elsevier Science Publishers, Amsterdam, Netherlands.

Ederington, J., A. Levinson, and J. Minier (2004), "Trade Liberalization and Pollution Havens", *Advances in Economic Analysis and Policy*, Vol. 4, No. 2, Article 6.

Elbers, C., and C. Withagen (2004), "Environmental Policy, Population Dynamics and Agglomeration," *Contributions to Economic Analysis and Policy*, Vol. 3, No. 2, Article 3.

Fredriksson, P. and M. Mani (2004), "Trade Integration and Political Turbulence: Environmental Policy Consequences", *Advances in Economic Analysis and Policy*, Vol. 4, No. 2, Article 3.

Golombek, R., and M. Hoel (2004), "Unilateral Emission Reductions and Cross-Country Technology Spillovers," *Contributions to Economic Analysis and Policy*, Vol. 3, No. 2, Article 4.

Grossman G.M., and A.B. Krueger (1993), "Environmental Impacts of a North American Free Trade Agreement," in *The US-Mexico Free Trade Agreement*, ed. P. Garber, MIT Press, Cambridge, MA.

Grossman G.M., and A.B. Krueger (1995), "Economic Growth and the Environment," *Quarterly Journal of Economics,* Vol. 110, No. 2, May, pp. 353-377.

Heal,G. (1994), "Formation of International Environmental Agreements", in *Trade, Innovation, Environment,* ed. C. Carraro, Kluwer Academic Publishers, Dordrecht, Netherlands.

Hultberg, P.T. and E.B. Barbier (2004), "Cross Country Policy Harmonization with Rent Seeking", *Contributions to Economic Analysis and Policy*, Vol. 3, No. 2, Article 4.

Javorcik, B.S., and S-J. Wei (2004), "Pollution Havens and Foreign Direct Investment: Dirty Secret or Popular Myth?", *Contributions to Economic Analysis and Policy,* Vol. 3, No. 2, Article 8.

Kahn, M. E. and Y. Yoshino (2004), "Testing the Pollution Havens Hypothesis Inside and Outside of Regional Trading Blocs," *Advances in Economic Analysis and Policy,* Vol. 4, No. 2, Article 4.

Levinson, A., Taylor M. S. (2003), "Unmasking the Pollution Haven Effect", National Bureau of Economic Research Working Paper No. 10629, Cambridge MA.

Levinson, A. (1996). "Environmental Regulations and Industry Location: International and Domestic Evidence." in Fair Trade and Harmonization: Prerequisites for Free Trade, ed. J.N. Bhagwati and R.E. Hudec. Cambridge: MIT Press.

List, J.A., D.L. Millimet, and W. McHone (2004), "The Unintended Disincentive in the Clean Air Act," *Advances in Economic Analysis and Policy*, Vol. 4, No. 2, Article 2.

McAusland, C. (2004), "Environmental Regulation as Export Promotion: Product Standards for Dirty Intermediate Goods," *Contributions to Economic Analysis and Policy*, Vol. 3, No. 2, Article 7.

Mulatu, A., R.J.G.M. Florax and C. Withagen (2004), "Environmental Regulation and International Trade: Empirical results for Germany, the Netherlands and the US, 1977-1992," *Contributions to Economic Analysis and Policy*, Vol. 3, No. 2, Article 5.

Rauscher, M. (1991), "National Environmental Policies and the Effects of Economic Integration", *European Journal of Political Economy,* Vol. 7, No. 3, pp. 313-329.

Regibeau, P.M. and A. Gallegos (2004), "Managed Trade, Liberalization and Local Pollution", *Advances in Economic Analysis and Policy*, Vol. 4, No. 2, Article 5.

Sigman, H. (2004), "Does Trade Promote Environmental Coordination?: Pollution in International Rivers", *Contributions to Economic Analysis and Policy*, Vol. 3, No. 2, Article 2.

Wu, X. (2004), "Pollution Havens and the Regulation of Multinationals with Asymmetric Information," *Contributions to Economic Analysis and Policy*, Vol. 3, No. 2, Article 1.

[2]

Environmental Policy, Population Dynamics and Agglomeration*

Chris Elbers and Cees Withagen

Abstract

We present and discuss a simple model of international trade in agricultural and manufactured commodities. Production of the latter takes places under monopolistic competition, and causes pollution. We incorporate mobility of skilled labor, the input in the manufacturing sector. One of the main findings is that pollution and environmental policy tend to countervail clustering that would occur in their absence.

KEYWORDS: Environmental policy, population dynamics, monopolistic competition

*The authors are indebted to two referees and Don Fullerton for their very helpful comments on an earlier draft of the paper. Elbers is at Department of Development Economics, Faculty of Economics and Business Administration, Vrije Universiteit Amsterdam, The Netherlands. Withagen is at Department of Spatial Economics, Faculty of Economics and Business Administration, Vrije Universiteit Amsterdam, De Boelelaan 1105, 1081 HV Amsterdam, The Netherlands; email cwithagen@feweb.vu.nl. He is also affiliated with Department of Economics, Faculty of Economics and Business Administration, Tilburg University.

One definition of environmental dumping is that a government sets environmental policy too lax so that producers "can dump their products in international markets at prices that do not reflect the true cost of production" (Rauscher, 1994, p.822). This definition is focused just on trade, however, whereas a broader definition might also consider a larger set of objectives, including plant location. Then environmental dumping occurs when non-cooperative governments set environmental taxes lower than cooperative ones (Wilson, 1996 and Pflüger, 2001). The alleged rationale for ecological dumping in the former definition is the fear of a deterioration of export performance, due to strict environmental policy, for example when conducted according to Pigouvian rules. The latter definition allows for a larger set of objectives including the prevention of delocation of domestic firms abroad or the enhancement of the attractiveness of a jurisdiction for foreign firms. In this paper we will take the broader perspective.

By now we have a vast theoretical and empirical literature on this topic, addressing the question under which circumstances ecological dumping is an equilibrium strategy and whether relocation or loss of competitiveness can actually be observed. In the present paper we aim at making a contribution to the theory of strategic environmental tax competition and choice of location of plants. We do not address plant location decisions as a consequence of exogenous environmental policy as treated by e.g., Markusen et al. (1993) and Motta and Thisse (1994), nor the theory of ecological dumping when plant locations are fixed (see e.g., Rauscher, 1994, Barrett, 1994, and Ulph, 1997, the latter for an excellent survey). The novelty of our contribution lies in bringing together the choice of plant location in a spatial model of monopolistic competition along the lines of the 'new trade theory', and labor mobility triggered not only by real wage differences but also by environmental considerations. We start by motivating this choice by means of a brief survey of the existing literature, thereby identifying the gap we aim to bridge.

The usual theoretical approach, followed by e.g., Markusen et al. (1995), and Hoel (1997) is to consider two countries and a single firm that has to decide where to locate. The two governments act strategically in their choice of pollution taxes. The government's objective in the Markusen et al. paper is to maximize consumer surplus, taking into account the externality caused by pollution. Producer surplus is not involved, because the firm is not owned by either of the two countries. In this respect Hoel's model is different: he takes ownership fully into account. In this literature it is found that when disutility from pollution is large, ecological dumping is not very likely: countries want firms to move out and pollute abroad rather than domestically. If pollution is not very harmful the result is reversed: countries have an incentive to undercut each other's emissions taxes. One possible extension of the models is to take explicit account of the time structure of the decisions by governments and firms, thereby investigating the role of commitment (see e.g., Ulph and Valentini, 2001 and 2002). We will not pursue this issue further. A second possible extension deals with capturing the so-called home market effect, absent in the former studies, saying that an inflow of firms may lower the general price index, implying an increase of consumer welfare. So, even with high disutility of pollution, jurisdictions might compete for firms to realize this effect. In order to model the home market effect, Pflüger (2001) considers the case of monopolistic competition in a new trade theory framework, but without labor mobility. In his set-up, ownership issues don't play a role because profits vanish. Each firm needs one unit of capital so that the total number of firms equals total available capital, which is given. Capital is mobile across the two countries. It is shown that in the case of identical countries that can only impose an emissions tax, ecological dumping occurs if the emission share in production is relatively low compared to transport costs and to the markup over marginal production costs.[1] In this comparison marginal disutility of pollution does not matter, contrary to the studies mentioned above. Hence, if in the social optimum the emissions tax is high, it will also be high in the non-cooperative equilibrium. In our paper we will stay close to Pflüger, but extend his analysis by allowing for mobility of labor.

Important contributions to the pollution-population mobility literature have been made by Wellisch (1994, 1995), Silva (1997), Haavio (2003a and 2003b), Sandmo and Wildasin (1999) and Hoel and Shapiro (2003). We discuss the last in some detail because it has the first three studies as special cases.[2] In a static framework the central idea is simple. In a set of different regions, welfare depends on the production of a commodity produced by homogeneous labor and pollution. The produced commodity is not traded but pollution is possibly transboundary. With an immobile population, emission levels are too high without coordination of environmental policy. But with perfect and costless migration welfare across all jurisdictions will be equal. It is then shown that social welfare maximization over all jurisdictions coincides with a Nash equilibrium of the game played between the individual regions. Hoel and Shapiro (2003) assert that other Nash equilibria may exist, implying that decentralized decision making does not necessarily result in the social optimum. However, the main result strongly suggests that population mobility is important in addressing transboundary pollution problems. In a series of papers Haavio (2003a and 2003b) also consider the mobility-pollution link. He demonstrates that the result obtained by Hoel and Shapiro (2003) and their predecessors no longer holds under alternative assumptions. He considers the case of imperfect (costly) mobility with a distinction between migrants and natives in the sense that the natives receive preferential treatment, and the case where pollution damage occurs as a stock, whose dynamics is explicitly modeled, possibly causing dynamic inconsistency in policy making. All these studies have in common that they assume no trade in commodities.

In the present paper we aim at bringing together two aspects that are lacking in the previous literature. We consider firm location choice in the case of monopolistic competition and trade, taking account of labor mobility, triggered not only by real wages but by environmental concerns as well.[3] We employ the insights from new

[1] Oates and Schwab (1988) also consider capital mobility and analyse the role of a capital tax in the location of capital and pollution. However, strategic interaction is not taken into account.

[2] Sandmo and Wildasin (1999) restrict themselves to a single country with immigration quotas.

[3] To our knowledge the only other paper that combines new geography and environmental policy issues is Calmette and Péchoux (2003), who study the case where clustering of firms increases the probability of

economic geography, so prominently brought to the fore by Fujita, Krugman and Venables (1999), henceforth FKV, in particular their chapters 4 and 5. The central idea in FKV is that monopolistic competition arising from internal scale economies at the firm level, together with costly transportation, may provide an incentive for clustering of activities in the case of mobility of production factors based on real factor reward differentials between regions or countries. The clustering effect comes about because concentration of activities and workers decreases the price index of the cost of living, thereby increasing the real wage. However, when economic activities cause pollution, and if potential migrants take into account not only the real wage but environmental amenities as well, the situation might change. It is the purpose of the present paper to investigate this effect. Our main finding is that strategic environmental policy tends to work against clustering by favoring the symmetric equilibrium. But the stability of an equilibrium with clustering cannot be ruled out.

The outline of the paper is as follows. In the next section we present the model. Section 2 describes a few preliminary theoretical results. Section 3 deals with optimal environmental policy. Finally, section 4 concludes.

1. A model of trade and the environment

The model closely resembles the Dixit-Stiglitz (1977) model of monopolistic competition, as extended by Fujita, Krugman and Venables (FKV, 1999, chapter 4)) in order to account for a spatial dimension. We confine ourselves to a world consisting of two countries or states, because that is the usual setting in the literature on the pollution haven hypothesis. The countries are called home and foreign. They are identical with regard to technology and preferences. Variables referring to the foreign region are denoted by asterisks. In the outline of the model we focus on the home country.

The economy produces two types of commodities: a homogeneous agricultural commodity, serving as the numeraire, and a continuum of manufactured commodities. The agricultural commodity is produced under perfect competition. For the manufactured commodities monopolistic competition rules. A variety is produced by one firm in one country only: if two firms would compete in the same market, then, given a continuum of potential varieties, it would be more profitable for one firm to specialize in a yet unexploited variety. We can therefore say that the home country produces varieties in the range $[0,n]$. The foreign country produces varieties in the range $[n,n+n^*]$. For the sake of brevity we say that the home country produces n varieties and the foreign country produces n^* varieties. Production of the manufactured commodities brings along pollution that causes local damage only. Emissions are subject to an emissions tax. We assume that within each country the emissions tax rates are equal, whereas across countries they may differ. Since the technologies to produce

environmental disasters. After the completion of the present paper we came across a very recent paper by Quaas and Lange (2004), analyzing the FKV model extended with environmental issues. However, strategic policy is not addressed.

varieties are identical, this implies that the producer prices of the actually produced manufacturing commodities within each country are identical. In both countries the revenues from the local emissions tax are redistributed to workers actually working in that country in a lump sum way. We also assume no transportation costs within each country. For domestically produced varieties, the consumer price therefore equals the producer price. Transportation costs between countries are modeled as "iceberg" transport costs: in order for one unit of a commodity produced in one country to arrive in the other country, $T(\geq 1)$ units have to be shipped. Transportation costs apply to manufactured commodities only.

We first describe the behavior of a representative domestic consumer. As in FKV, preferences for commodities are described by a Cobb-Douglas utility function. Following Markusen et al. (1995), Hoel (1997) and Pflüger (2001), it is assumed that marginal damage from pollution φ is constant ($\varphi > 0$). Total welfare of the consumer is given by

$$U(A, M, E) = M^{\mu} A^{1-\mu} - \varphi E$$

where A denotes consumption of the agricultural commodity, E stands for total emissions by the domestic firms, μ is constant with $0 < \mu < 1$, and

$$M \equiv \left[\int_0^{\bar{n}} m(i)^{\rho} di \right]^{1/\rho}$$

where $\bar{n} = n + n^*$ denotes the range of available varieties, $m(i)$ is the consumption of variety i of manufactured goods, and ρ is a constant ($0 < \rho < 1$). The parameter ρ is a measure of substitutability of the manufactured commodities. If it approaches unity, the manufactured commodities become perfect substitutes; if it approaches zero, the manufactured commodities are complements. The consumer's budget constraint reads

$$Y \geq A + \int_0^{\bar{n}} p(i) m(i) di$$

where Y is individual income and $p(i)$ is the consumer price of variety i with $i \in [0, \bar{n}]$. Define

$$G = \left[\int_0^{\bar{n}} p(i)^{\frac{\rho}{\rho-1}} di \right]^{\frac{\rho-1}{\rho}}$$

It is then straightforward to derive the individual demand functions (see e.g., FKV p. 47).

$$A = (1-\mu)Y$$

$$m(i) = \mu Y \frac{p(i)^{-\sigma}}{G^{1-\sigma}}, \quad i \in [0, \bar{n}]$$

with $\sigma = 1/(1-\rho) > 1$, the elasticity of substitution between two varieties. The indirect utility is therefore $(1-\mu)^{1-\mu}\mu^{\mu}YG^{-\mu}$. Following FKV, we define the function G as the price index. The rationale for this definition is that total expenditures on manufactured commodities equal the price index times the quantity composite M. If $p(i) = p$ for all $i \in [0, \bar{n}]$ then the price index G reduces to $p\bar{n}^{\frac{1}{1-\sigma}}$. Since $\sigma > 1$, more available varieties decrease the price index, thereby increasing utility from private consumption. This exactly represents the benefits of agglomeration in the original FKV model.

Total demand for a variety produced in the home country consists of demand from the home country and demand from the foreign country. In the absence of transportation costs within each country, the consumer price of domestically produced varieties equals the producer price. But when manufactured commodities are imported, the consumer price in increased. This yields the free trade price indexes G and G^* for the manufactured commodities:

$$\text{(1a)} \quad G = \left[np^{1-\sigma} + n^*(p^*T)^{1-\sigma}\right]^{1/(1-\sigma)}$$

$$\text{(1b)} \quad G^* = \left[n^*(p^*)^{1-\sigma} + n(pT)^{1-\sigma}\right]^{1/(1-\sigma)}$$

Demand for a manufactured commodity produced in the home country equals

$$m = \mu Y p^{-\sigma} G^{\sigma-1} + \mu Y^*(pT)^{-\sigma}(G^*)^{\sigma-1}T$$

where Y and Y^* now denote aggregate domestic and foreign income, respectively. The first term on the right-hand side needs no explanation. The second term includes the transportation loss.

Let us turn now to the supply side of the economy. The agricultural sector exclusively employs unskilled labor. Production of agricultural commodities is described by a linear technology:

$$x = l_u$$

where x denotes production and l_u is the input of unskilled labor. The agricultural commodity is produced under perfect competition. Since it serves as the numeraire, the wage rate of unskilled labor is equal to unity. The agricultural sector is clean: it creates no pollution externality.

The manufacturing sector employs skilled labor only. Consider an arbitrary variety. Production y requires skilled labor input l_s given by

$$l_s = F + \gamma y \text{ if } y > 0$$

where F denotes the constant fixed cost and γ is the constant marginal production cost. Production brings along pollution, proportional to labor input. A unit of emissions is defined as the amount from one unit of labor input. We assume no abatement possibilities, so net emissions are

$$e = l_s$$

Profits amount to

$$\pi(y, l_s) = py - (w+t)l_s$$

where p is the producer price, w is the skilled labor wage rate, and t is the emissions tax. Demand for the variety was derived above. It will be assumed that all individual producers take the price indexes G and G^* as well as incomes as given. It then follows from monopolistic profit maximization that

$$p = \frac{(w+t)\gamma\sigma}{\sigma - 1} \tag{2a}$$

For later reference, we also give the producer price of goods produced in the foreign country:

$$p^* = \frac{(w^* + t^*)\gamma\sigma}{\sigma - 1} \tag{2b}$$

Equations (2) give the constant markup over marginal cost, not depending on market size. This result crucially depends on the assumption of iso-elastic demand. It is discussed extensively in FKV (1999, p.52). Equilibrium profits are zero, implying

$$y = F(\sigma - 1)/\gamma \text{ and } l_s = F\sigma$$

A country's income is composed of the wage income of the workers *within* that region, augmented with the lump sum redistribution of emissions taxes. The immobile unskilled agricultural labor force in the home country is l_u. Since the home country produces n varieties and each variety requires $l_s = F\sigma$ skilled workers, the skilled labor force is $nF\sigma$. With a wage rate w for skilled labor and an emissions tax of t per unit of emissions (equal to labor input), we have for total income in the home country

$$(3a) \qquad Y = l_u + wnF\sigma + tnF\sigma$$

Similarly, total income in the foreign country is

$$(3b) \qquad Y^* = l_u^* + w^*n^*F\sigma + t^*n^*F\sigma$$

Market equilibrium conditions for final manufactured products and skilled labor are

$$(4a) \qquad \mu Yp^{-\sigma}G^{\sigma-1} + \mu Y^*(pT)^{-\sigma}(G^*)^{\sigma-1}T = (\sigma-1)F/\gamma$$

$$(4b) \qquad \mu Y^*(p^*)^{-\sigma}(G^*)^{\sigma-1} + \mu Y(p^*T)^{-\sigma}G^{\sigma-1}T = (\sigma-1)F/\gamma$$

$$(5) \qquad (n+n^*)F\sigma = \bar{l}_s$$

where $\bar{l}_s$ is total endowment of skilled labor, supplied inelastically.

The main difference between our model and the original FKV model lies in the motives for migration of skilled workers. Unskilled labor is immobile across countries, for example because agricultural workers are tied to their land. One reason for skilled workers to migrate can be real wage differentials of skilled labor between the two regions, as in the original FKV model.[4] In the absence of environmental amenities this makes good sense. In the case at hand we assume that skilled workers take a broader perspective and move to the region with the higher per capita welfare of skilled workers, where welfare takes account of inflicted pollution damage. In order to derive a mathematical expression for welfare, we define total skilled labor income as Y_s. It consists of total wage income $wnF\sigma$, to which is added the share in tax redistribution. Regarding the latter it is assumed that all workers, including the unskilled, are treated equally. Hence, taking into account that $l_s = l_s^* = F\sigma$,

$$(6a) \qquad Y_s = wnF\sigma + tnF\sigma\frac{nF\sigma}{l_u + nF\sigma}$$

[4] The literature on migration of skilled labor is scarce. One reference is Lowell and Findlay (2001).

(6b) $$Y_s^* = w^* n^* F\sigma + t^* n^* F\sigma \frac{n^* F\sigma}{l_u^* + n^* F\sigma}$$

Per capita welfare of skilled labor in each country is then:

(7a) $$W_s = \frac{\mu^\mu (1-\mu)^{1-\mu} Y_s G^{-\mu}}{nF\sigma} - \varphi nF\sigma$$

and

(7b) $$W_s^* = \frac{\mu^\mu (1-\mu)^{1-\mu} Y_s^* (G^*)^{-\mu}}{n^* F\sigma} - \varphi n^* F\sigma$$

where use has been made of the fact that the indirect utility from consumption equals $(1-\mu)^{1-\mu}\mu^\mu YG^{-\mu}$. The number of skilled workers in the home country is nl_s. Labor input in an active sector (l_s) is fixed and equals $F\sigma$. The migration flow can therefore be described by the change in the number of varieties n only. This is specified by the following differential equation:

(8) $$\dot{n} = \vartheta(W_s - W_s^*)$$

Here $\dot{n}$ denotes the change in n over time and ϑ is a measure of the speed of adjustment. No migration occurs if welfare of skilled labor is the same in both countries.

Finally, we provide an expression for aggregate welfare in both countries:

(9a) $$W = \mu^\mu (1-\mu)^{1-\mu} YG^{-\mu} - \varphi nF\sigma[l_u + nl_s]$$

(9b) $$W^* = \mu^\mu (1-\mu)^{1-\mu} Y^* (G^*)^{-\mu} - \varphi n^* F\sigma[l_u + n^* l_s]$$

So, total welfare consists of total utility derived from consumer commodities and the negative externality from pollution inflicted upon all residents. Given the number of varieties produced in the foreign country, home country's welfare from consumption increases when the number of varieties increases: it boosts income (see equation (6)) and decreases the price index. However, pollution damage increases when more manufacturing production takes place.

This completes the outline of the model.

2. Normalization and preliminary results

In this section we follow FKV and make several normalizations, in order to make the subsequent analysis tractable. As explained in FKV, chapters 4 and 5, these normalizations do not cause loss of generality. Units are chosen such that $\gamma = \rho$, implying that $p = w + t$ and $p^* = w^* + t^*$. We also take $F = \mu / \sigma$, implying that $l_s = l_s^* = \mu$. Moreover, it will be assumed that total available labor equals unity and that labor available for agriculture equals $(1-\mu)$, uniformly distributed over the two countries: $l_u = l_u^* = \frac{1-\mu}{2}$. Then, with full employment, we have $(1-\mu) + n\mu + n^*\mu = 1$, so that $n + n^* = 1$. Finally, we assume $\vartheta = 1$.

The set of equilibrium conditions becomes:

$$\text{(10a)} \quad G = \left[np^{1-\sigma} + n^*(p^*T)^{1-\sigma}\right]^{1/(1-\sigma)}$$

$$\text{(10b)} \quad G^* = \left[n^*(p^*)^{1-\sigma} + n(pT)^{1-\sigma}\right]^{1/(1-\sigma)}$$

$$\text{(11a)} \quad Y = \frac{1-\mu}{2} + np\mu$$

$$\text{(11b)} \quad Y^* = \frac{1-\mu}{2} + n^*p^*\mu$$

$$\text{(12a)} \quad p^\sigma = YG^{\sigma-1} + Y^*(G^*)^{\sigma-1}T^{1-\sigma}$$

$$\text{(12b)} \quad (p^*)^\sigma = YG^{\sigma-1}T^{1-\sigma} + Y^*(G^*)^{\sigma-1}$$

$$\text{(13)} \quad n + n^* = 1$$

The number of varieties in the home country depends on available labor, which is governed by migration. From (1) and (6) we have for the per capita income of skilled workers:

$$\text{(14a)} \quad Y_s / nl_s = w + \frac{2n\mu t}{1-\mu+2n\mu} = p - \frac{(1-\mu)t}{1-\mu+2n\mu}$$

$$\text{(14b)} \quad Y_s^* / n^*l_s^* = w^* + \frac{2n^*\mu t^*}{1-\mu+2n^*\mu} = p^* - \frac{(1-\mu)t^*}{1-\mu+2n^*\mu}$$

For the sake of notation we define $\psi = \mu^{\mu}(1-\mu)^{1-\mu}$. Then per capita welfare of skilled workers is

(15a) $$W_s = \psi\left(p - \frac{(1-\mu)t}{1-\mu+2n\mu}\right)G^{-\mu} - \varphi n\mu$$

(15b) $$W_s^* = \psi\left(p^* - \frac{(1-\mu)t^*}{1-\mu+2n^*\mu}\right)(G^*)^{-\mu} - \varphi n^*\mu$$

For any n, we can solve the endogenous variables $(Y, Y^*, G, G^*, p, p^*, n^*)$ from equations (10)-(13). This allows us to write welfare in each country as an (implicit) function of the number of varieties produced:

(16a) $$W(n) = [\frac{1-\mu}{2} + np\mu]\psi G^{-\mu} - [\frac{1-\mu}{2} + n\mu]\varphi n\mu$$

(16b) $$W^*(n^*) = [\frac{1-\mu}{2} + n^* p^* \mu]\psi(G^*)^{-\mu} - [\frac{1-\mu}{2} + n^*\mu]\varphi n^*\mu$$

where it is to be understood prices and the price indexes depend on the number of varieties.

Before embarking on a policy analysis we illustrate the factors that have an impact on migration. In the steady state symmetric equilibrium, where $n = ½$ and $\dot{n} = 0$, no welfare difference arises between the skilled workers in the different countries. In case of complete specialization in the home country, where $n = 1$, and under the assumption of no emissions taxes in the (clean) foreign country, we have

(17a) $$G = 1, Y = \frac{1+\mu}{2}, p = 1$$

(17b) $$G^* = T,\ Y^* = \frac{1-\mu}{2},\ p^* = \left[\frac{1+\mu}{2}T^{1-\sigma} + \frac{1-\mu}{2}T^{\sigma-1}\right]^{1/\sigma}$$

The expressions for welfare become

(18a) $$W(1) = \frac{1+\mu}{2}[\psi - \varphi\mu]$$

(18b) $$W^*(0) = \frac{1-\mu}{2}\psi T^{-\mu}$$

The welfare comparison of skilled workers boils down to the comparison of

$$\text{(19a)} \quad W_s(1) = \psi(1 - \frac{(1-\mu)t}{1+\mu}) - \varphi\mu$$

and

$$\text{(19b)} \quad W_s^*(0) = \psi\left[\frac{1+\mu}{2}T^{1-\sigma} + \frac{1-\mu}{2}T^{\sigma-1}\right]^{1/\sigma} T^{-\mu}$$

One can identify four factors that influence the sustainability of concentration in the home country. The first two are found in the expression for W_s and are related to pollution and environmental policy. The final two appear in W_s^* and are related to transport cost. These effects are identical to the effects discussed in FKV.

The first factor is the emissions tax in the expression for W_s. Part of the emissions-tax revenue is recycled to the skilled workers, but not all of it. Therefore, high emissions taxes reduce skilled workers' real income (recall that $p = 1$), which reduces the incentive to move into the home country.

The second factor is pollution (the term $-\varphi\mu$). It has a negative impact: workers prefer a clean environment.

The third factor (captured by $G^{-\mu} = T^{-\mu}$) concerns the price index that is higher in the foreign country because it has to import all manufacturing commodities, which makes them more expensive. Hence, it is less attractive for skilled workers to move out of the home country.

The fourth factor is captured by the term in brackets. Essentially, it is the wage rate a foreign firm would offer, if it became active. In the absence of transport costs ($T = 1$) this factor doesn't play a role. High transport costs are beneficial to skilled labor in the sense that the firm can afford to pay a higher nominal wage because domestic consumers want to buy more domestically. This outweighs lower demand from abroad.

What matters is the total effect of transportation costs. Under the so-called no-black-hole assumption,[5] the expression $W_s^*(0)$ is decreasing in T for T sufficiently small, and increasing in T without bound afterwards.

As explained by FKV, in the absence of environmental amenities and policy, a symmetric equilibrium with $n = n^* = ½$ is stable for T sufficiently high, meaning that from any initial state (characterized by n) the system converges to $n = n^* = ½$. For low

[5] This is $\sigma - 1 > \sigma\mu$, see FKV, p. 59.

transportation costs two stable equilibria arise: $n=1$ and $n=0$: clustering occurs. For intermediate values of transportation costs the clustering equilibria as well as the symmetric equilibrium are locally stable. However, with environmental externalities and environmental policy this is no longer true. If transport costs are low, pollution and pollution tax policy might make skilled workers move to the 'clean' country.

3. Policy

In the present section, we analyze environmental policy. We compare the coordinated optimum with the Nash equilibrium of the emissions tax competition game played by the two governments. We start by making some assumptions on the policy game.

Initially the two economies find themselves in a certain state, characterized by n. Consider the home country. It takes the emissions tax t^{*} of the foreign country as given and determines its own optimal tax. We assume that the tax rates are determined at the outset of the game, and, moreover, that all transition effects are neglected. The first assumption implies an open-loop setting with full commitment. The issue of time consistency therefore does not arise. The second assumption neglects the fact that the transition might be costly, and says that the countries only care about the welfare evaluation in the steady state equilibrium.

The strategies consist of the emissions tax rates, which we assume are bounded from below by zero. In some cases governments might want to subsidize pollution. We rule out this possibility; pressure groups might be fiercely against subsidizing pollution. Taxes are bounded from above by the requirement that skilled workers' income cannot be negative. In view of (14) this might imply a negative wage rate. An alternative upper bound might be induced by the requirement that the wage rate is non-negative. In view of the fact that $p=w+t$ the upper bound on the emissions tax is then equal to p. With the former upper bound the expression for welfare is more tractable than with the latter. In the sequel we use this. The consequences of the alternative upper bound will be discussed later.

Each country's payoff consists of welfare aggregated over all unskilled and skilled workers within the country. Therefore, the governments also care for foreign skilled workers working within their jurisdiction. Alternatively, we could identify the initial citizenship of skilled workers and assume that governments only care about their own citizens. If the redistribution scheme is another policy instrument, then it becomes easier for governments to make their country less attractive to foreign skilled workers. We leave this issue for future research (see also Haavio, 2003b). It might be interesting to include other objectives as well. For example, insights from political economy suggest that lobbying for low emissions taxes by exposed sectors plays a role in policy design. We will not discuss such issues here.

In the sequel of this section we analyze optimal environmental policy, from the point of view of the individual countries as well as from the perspective of overall welfare. Social welfare over both countries is defined as the sum of the individual welfare functions. We first focus on the case of no transportation costs. Subsequently transportation costs are taken into account.

No transportation costs ($T = 1$)

Complexity is reduced drastically when $T = 1$. It follows from (10)-(13) that $p = p^* = G = G^* = Y + Y^* = 1$, implying

(20a) $$W(n) = [\frac{1-\mu}{2} + n\mu][\psi - \varphi n\mu]$$

(20b) $$W(n^*) = [\frac{1-\mu}{2} + n^*\mu][\psi - \varphi n^*\mu]$$

Hence, welfare is quadratic in the number of varieties produced. Over the relevant range of n (the unit interval) it may be monotonically increasing, monotonically decreasing or inverted-U shaped. If $\psi - \frac{1}{2}\varphi(1-\mu) < 0$ then the individually optimal number of varieties $\hat{n}$ equals zero. If $\frac{\psi - \frac{1}{2}\varphi(1-\mu)}{2\mu\varphi} > 1$, then $\hat{n} = 1$. Otherwise we have

(21) $$\hat{n} = \frac{\psi - \frac{1}{2}\varphi(1-\mu)}{2\mu\varphi}$$

Social welfare is defined as

(22) $$W(n) + W(n^*) = [\frac{1-\mu}{2} + n\mu][\psi - \varphi n\mu] + [\frac{1-\mu}{2} + n^*\mu][\psi - \varphi n^*\mu]$$

Due to strict concavity, maximization of this expression, subject to $n + n^* = 1$, yields $n = \frac{1}{2}$: from an aggregate welfare point of view the symmetric solution is better than the clustering solution. However, what equilibrium eventually prevails, is the result of the policy game played by the two jurisdictions, and depends crucially on marginal damage from pollution (φ). It follows from (21) that the individually optimal number of varieties for the home country is smaller than the socially optimal one if and only if $\varphi(1+\mu) > 2\psi$.

Suppose the initial state is a symmetric equilibrium ($n = n^* = \frac{1}{2}$). With high marginal damage, implying $\hat{n} < n = \frac{1}{2}$, each country prefers to have a smaller manufacturing sector. In principle, this is accomplished by making it unattractive to skilled workers to move into the country. To see how this can be accomplished, consider the welfare differential of skilled workers (see equations (14) and (15)):

$$\dot{n} = W_s - W_s^* = \frac{Y_s}{n\mu} - \frac{Y_s^*}{n^*\mu} - \varphi\mu(2n-1) =$$

(23)

$$\psi[1 - \frac{(1-\mu)t}{1-\mu+2n\mu}] - \psi[1 - \frac{(1-\mu)t^*}{1-\mu+2(1-n)\mu}] - \varphi\mu(2n-1)$$

Given that emissions taxes are the only instrument, they will be used in a 'race to the top'. Since the upper bound on emissions taxes is determined by the requirement that $Y_s = Y_s^* = 0$, the Nash equilibrium is attained when $t = t^* = \frac{1}{1-\mu}$ (see (14) and take into account that $p = p^* = 1$ and $n = n^* = \frac{1}{2}$). Consequently, both economies stay in the symmetric equilibrium. The wage rate is negative, but to the producers only the gross labor costs, consisting of the wage and the pollution tax per unit of labor, matters. For the workers too, the wage rate per se is irrelevant: they care about income. In order for the proposed policy to be effective, it has to be assumed that skilled workers are entitled to recycled emissions taxes only if they actually work.

For low marginal damage, implying $\hat{n} > n = \frac{1}{2}$, countries engage in a 'race to the bottom'. In fact, governments would even like to subsidize emissions. Since emissions taxes are bound to be non-negative, this is ruled out, and the Nash equilibrium strategy is to have zero taxation.

Therefore, in both situations the symmetric equilibrium is stable in the sense that no country will deviate from a symmetric policy when it finds itself in this state.

What happens if the initial equilibrium, characterized by n, is not symmetric? Several cases can be distinguished. We elaborate on three of them. The other cases are equivalent mirror images, and are not discussed here.

- Suppose $\hat{n} < n < \frac{1}{2}$. The home country wants to decrease the number of varieties it produces, and the foreign country wants this as well, because $n^* > \frac{1}{2} > \hat{n}^* = \hat{n}$. Both countries observe the number of varieties and the tax rate set by the other country. Denote time by $\tau(\geq 0)$. Suppose the foreign country sets

(24) $$t^*(\tau) = \frac{1-\mu+2n^*(\tau)\mu}{1-\mu}$$

for all $\tau \geq 0$. This implies that $Y^*(\tau) = 0$ for all $\tau \geq 0$. Given the trajectory of t^* and given that the home country is interested in the new steady state only, the best the home country can do is to set

$$(25) \quad t(\tau) = \frac{1-\mu+2n(\tau)\mu}{1-\mu}$$

because then the solution of $\dot{n}(\tau) = -\varphi\mu[2n(\tau)-1]$ converges to $n = \frac{1}{2}$, which is the best the home country can obtain. Therefore, in the Nash equilibrium both set their taxes according to (24) and (25) implying that $Y_s = Y_s^* = 0$. Hence, eventually the symmetric equilibrium arises. The home country cannot prevent the foreign country from getting at least half of the varieties. For the skilled workers this is a bad outcome, since they have no income.

- Suppose $n < \hat{n} < \frac{1}{2}$. The home country wants to increase the number of varieties produced, and the foreign country wants to decrease its number of varieties, because $n^* > \frac{1}{2} > \hat{n}^* = \hat{n}$. Suppose the foreign country sets the emissions tax according to (24) such that $Y_s^*(\tau) = 0$ for all $\tau \geq 0$. The home country can enforce a steady state equilibrium with $n = \hat{n}$ by setting its tax according to (23) for an initial period of time, implying $Y_s = 0$ initially, so that n reaches $\hat{n}$ within finite time (see (23)) and by choosing t thereafter such that it solves

$$\dot{n} = \psi[1 - \frac{(1-\mu)t}{1-\mu+2\hat{n}\mu}] - \varphi\mu(2\hat{n}-1) = 0$$

Moreover, given this policy by the home country, the foreign country cannot do better. It ends up in the best possible state, closest to $\hat{n}^*$. So, it does not pay to prevent the home country from reaching $\hat{n}$ nor to prevent the home country from staying there.

- Suppose $\hat{n} < \frac{1}{2} < n$. The home country wants to reduce the number of varieties produced. Suppose it sets the emissions tax such that $Y_s(\tau) = 0$ for all $\tau \geq 0$. If $n^* > \hat{n}$, then the foreign country wants to reduce its number of varieties too. Given home country's policy it is optimal for the foreign country to set the emissions tax according to (24) and $Y_s^*(\tau) = 0$ for all $\tau \geq 0$. Eventually a symmetric equilibrium results. This is also the Nash equilibrium. If $n^* < \hat{n}$, the foreign country wants to have more varieties, and it can find a tax schedule leading to the individual optimum, along the lines set out before.

In the case at hand, without transportation costs, equilibrium prices are unity. Hence, a requirement of a non-negative wage rate would imply that the emissions taxes are uniformly bounded above by unity. Therefore, the results derived above still hold if this upper bound would be employed instead of an upper bound leading to zero income for skilled workers.

The conclusion we can draw from this analysis is that the socially optimal symmetric equilibrium is stable if the initial state is symmetric. It will also be reached from any other initial equilibrium, if both countries want to reduce or both countries want to enlarge their manufacturing sector. When the interests are opposite, one of the countries will reach its individual optimum. In this stylized model we also see that chosen emissions taxes are not directly related to marginal damage, but do play a role in making countries attractive or unattractive to potential skilled migrants. Environmental dumping does not occur because the social optimum can be reached without any emissions tax. Therefore, given the lower bound on emissions taxes, individual countries will not engage in environmental dumping in the sense discussed in the introduction. But they are inclined to subsidize pollution when their individually optimal number of varieties is higher than $\frac{1}{2}$.

Transportation costs ($T > 1$)

Since both countries have identical welfare functions, the allocation that maximizes the sum of the welfare of the individual countries corresponds either with complete specialization or with symmetry. From an environmental perspective it is optimal to have symmetry: this minimizes total damage from pollution, as we have seen in the previous case of no transportation costs. Moreover, in the case of transportation cost, equalization of the price indexes G and G^* also works in the direction of symmetry. However, the symmetric equilibrium is not necessarily socially optimal in the case of transportation cost, as the following example shows. In a symmetric equilibrium, with $n = n^* = \frac{1}{2}$, we have from (10)-(13) that $G = G^* = \left(\frac{1}{2}(1+T^{1-\sigma})\right)^{\frac{1}{1-\sigma}}$, $Y = Y^* = \frac{1}{2}$ and $p = p^* = 1$. Hence

$$(26)\qquad W(\tfrac{1}{2}) = W^*(\tfrac{1}{2}) = \tfrac{1}{2}\psi[\tfrac{1}{2}(1+T^{1-\sigma})]^{\frac{-\mu}{1-\sigma}} - \tfrac{1}{4}\varphi\mu$$

If $n = 1$, then (17) and (18) hold. Take $1-\sigma = -\mu$ and use this in comparing welfare in the two states. It follows that $W(\frac{1}{2}) + W^*(\frac{1}{2}) < W(1) + W^*(0)$ if and only if

$$\varphi\mu < \psi(1 - T^{-\mu})$$

Hence, if marginal damage is low relative to transportation costs, specialization is socially optimal. If marginal damage is high, symmetry is optimal. This latter statement obviously holds in general (also when $1-\sigma \neq -\mu$). Of course it matters that the social welfare function is just the sum of the individual welfare functions. If the social welfare function would display inequality aversion, then clustering would be considered bad.

For the individual country aggregate welfare as a function of the number of varieties ($W(n)$) can have different shapes. It may be monotonically increasing, monotonically decreasing or inverted-U shaped. Bearing this in mind, we now consider both social optima. In the first case clustering is optimal. In the second, symmetry is optimal.

Socially optimal clustering

First we characterize welfare in a situation of complete specialization of manufacturing in the home country: $n=1$. The system reduces to (17) and (18), and the initial welfare comparison made by an individual skilled worker boils down to considering

$$W_s(1) - W_s^*(0) =$$

(27)

$$\psi[1-\frac{1-\mu}{1+\mu}t] - \varphi\mu - \psi\left\{\left[\frac{1+\mu}{2}T^{1-\sigma} + \frac{1-\mu}{2}T^{\sigma-1}\right]^{1/\sigma} - t^*\right\}T^{-\mu}$$

We distinguish between three cases. As before, $\hat{n}$ is the number of varieties produced in the home country that maximizes that country's individual welfare. Initially, we are in the social optimum: $n=1$.

- $\hat{n}=1$ and $W(1)=W(0)$. It follows from (17) that a necessary and sufficient condition for $W(1)=W(0)$ to hold is:

$$\psi - \psi\frac{1-\mu}{1+\mu}T^{-\mu} = \varphi\mu$$

If $\sigma-1=\mu$ the requirement $\varphi\mu < \psi - \psi T^{-\mu}$ is not met. But for σ sufficiently large the social optimum and the individual optima coincide at complete specialization. The commodities become close substitutes, so that the need to import gets negligible. In this case, clearly no environmental policy is needed to sustain the equilibrium.

- $\hat{n}=1$ and $W(1)>W(0)$. The foreign country wants to attract as much manufacturing as possible. It sets $t^*=0$. The home country wants to keep its entire manufacturing industry and sets $t=0$ as well. Which equilibrium prevails eventually depends on transportation costs and on marginal damage. The case at hand reduces to the original FKV model, discussed at the end of Section 2, except for the damage playing a role in the welfare comparison of skilled workers. Therefore, if transportation costs are high enough and if the no-black-hole assumption holds (see footnote 5), skilled workers will move out of the home country into the foreign country, since then transportation costs are dominant in (27). This is enhanced by the fact that pollution damage plays a role now. For very low transportation costs, clustering is not socially optimal in the first place, as we have seen above. But for intermediate values of marginal environmental damage and transportation costs, workers may want to stay in the home country. However, the presence of pollution damage makes this less likely.

- $0 < \hat{n} < 1$. This holds for marginal pollution damage sufficiently large as well as for transportation costs sufficiently small. The home country wants to dispose of part of the manufacturing sector. The foreign country wants to acquire some manufacturing industry.

Suppose $\hat{n} < ½$. The Nash equilibrium taxes are such that $Y_s = 0$ with $n = ½$ eventually. This can be seen as follows. Given that $Y_s(\tau) = 0$ for all $\tau \geq 0$, the foreign country attain any $n \leq ½$ or $n^* \geq ½$. The best the foreign country can do is to set the emissions tax equal to zero initially and such that such that $Y_s^* = 0$ thereafter. This strategy yields $n = n^* = ½$, whereas $Y_s^* > 0$ after $n = ½$ has been reached, gives smaller n and therefore larger n^*. The same reasoning applies to the home country.

Suppose $\hat{n} > ½$. The home country can always reach $\hat{n}$ starting from $n = 1$, by setting the emissions prohibitively high ($Y_s = 0$) for an initial period of time. So, we can do as if we start from $n = \hat{n}$. The foreign country still wants to increase the number of varieties it produces. Suppose it puts $t^*(\tau) = 0$ for all $\tau \geq 0$. If transportation costs are sufficiently high, the resulting equilibrium is the symmetric equilibrium because even with a zero emissions tax, n decreases, as explained above in the case of $\hat{n} = 1$. For intermediate values of T the situation is more complex, because with zero emissions taxes two equilibria are locally stable: $n = ½$ and $n = 1$. It could be that from $\hat{n}$ the state of the system converges to $n = ½$, which might be worse for the home country than $n = 1$. If that is the case it would have been better to stay in the state of perfect specialization. Alternatively, starting from $\hat{n}$ the system converges to $n = 1$ if emissions taxes are zero. The home country can then stay in $\hat{n}$ by imposing a positive emissions tax.

We conclude that if clustering is socially optimal it may be realized as a Nash equilibrium; but high marginal damage costs (φ) tend to hamper the stability of this social optimum as well as the likelihood of clustering being the social optimum. The alternative upper bound on emissions taxes complicates the analysis. When the Nash equilibrium strategy for each country is to set taxes equal to zero, the upper bound is irrelevant. But races to the top require a tax rate equal to the producer price. The price and the number of varieties change during the transition to the new equilibrium. When clustering is socially optimal, a problem might therefore arise when $0 < \hat{n} < 1$. But since this occurs for relatively high marginal pollution damage, the home country can still make sure that its number of varieties decreases.

Socially optimal symmetry

Finally, consider the case where the symmetric equilibrium is socially optimal, and assume that the two economies find themselves in this equilibrium initially. If individual welfare is monotonically increasing in the number of varieties, a race to the bottom will result with zero taxation in both countries. The countries will not leave the symmetric equilibrium. If individual welfare is monotonically decreasing in the number of varieties, then clearly governments will race to the top, which in the case of a uniform upper bound on taxes results in the symmetric equilibrium again. If individual welfare is an inverted-U shaped in the number of varieties both scenarios are possible. Suppose welfare of the home country reaches a maximum for $\hat{n} < \frac{1}{2}$. Starting in the symmetric equilibrium, the home country as well as the foreign country wants to reduce the number of varieties produced in their country. Consequently, taxes will be set as high as possible: a race to the top. If $\hat{n} > \frac{1}{2}$, then a race to the bottom is to be expected. In both cases, with uniform lower bounds as well as uniform upper bounds on the emissions taxes, the symmetric equilibrium is stable. The symmetric equilibrium is a Nash equilibrium.

It is interesting to see how optimal policies are affected by changes in the pollution and transportation parameters. Clearly, for higher marginal pollution damage (φ), individual countries find it less attractive to engage in manufacturing production. Therefore, the optimal amount of varieties decreases as marginal damage increases. And this enlarges the parameter set for which races to the top occur. If transportation costs get higher, it becomes more profitable to have manufacturing at home. This will enhance races to the bottom. An illustration is the following numerical example, where $\mu = 0.4$, $\sigma = 1.7$ and $\varphi = 0.9$. If $T = 1.5$ the social optimum is attained for $n = \frac{1}{2}$ and the individual optimum is reached at $n = 0.4$ approximately. For $T = 3.5$ we still have a symmetric social optimum but the individual optimum is at $n = 0.6$ and $n^* = 0.6$.

Given that the initial equilibrium is symmetric, imposing the alternative upper bound on the tax rate, implying a zero wage rate, does not offer new insights.

When the initial state does not coincide with the social optimum, be it clustered or symmetric, many scenarios are possible again. We do not provide an exhaustive analysis or a complete taxonomy here, because too many cases must be considered. Moreover, a full account of all possibilities seems insurmountable analytically. Numerical simulations might provide further insight. This is left for future research. One result, however, is that pollution plays a stabilizing role when the original state of the economy is in the neighborhood of the symmetric equilibrium.

4. Conclusions

Our model emphasizes the importance of labor mobility in the context of ecological dumping. Labor mobility is triggered not only by real wages but by environmental quality as well. It was found in many cases that this enhances the stability of a cooperative equilibrium, which is in line with the earlier results on environmental policy and mobility. But our approach offers a much broader perspective than that literature because we allow for trade in a world with monopolistic competition that may be favorable to clustering by itself. With regard to ecological dumping, we have seen that races to the bottom occur, but this does not imply ecological dumping, defined as emissions taxes that are too low from the point of view of social welfare.

Many aspects of the present paper call for further research. To start with, our production structure is rather rigid. Labor is the only factor of production and therefore substitution possibilities are lacking. Also, we don't allow for abatement opportunities. Moreover, the payoffs of the countries are evaluated in the eventual equilibrium only, whereas competition will actually evolve over time. Therefore, a genuine differential game approach would be more appropriate. Finally, it is important to consider the dynamics of the system when policy options are different across countries.

References

Barrett, S. "Strategic environmental policy and international trade." *Journal of Public Economics.* 54, 1994: 325-338.

Calmette, M.F. and I. Péchoux. "Agglomeration, major risky accidents and environmental policies." Mimeo University of Toulouse, 2003.

Dixit, A. and J. Stiglitz. "Monopolistic competition and optimum product diversity." *American Economic Review.* 67, 1977: 297-308.

Elbers, C. and C. Withagen. "Environmental regulation and international trade: a general equilibrium approach.", in J. List and A. de Zeeuw, eds. *Recent Advances in Environmental Economics*. Cheltenham: Edward Elgar, 2002a: 98-118.

________________________. "Environmental policy and Bertrand competition: a general equilibrium approach." in C. Hommes, R. Ramer and C. Withagen, eds. *Equilibrium, Markets and Dynamics*. Berlin: Springer, 2002b: 183-196.

Fujita, M., P. Krugman, and A. Venables. *The Spatial Economy: Cities, Regions, and International Trade.* Boston: MIT Press, 1999.

Haavio, M. "Transboundary pollution and household mobility: Are they equivalent?" Mimeo University of Helsinki, 2003a.

Haavio, M. "Migration and the environment: Instrument choice matters." Mimeo University of Helsinki, 2003b.

Hoel, M. "Environmental policy with endogenous plant locations." *Scandinavian Journal of Economics*. 99, 1997: 241-259.

Hoel, M. and P. Shapiro. "Population mobility and transboundary environmental problems." *Journal of Public Economics.* 87, 2003: 1013-1024.

Lowell, B. and A. Findlay. "Migration of highly skilled persons from developing countries: Impact and policy responses." Geneva: International Labour Office, 2001.

Markusen, J., E. Morey, and N. Olewiler. "Environmental policy when market structure and plant location are endogenous." *Journal of Environmental Economics and Management.* 24, 1993: 69-86.

Markusen, J., E. Morey, and N. Olewiler. "Noncooperative equilibria in regional environmental policies when plant locations are endogenous." *Journal of Public Economics.* 56, 1995: 55-77.

Motta, M. and J.F. Thisse. "Does environmental dumping lead to delocation?" *European Economic Review*. 38, 1994: 563-576.

Oates, W. and R. Schwab. "Economic competition among jurisdictions: Efficiency enhancing or distortion inducing?" *Journal of Public Economics.* 35, 1988: 333-354.

Pflüger, M. "Ecological dumping under monopolistic competition." *Scandinavian Journal of Economics.* 103, 2001: 689-706.

Quaas, M. and A. Lange. "Economic geography and urban environmental pollution." Mimeo University of Heidelberg, 2004.

Rauscher, M. "On ecological dumping." *Oxford Economic Papers.* 46, 1994: 822-840.

Sandmo, A. and D. Wildasin. "Taxation, migration and pollution." *International Tax and Public Finance.* 6, 1999: 39-61.

Silva, E. "Decentralized and efficient control of transboundary pollution in federal systems." *Journal of Environmental Economics and Management.* 32, 1997: 95-108.

Ulph, A. "International trade and the environment: a survey of recent economic analysis." in H. Folmer and T. Tietenberg, eds. *The International Yearbook of Environmental and Resource Economics 1997/1998.* Cheltenham: Edward Elgar, 1997: 205-242.

Ulph, A. and L. Valentini. "Is environmental dumping greater when plants are footloose?" *Scandinavian Journal of Economics.* 103, 2001: 673-688.

Ulph, A. and L. Valentini. "Modelling commitment in multi-stage models of location, trade and environment." in L. Marsiliani, M. Rauscher and C. Withagen, eds. *Environmental Economics and the International Economy.* Dordrecht: Kluwer, 2002: 55-88.

Wellisch, D. "Can household mobility solve basic environmental problems?" *International Tax and Public Finance.* 2, 1995: 245-260.

Wellisch, D. "Interregional spillovers in the presence of perfect and imperfect household mobility." *Journal of Public Economics.* 55, 1994: 167-184.

Wilson, J. "Capital mobility and environmental standards: Is there a theoretical basis for a race to the bottom?" in J. Bhagwati and R. Hudec, eds. *Fair Trade and Harmonization, Prerequisites for Free Trade?, Vol. 1 Economic Analysis.* Cambridge MA: MIT Press, 1996: 393-427.

[3]

Managed Trade, Trade Liberalisation and Local Pollution*

Pierre M. Regibeau and Alberto Gallegos

Abstract

Abstract The current paper addresses the relationship between trade and endogenous pollution levels, with a focus different from the previous literature. The mechanism linking pollution and trade here is that trade policy provides the home government with a credible threat that helps motivate domestic firms to adopt cleaner technologies. This credible threat comes from the fact that the government has a greater incentive to protect a clean industry than to protect a very polluting one. In that sense, the existence of trade helps reduce domestic pollution compared to what would prevail in a situation of autarky. On the other hand, a commitment to free trade would be counterproductive: it removes the government's ability to credibly threaten lower levels of protection. In fact we show that any trade liberalization hurts the welfare of the home country. In terms of world welfare, moderate trade liberalization is helpful, but only as long as it does not affect the technology choices of the firms. Because committing to lower 'bounded' tariffs limits a government's ability to enforce strict environmental standards, a country that has agreed to tighter tariff limits under the World Trade Organization would, other things equal, be a more likely "pollution haven" than a country with weaker WTO commitments.

KEYWORDS: pollution, trade, environment

*Pierre Regibeau is at the Department of Economics, University of Essex, Wivenhoe Park, Colchester, CO4 3SQ, UK, and Alberto Gallegos is at the Instituto Tecnologico de Monterrey, Campus Ciudad de Mexico, Calle del Puente 222, Col. Ejidos de Huipulco, Tlalpin, Mexico D.F., CP 14380, Mexico. We would like to thank the referees and the Editor for their many helpful comments. We also benefited from the comments of C. Montagna and K. E. Rockett.

Over the last two decades economists and environmentalists have tried to disentangle the complex linkages between environmental and trade policies to assess the likely effects of freer trade on the environment. The environmentalists' great fear is that an expansion of international trade might lead to a systematic increase in pollution through an increase in economic activity, the location choices of multinational corporations and/or the weakening of national environmental policies. These concerns are the subject of a large economics literature.[1]

Our paper also addresses the relationship between trade and endogenous pollution levels, but we take a different perspective. Rather than ask how more trade is likely to affect the environment, we ask what the effects of a *commitment* to free trade would be. We do this in a partial equilibrium model with local pollution. As several of the more polluting industries (e.g. mining, chemicals) are best characterized as oligopolies, models of trade that allow for market power appear especially well-suited to this kind of analysis.

Many papers examine the strategic interaction between oligopolistic firms and governments in an international context. Spencer and Brander (1983), Brander and Spencer(1983) and Eaton and Grossman (1986) show how governments can improve national welfare by committing to trade policies that modify the strategic interaction between firms. In such models, where policies are set before firms make their own choices, governments essentially 'lend' their commitment power to their home firms, helping them to ensure a more advantageous position. This type of model has also been used to discuss how environmental policies can affect the competitiveness of home firms (see Conrad (1993), Barrett(1994), Kennedy (1994) and Ulph (1996)). We depart from this traditional approach by assuming that governments have limited commitment power. More precisely, as is traditional, we consider a simple game where governments can choose their trade policies before firms choose their output levels, but we also allow the firms to choose the type of production technology that they will use before trade policies are set. The main rationale for this is that technology adoption can be argued to be less easily reversible than trade policy choices. In that sense, our timing is similar to the model of Brander and Spencer (1987), where multinational firms choose whether to serve a market through exports or foreign direct investment before the host government can commit to levels of trade protection or local taxation.

[1]Papers on the topic differ mostly according to whether pollution is strictly local or spills across borders, whether firms are allowed to choose an abatement technology, whether the location of production is endogenous, and according to the set of policy tools considered. See Copeland and Taylor (1994 and 2001) and Grossman and Krueger (1991) for seminal contributions in a general equilibrium framework. For partial equilibrium analysis, see Markusen et al. (1995), who examine the location decision of (polluting) firms as well as Conrad (1993), Barret (1994) and Kennedy (1994), where environmental policies are used to affect the international competitiveness of local firms. See Karp et al. (2001) for a recent review of the literature.

In our model, the mechanism linking pollution and trade is that trade provides the home government with a credible threat that helps motivate domestic firms to adopt cleaner technologies.[2] This credible threat comes from the fact that the government has a greater incentive to protect a clean industry than to protect a polluting one. Unless the government controls other perfectly efficient instruments to affect the firms' technological choices, this threat is useful. In that sense, the existence of trade helps reduce domestic pollution compared to what would prevail in a situation of autarky. On the other hand, a commitment to free trade would be counter-productive: it removes the government's credibility to threaten lower levels of protection and, therefore, decreases the domestic firms' incentives to invest in cleaner production methods.

Extending our analysis to allow for emission taxes, we find that a commitment to free trade can again limit a country's ability to enforce tough pollution standards. Under free trade, imposing high emission taxes on the domestic firms puts them at a competitive disadvantage with respect to their foreign rivals. Under managed trade, this competitive handicap can be alleviated by taxing imports of foreign goods.

We therefore conclude that committing to lower tariff bounds limits a country's ability to enforce tough standards for local pollution. In that sense, we would expect countries with stricter WTO commitments to be more likely *pollution havens* than countries that retain considerable discretion in setting up their trade barriers.

This basic mechanism described above is analysed in the simplest possible two-country partial equilibrium game. It is presented in section 1, where we show how *ex post* trade policy provides incentives to invest in pollution-reducing technologies. The effect of trade liberalization on home and world welfare is analysed in sections 2 and 3. Section 4 discusses the robustness of our results and a few extensions. Section 5 concludes.

1. The Model

We consider two countries, 'Home' and 'Foreign'. Each country has one firm that produces in its country of origin. The home firm is indexed as firm 1, while firm 2 is the foreign firm. These firms sell a homogenous good that can be produced at a constant marginal (and average) cost of c, which we will set equal to zero. Firms compete in quantities. Two production technologies are available, one clean, one dirty. The dirty technology generates local pollution, while the clean technology does not. The welfare cost of this pollution is proportional to local output, i.e. the social cost pollution at Home is:

θq_1 if the home firm uses the dirty technology
0 if the home firm adopts the clean technology,

[2] The basic idea is therefore related to Matsuyama (1990), where the government uses the threat of trade liberalization to induce efficiency-enhancing investments by domestic firms. However, Matsuyama's set-up is more complex, as the timing is endogenous for both the government and firms' decisions.

where q_1 is the output of the home firm and $\theta \geq 0$. We assume that the home firm is initially endowed with the dirty technology but can adopt the clean technology at a fixed cost of F.

In order to concentrate on the new effect presented in this paper, we assume that the choice of technology does not affect the marginal cost of production. If it did, then policies aimed at reducing production would also affect the reaction function of the domestic firm. Such well-known 'strategic trade' effects would make computations messier without adding anything new to the analysis.

We will focus on the home market where the inverse demand for the homogenous good takes the simplest possible linear form:[3]

$$P = 1 - Q,$$

where $Q = q_1 + q_2$ is total industry output and P is the price in the domestic market. Trade policy takes the form of a unit tariff t. We assume that the home country is the only one with an active trade policy.[4]

The benchmark is a situation where the home country is free to set its tariff at any level that it pleases. In such a 'freely managed trade' regime, the timing of the game is as follows. In the first stage the domestic firm must decide whether to keep its current polluting technology or to adopt the new, cleaner technology. Once the adoption decision is made, the home government chooses the level of its import tariff. In a third stage, firms simultaneously choose their levels of output. As discussed in the introduction, the order of the three stages reflects the assumption that the government has only limited commitment power: while it might be able to commit to a specific level of tariff before firms actually compete in the product market, the choice of tariff is more easily reversible than the firm's choice of technology.[5] This timing also rules out 'contingent' trade policies, whereby the government commits to tariff levels that explicitly depend on the adoption decisions of the domestic firms. The game is solved for its unique sub-game perfect equilibrium.

Starting from the managed trade benchmark, we then consider the effects of 'trade liberalization', which is modelled as a legally binding upper limit on the level of tariff that the government can choose in stage 2. We also consider the extreme case of a commitment to free trade, where the home government completely forgo the use of trade policy. This means that $t = 0$ even if the optimal tariff t^* were to be negative

[3] We can either assume that the firms are only active in the Home market or that they are active in both markets but hold 'segmented markets' perceptions, i.e. each firm perceives each country as a separate market and makes distinct quantity decisions for each. Together with our assumption of constant unit costs, the segmented perception assumption would allow us to treat the two markets independently.

[4] See Gallegos (2003) for a discussion of the case with two active governments. As this extension does not add any significant new insight, it is omitted here.

[5] A similar timing can be found in Brander and Spencer (1987)'s analysis of foreign direct investment.

(which would occur for $\theta > 1$). In this case, the second stage of the game becomes irrelevant.[6]

For both versions of the model, we solve the game by starting with the final stage (given t of stage 2). The equilibrium quantities of the two firms in the third stage of the game are easily obtained as:

$$q_1^{\bullet} = \frac{1+t}{3}$$

$$q_2^{\bullet} = \frac{1-2t}{3},$$

where t is the per unit import tariff chosen by the home government in stage 2. To ensure an interior solution, we need $-1 < t < ½$. The corresponding tariff revenue is:

$$T = tq_2^{\bullet} = \frac{t(1-2t)}{3}$$

The equilibrium profit of the domestic firm, π_1, domestic consumer surplus CS, and domestic welfare W are:

$$\pi_1^{\bullet} = \frac{(1+t)^2}{9},$$

$$CS = \frac{(Q^{\bullet})^2}{2} = \frac{(2-t)^2}{18} \text{ and}$$

$$W = CS + T + \pi_1 - (\theta q_1 / 6)[2(1+t)(1-\theta) - 3t^2]$$

We can now consider the second stage of the game. Under freely managed trade, the optimal tariff is obtained by maximizing W with respect to t. This yields:

$$t^*(\theta) = \frac{1-\theta}{3}$$

In the absence of pollution, this tariff is positive.[7] Crucially for our argument, the chosen tariff is inversely related to the pollution parameter θ. For the domestic

[6] A commitment to free trade is therefore a special case of trade liberalization, where the upper bound on t is smaller than or equal to $Min[0,(1-\theta)/3]$.

[7] It also must be smaller than ½ for all possible values of θ, so that the output of the foreign firm will always be positive in equilibrium.

firm, this means that adopting a cleaner technology is "rewarded" by stronger *ex post* protection. As the domestic firm anticipates this reward, the home country's ability to choose its trade policy *ex post* increases the domestic firm's incentive to invest in the new technology in stage 1. This is confirmed by substituting the expression for the optimal tariff into the equilibrium profits of the home firm:

$$\pi_1^* = \frac{(4-\theta)^2}{81}$$

The domestic firm's incentives to invest in the non-polluting technology are therefore given by:

$$B_1 \equiv \pi_1^*(0) - \pi_1^*(\theta) = \frac{\theta(8-\theta)}{81},$$

where $\pi(0)$ is the firm's profit when it has chosen the clean technology. It is useful to replace t by its optimal value in a number of other variables. We get:

$$q_1^* = \frac{4-\theta}{9}$$

$$q_2^* = \frac{1+2\theta}{9}.$$

To ensure interior solutions, we assume that $\theta < 4$. Notice that, under this assumption, B_1 is always positive. Hence, the *ex post* tariff policy that emerges under 'managed trade' provides the domestic firm with greater incentives to invest in the cleaner technology than if the home government had committed to free trade. Under free trade, the domestic firm has no incentive to adopt the less polluting technology: adoption is costly and offers no benefit since the *ex post* tariff is set equal to zero anyway.

The range of investment costs for which the technology would be adopted under managed trade corresponds to all positive values lower than B_1, as shown on Figure 1.

Let us now consider the effect of less extreme trade liberalization. Define the maximum tariff allowed as $\bar{t}$. Clearly, liberalization has no effect if it is not binding, i.e. if $\bar{t} \geq t^*(\theta = 0) = 1/3$. On the other hand, any trade liberalization that lowers the maximum allowed tariff below 1/3 decreases the domestic firm's incentive to adopt the clean technology. For $\bar{t} \in [\max(0,(1-\theta)/3), 1/3]$, liberalization reduces the protection granted after adoption, while it leaves the protection level without adoption unchanged. If pollution is not too intense (i.e. $\theta < 1$), then further liberalization in the range $\bar{t} \in [0,(1-\theta)/3]$ eliminates all incentives to switch to the cleaner technology. We can therefore state the following proposition:

Proposition 1: *In this model, any binding trade liberalization lowers the domestic firm's incentives to invest in cleaner technology. In particular, incentives to invest in cleaner technology are higher under managed trade than under free trade.*

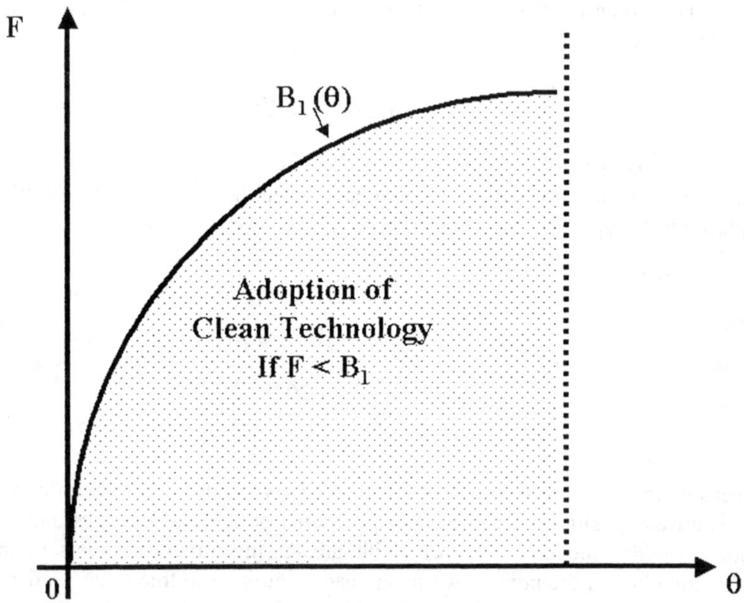

Figure 1: Private Incentives to Adopt the Clean Technology

2. Home Country's Welfare

The fact that private incentives to choose cleaner technologies are enhanced by managed trade does not of course guarantee that these technological choices are optimal from the point of view of the home country. To compare private and social incentives to adopt the cleaner technology, we must compare B_l to the change in the home country's welfare that is triggered by a move to the new technology. This change is given by:

$$\Delta W = \Delta CS^{*} + \Delta T^{*} + \Delta \pi_1^{*} - F + \theta q_1^{*}(\theta),$$

where Δx refers to the difference between the value of x when the clean technology is adopted – and the tariff set accordingly – and the value of x when the firm chooses to use the dirty technology. The first term represents the effect on consumer surplus. This is unambiguously negative as the adoption of a cleaner technology leads to greater levels of protection and, therefore, a higher domestic price. The last term corresponds to the decrease in pollution levels and is therefore positive. The third term represents the firm's private incentives to adopt the cleaner technology (i.e. it is equal to B_1). As we have just seen, this is positive. Anyway, as $\Delta\pi - F$ is fully taken into account by the domestic firm, these terms do not affect a comparison of private and social incentives. Finally, the change in tariff revenues can *a priori* be either positive or negative.

Our interest lies in the difference between ΔW and B_1 :

$$\Delta W - (B_1 - F) = \Delta CS^{*} + \Delta T^{*} + \theta q_1^{*}(\theta) = \frac{7\theta}{162}[8 - \theta],$$

which is positive for all relevant values of the pollution parameter (i.e. $\theta < 4$), and negative for high values of θ. This means that F has values for which adoption of the clean technology is desirable but does not happen, but no values for which socially undesirable adoption is observed. This is shown in Figure 2 and summarised in proposition 2.

Proposition 2: *Freely managed trade provides incentives for the adoption of the clean technology, but these incentives are insufficient from the point of view of the home country overall welfare.*

This result immediately suggests that managed trade always yields higher domestic welfare than free trade or, indeed, than any binding level of trade liberalization (i.e. $\bar{t} < 1/3$). If trade liberalization does not change the adoption behaviour of the domestic firm, then it cannot increase the home country's welfare. If the domestic firm invests in the clean technology anyway, any value of $\bar{t}$ below 1/3 simply prevents the domestic government from setting its *ex post* optimal tariff. If the home firm sticks with the dirty technology anyway then any $\bar{t}$ above $(1-\theta)/3$ does not affect welfare, since the *ex post* optimal tariff can still be chosen. On the other hand, further liberalization bringing $\bar{t}$ below $(1-\theta)/3$, would force the country to set a sub-optimal tariff *ex post*, lowering its welfare.

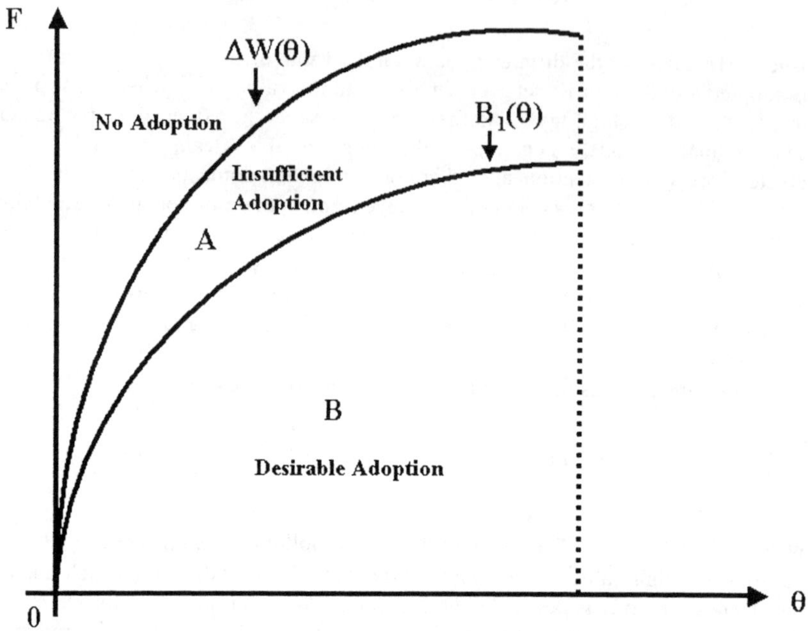

Figure 2: Private Incentives to Adopt and Home Country Welfare

We must now examine whether trade liberalization could help by changing the technological choice of the domestic firm. From proposition 1 we know that trade liberalization cannot *increase* the firm's incentives to adopt the clean technology. Hence we only need to analyse the case where the firm would adopt under freely managed trade but would not after trade has been liberalized. Define the level of tariff below which the firm ceases to invest in the clean technology as $\bar{t}_c$. Let us then assume that $\bar{t}_c < t^*(\theta)$ so that freely managed trade would ensure adoption of the clean technology and consider a trade liberalization that would reverse this technological choice, i.e. $\bar{t} < \bar{t}_c$. Since the firm already has insufficient incentives to invest under freely managed trade, such liberalization can only be welfare decreasing: the country reacts to any tariff bound just below the critical level $\bar{t}_c$ by choosing its optimal 'no adoption' *ex post* tariff $t = (1-\theta)/3$. As $\Delta W - B_1 > 0$, we know that this yields a level of welfare that is lower than when the country can induce adoption and set its *ex post* tariff equal to 1/3. Hence we can state:

Proposition 3: *In this model, trade liberalization can never increase domestic welfare.*

The evolution of domestic welfare as the maximum allowed tariff is lowered is summarized in Figure 3.

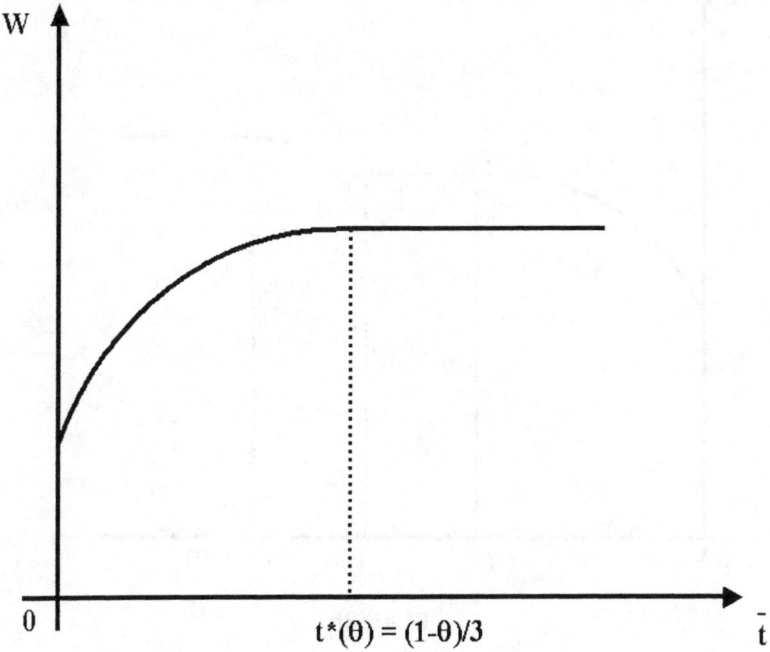

Figure 3a

Figure 3a corresponds to an initial situation of insufficient adoption (region A in Figure 2). As explained above, trade liberalization has no effect on the home country's welfare until the maximum tariff reaches $(1-\theta)/3$. Further liberalization beyond this point decreases welfare. Figure 3b is drawn for an initial point in region B of Figure 2. As soon as liberalization becomes binding, it decreases home welfare. This decrease continues up to the point where the firm no longer adopts the clean technology. At this point, welfare can in principle jump up or down. This depends on whether welfare without adoption and an optimal *ex post* tariff of $(1-\theta)/3$ is higher or lower than welfare with adoption an a constrained tariff equal to $\bar{t}_c$. If welfare jumps up, we still know that it remains lower than without trade liberalization: in region B adoption with an unconstrained tariff is better than no adoption with an unconstrained tariff. Further liberalization does not affect welfare until $\bar{t}$ reaches $(1-\theta)/3$, at which point further decreases in $\bar{t}$ decrease welfare.

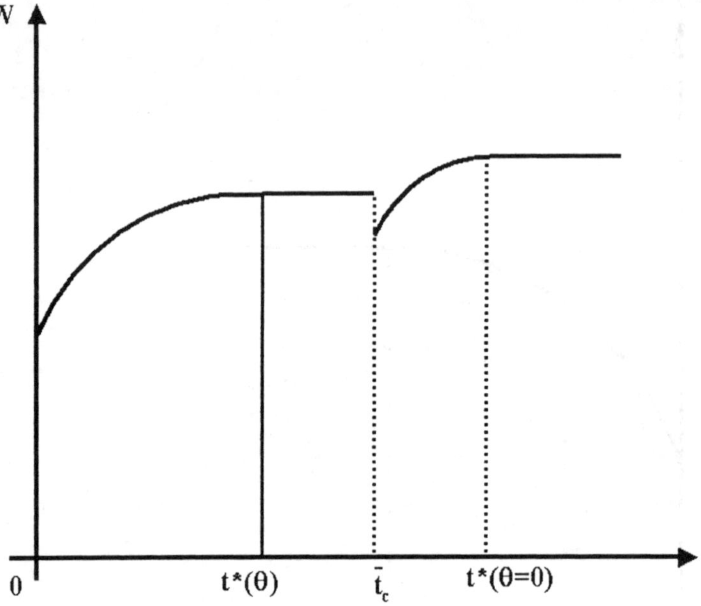

Figure 3b

While we have focused on the issue of trade liberalization, it should be clear that, in our model, autarky is never optimal either. To see this, notice that autarky, just like trade with $\bar{t} \leq \bar{t}_c$, never induces adoption of the clean technology. With no adoption, the home country does best by setting its *ex post* optimal tariff of $(1-\theta)/3$. As this tariff is not prohibitive, it must, by definition, yield higher home welfare than does autarky. Hence, some amount of trade liberalization is always preferable to autarky. A similar argument holds when adoption would occur under freely managed trade and would be socially optimal (region B in Figure 2): if managed trade is better than the best possible regime without adoption, and adoption is desirable, then managed trade must be better than autarky.

Proposition 4: *A degree of trade liberalization always exists for which the home country's welfare is greater than under autarky. Autarky is also dominated by freely managed trade.*

3. World Welfare

World welfare differs from the home country's welfare in two respects: it includes the profits of the foreign firm as well as the pollution associated with foreign production. Because of this last term, we cannot evaluate world welfare without making some assumption about the technology used by the foreign firm. In the first part of this section, we simply assume that the foreign firm is using the clean technology, so that foreign pollution is not an issue. In the second part, we look at the opposite case where the foreign firm uses the dirty technology, so that foreign pollution further complicates the analysis.

3.1 'Clean' Foreign Firm

Adoption of the clean technology by the home firm leads to higher tariff levels *ex post*, thereby reducing the equilibrium profits of the foreign firm. Since these profits are the difference between world welfare and home welfare, adoption must be less desirable from the world point of view than from the viewpoint of the home country alone. Therefore we cannot *a priori* rule out values of the parameters for which the domestic firm's private decision to adopt the clean technology might actually reduce world welfare, defined as W^w. We are interested in the difference between the effect of the adoption of the clean technology by domestic firm on world welfare, ΔW^w, and its effect on the domestic firm' s profit, B_1 :

$$\Delta W^w - B_1 = \Delta W - B_1 - \Delta\pi_2 = \frac{3\theta(14-\theta)}{162},$$

where $\Delta\pi_2 \equiv \pi_2(t^*(\theta = 0)) - \pi_2(t^*(\theta))$. Since this expression is positive for all $\theta \leq 4$, the firm's incentives to adopt cannot ever be excessive from the point of view of world welfare.

3.2 'Dirty' Foreign Firm

If the foreign firm relies on a dirty technology, then adoption of the clean technology by the home firm has *an additional positive effect* on world welfare. Adoption results in higher tariffs. These higher tariffs reduce the output of the foreign firm and the corresponding foreign pollution. Since incentives to adopt were already insufficient in the presence of a 'clean' foreign firm, they still are when the foreign firm itself contributes to overall pollution.

Proposition 5: *Freely managed trade provides incentives for the adoption of the clean technology, but these incentives are insufficient from the point of view of the world welfare.*

The qualitative effect of trade liberalization on world welfare does not depend on whether the foreign firm pollutes. The general pattern is shown in Figure 4. We focus on the non-trivial case where the domestic firm would adopt the clean technology under freely managed trade. That is, we assume that the adoption cost F is small enough that $\bar{t}_c \leq t^*(\theta = 0)$.

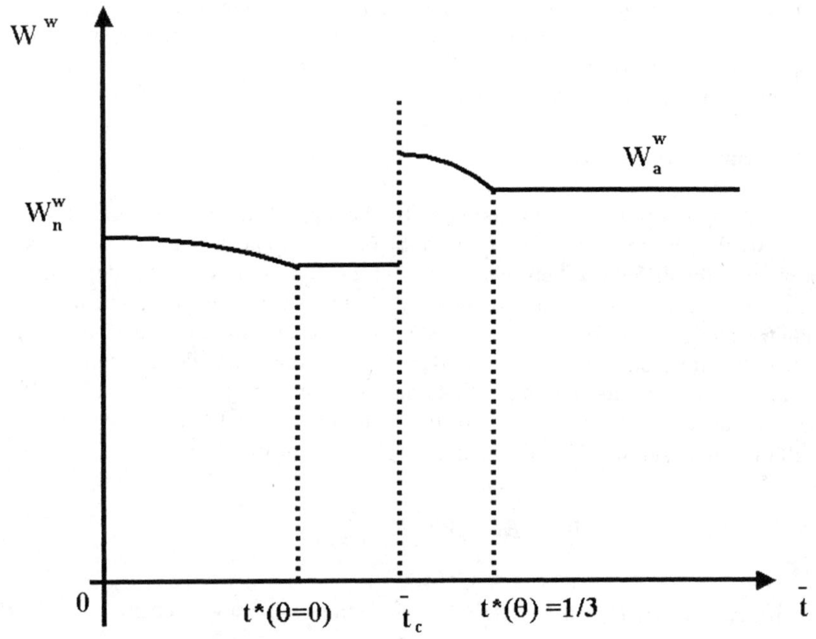

Figure 4: Trade Liberalization and World Welfare

For $\bar{t} \geq t^*(\theta = 0) = 1/3$, the home country would choose its *ex post* optimal tariff $t^*(\theta = 0)$ if the firm were to adopt. Since this tariff is larger than $\bar{t}_c$, it is enough to trigger adoption. As the *ex post* tariff is independent of $\bar{t}$ over this range, world welfare is also independent of $\bar{t}$. For all $\bar{t} \in [\bar{t}_c, 1/3[$, the home country will set the maximum *ex post* tariff allowed, $\bar{t}$, if the home firm adopts the clean technology.

Since $\bar{t} \geq \bar{t}_c$, this ensures adoption. As, for a given choice of technology, world welfare is a decreasing function of the home tariff, then W^w increases as $\bar{t}$ decreases. To the left of the critical tariff $\bar{t}_c$, the domestic firm sticks to its dirty technology, as the government cannot 'reward' adoption with a large enough tariff. For all values of $\bar{t}$ between $(1-\theta)/3$ and $\bar{t}_c$, the home country then chooses the *ex post* optimal tariff $t = (1-\theta)/3$ and world welfare is independent of $\bar{t}$ over this range. With further liberalization, the home country's *ex post* tariff is now constrained by $\bar{t}$. Hence any decrease in this upper bound translates into a decrease in the equilibrium tariff, increasing world welfare.

We must still discuss the relative 'height' of the segments of the welfare function to the right and to the left of $\bar{t}_c$ in Figure 4. From proposition 5, we know that whenever adoption occurs and the home country's optimal *ex post* tariff is chosen (i.e. $\bar{t} \geq t^*(\theta = 0)$), world welfare is higher than when the clean technology is not adopted and the home country chooses its *ex post* optimal tariff (i.e. $\bar{t} \in [t^*(\theta), \bar{t}_c[$). This means that the horizontal segment to the right of $\bar{t}_c$ must lie higher than the horizontal segment to the left. On the other hand, the relative position of the rightmost horizontal segment and the level of welfare under free trade is not clear *a priori*. To rank these, we take the technology of the foreign firm as given. Hence, the only relevant fixed costs of adoption are those relating to the technological choice of the home firm.[8]

Suppose first that the foreign firm does not pollute. World welfare under free trade (and, therefore no domestic adoption) is:

$$W_f^w(\theta_2 = 0) = \frac{2-3\theta}{9}$$

Under freely managed trade (and adoption), we have:

$$W_m^w(\theta_2 = 0) = \frac{20}{9} - F,$$

[8] Notice that the technological choice of the foreign firm would not in any case depend on the trade policy of the home country. Home welfare is independent of foreign pollution both directly (pollution is local) and indirectly (because the technology does not affect marginal costs) so that home's *ex post* tariff is not affected by the technological choice of the foreign firm. Hence the profitability of technological adoption by the foreign firm is independent of home trade policy.

which is higher than $W_f^w(\theta_2 = 0)$ for all $\theta \geq 0$.[9] Similarly, if the foreign firm pollutes, we have:

$$W_f^w(\theta_2 = \theta) = \frac{2(1-3\theta)}{9}$$

and

$$W_a^w(\theta_2 = \theta) = \frac{20-\theta}{9} - F$$

Again, welfare is higher under freely managed trade. Having shown that Figure 4 can actually be drawn as presented, we see that welfare is maximised at $\bar{t} = \bar{t}_c$. Hence we can state:

Proposition 6: *World welfare is higher under freely managed trade than under free trade. Still a moderate trade liberalization that constrains the home country tariff without preventing the adoption of the clean technology by the home firm improves world welfare.*

Our analysis of 'world welfare' relies on a very partial equilibrium exercise: there is only one active market and one active government. Allowing for two active governments and two active markets does not change matters very much. Under our assumptions of constant cost, segmented market perception and local pollution, the *ex post* optimal tariffs of the two active governments are independent of each other and are exactly the same as the optimal tariff in our 'one market/one government framework'. Furthermore, the adoption decision of a given firm is itself independent of the policy of the foreign government and of the adoption choice of its foreign rival. The only significant difference concerns the impact of adoption on world welfare. The presence of two markets increases output levels, making pollution reduction more socially desirable for any level of fixed investment cost *F*. As private incentives to invest do not change, this means that they are even more insufficient in the full two-country framework. A more complete discussion of the case with two active governments and two markets can be found in the Appendix.

4. Discussion and Robustness

Our analysis suggests that governments might actually gain from retaining some discretion over their trade policies. This contrasts with a significant body of work that

[9] Comparing the two expressions, we see that welfare under freely managed trade dominates welfare under free trade iff $F < ((18+3\theta)/9$. We then use the fact that the home firm adopts under freely managed trade so that $F < B_l = \theta(8-\theta)/81$. As $(18+3\theta)/9 > \theta(8-\theta)/81$ for all $\theta \geq 0$, $F < B_l$ ensures that $W^w_a > W^w_f$.

shows that international trade agreements can be useful precisely because they remove discretion from individual governments.[10] This striking difference arises because of a restriction that we impose on the class of agreements that can be reached. In our framework, trade liberalization simply imposes an upper bound on tariffs. We do not allow these upper bounds to be conditional on the technology choices of home firms. If we did, then *ex ante* agreements to liberalize trade would be at least as effective as freely managed trade. While trade agreements could in principle accommodate such contingent tariff bounds, they do not seem to do so in practice, suggesting that such 'contracts' might be prohibitively costly to enforce.

In the rest of the section, we briefly discuss the robustness of our results to four kinds of extensions: more firms, more general demands, other models of oligopoly, and the availability of other policy instruments.

4.1 More Firms, More General Demands, and Other Models of Oligopoly

The conclusion that the domestic firms' incentives to adopt the clean technology are larger under freely managed trade than under free trade (or partial liberalization) holds under quite general conditions. Define θ_i as firm *i*'s pollution parameter. Firm i's incentives to clean up its technology is determined by:

$$d\pi_i / d\theta_i = (d\pi_i / dt^*)(dt^* / d\theta_i)$$

reflecting the fact that firm *i*'s technological choice only affects its profits through its effect on the *ex post* optimal tariff. As we can see, our main result depends on two effects.

Firstly we need an increase in the cleanliness of firm *i*'s technology to result in a higher *ex post* tariff (i.e. $dt^*/d\theta_i < 0$). Let us write the country's welfare as

$$W = CS(t) + tn^f q^*(t) + n\pi(t) - \theta_i q_i(t) - \sum_{j \neq i}^{n} \theta_j q_j(t),$$

where n^f is the number of foreign firms, *n* is the number of domestic firms and θ_i is the degree of dirtiness of firm *i*'s technology. The value of t^* is therefore determined by

$$\frac{dW}{dt} = CS'(t) + n^f q^*(t) + tn^f q^{*\prime}(t) + n\pi'(t) - \theta_i q_i{}'(t) - \sum_{j \neq i}^{n} \theta_j q_j{}'(t),$$

where a prime refers to a first order derivative. Hence,

[10] See, for example, Staiger and Tabellini (1989), where trade agreements help the policy-maker solve his own time inconsistency problem, and Maggi and Rodriguez-Clare (1998), where trade agreements 'save' the policy-maker from domestic lobbies.

$$dt^* / d\theta_i = -q_i'(t)/(d^2W / dt^2)$$

which is negative in all models of oligopoly where a firm's output expands when some of its rivals are hit by a per unit tax.[11]

The second effect that we need in order to reach our conclusion is simply that the domestic firm benefits from a higher tariff (i.e. $d\pi_i/dt^* > 0$). As a tariff just shifts the residual demand of the domestic industry out, this effect arises in every common model of oligopolistic competition.

The precise welfare results that we obtain are much less robust. They might well change if different functional forms were used. Crucially we cannot be certain that the domestic firm's incentives to adopt the new technology will always fall short of the associated benefits for its home country or for the world as a whole. If they did not, then a range of parameters exists for which adoption would occur even though it would not be welfare maximising. In such a case one can show that trade liberalization that is drastic enough to discourage adoption would in fact increase welfare. Moreover such a 'drastic' level of trade liberalization would always exist. This argument is sketched briefly below.

Suppose that, with different functional forms, we obtained a range of parameters for which the domestic firm would adopt the clean technology even though home welfare would be higher if it stuck with the dirty technology. The effect of trade liberalization on the welfare of the home country could then be represented in Figure 5. The graph is similar to Figure 3b, so is the reasoning involved.

Starting from unrestricted managed trade, a progressive tightening of $\bar{t}$, leaves home welfare unchanged until it reaches the *ex post* optimal tariff given that the cleaner technology has been adopted, defined as t_n^* (i.e. 1/3 in our linear demand example). For $\bar{t}$ between $\bar{t}_c$ and t_n^*, adoption still takes place but the home country cannot set its preferred tariff *ex post*. Hence home welfare decreases. As in Figure 4b. trade liberalization that moves $\bar{t}$ below $\bar{t}_c$ prevents the adoption of the new technology. Since technology adoption was undesirable *even if the tariff could be set at* t_n^*, avoiding adoption when the tariff is capped below t_n^* must increase domestic welfare. Moreover, domestic welfare must jump to a level that is higher than when $\bar{t} \geq t_n^*$. Domestic welfare remains at this level as long as $\bar{t}$ remains higher than the *ex post* optimal tariff in the absence of adoption, defined as t_o^* (equal to

[11] This will be true in most of the oligopoly models used in trade theory. In any model of quantity competition with downward sloping reaction functions, the tariff shifts the reaction functions of foreign firms in, resulting in larger equilibrium output for all (symmetric) domestic firms. In a model of price competition with differentiated products and upward sloping reaction functions, the tariff shifts the reaction functions of the foreign firms out. This results in higher equilibrium prices for all firms but the prices of domestic firm increases less than the prices of foreign firm. Using stability conditions, one can show that the output of each (symmetric) domestic firm actually increases (see Vives (1999)).

$(1-\theta)/3$ in our linear example). Tightening the tariff cap beyond this level is again welfare decreasing. Overall then, the home country' s welfare is maximised by committing to a tariff cap that lies between t_o* and $\bar{t}_c$. We can then conclude that *when freely managed trade results in undesirable adoption of the new technology, moderate trade liberalization – sufficient to prevent adoption but leaving the country free to set its* ex post *optimal tariff – unambiguously increases the home country's welfare.*

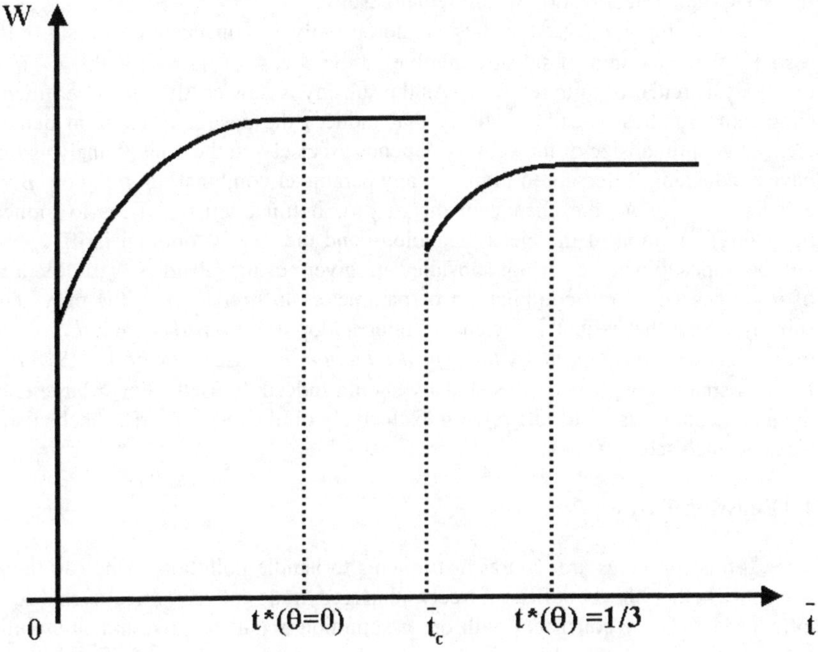

Figure 5

Essentially, trade liberalization helps the home country credibly commit not to 'reward' the home firm for the adoption of the new technology. Incentives to adopt disappear at a tariff rate that is above the *ex post* optimal tariff rate without adoption, so a suitable tariff cap can eliminate undesirable adoption without affecting the level of domestic welfare in the non-adoption outcome.

4.2 Subsidizing Adoption of the Clean Technology

As presented, the analysis is obviously vulnerable to the introduction of other policy instruments. In fact, a government who could commit to paying a lump sum subsidy S to domestic firms adopting the clean technology and who could finance this costlessly would always be able to induce adoption of the clean technology whenever this is socially desirable. On the other hand, unless lump sum *taxes* on pollution-reducing investments were also contemplated (which seems politically difficult), the government would not be able to avoid excessive adoption. In that case, as we have just seen, trade liberalization would remain useful.

More importantly, subsidies cannot usually be financed costlessly. If one assumes that each unit of subsidy involves a social cost of $(1+\alpha)$, with $\alpha \geq 0$, then our analysis remains quite relevant. As the subsidy is now costly, the government will choose the smallest possible subsidy that induces the desirable adoption behaviour. Hence the optimal size of the subsidy depends precisely on the type of analysis that we have conducted. Referring to Figure 2, any parameter combination in region B would call for $S = 0$. As the change in the *ex post* optimal tariff suffices to induce the (desirable) adoption of the clean technology and the *ex post* optimal tariff $t_n{}^* = 1/3$ will be imposed whether or not subsidies are given, using subsidies would be a waste of resources. For any combination of parameters in region A , the only sensible subsidy is one that is just sufficient to induce adoption *given the reward that will be granted anyway through the optimal post-adoption increase in tariff,* i.e. $S = F - B_1$ If α is small enough, then this subsidy should indeed be used. For α large enough, the home country is better off relying exclusively on the *ex post* tariff mechanism that we have analysed.

4.3 Emission Taxes

Emission taxes are natural instruments to handle pollution. One can therefore wonder whether the benefits of freely managed trade survive once such taxes are introduced.[12] To be consistent with our assumption about the government's ability to commit, we assume that the emission taxes are set *after* the domestic firm has chosen its technology. The tax is simply proportional to the firm's emissions, i.e. tax revenues E are given by:

$$E = \lambda \theta q_1,$$

where λ is the tax rate.

[12] Even if emission taxes eliminated the benefits of freely managed trade, one could still argue that implementing such taxes might be significantly more costly than relying on *ex post* trade policy. In particular, monitoring costs might be significant. If so, then a commitment to free trade might still be undesirable as it sacrifice the more efficient of the two instruments.

To assess the effect of freely managed trade on the domestic firm's incentives to adopt the clean technology, we must obtain the value of $B_l = \pi_l(\theta=0) - \pi_l(\theta)$ when both instruments can be used an compare it to the value of B_l when the government has committed to free trade. The profit functions of the two firms are given by:

$$\pi_1 = (1 - q_1 - q_2)q_1 - \lambda\theta q_1$$

$$\pi_2 = (1 - q_1 - q_2)q_2 - tq_2$$

In the third stage of the game, the firms choose their respective output simultaneously, yielding the following equilibrium quantities;

$$q_1 = \frac{1 + t - 2\lambda\theta}{3}$$

$$q_2 = \frac{1 + \lambda\theta - 2t}{3}$$

As emission taxes are a transfer between domestic firms and the home government, domestic welfare can be written as:

$$W = CS + T + Pq_1 - \theta q_1$$

Using the equilibrium values of the firms' outputs and maximizing W with respect to both t and λ, we obtain the following first order conditions:

$$\frac{\partial W}{\partial \lambda} = 2\theta + t - \lambda\theta = 0$$

$$\frac{\partial W}{\partial t} = (1 - \theta) - 3t + \lambda\theta$$

One can see immediately that the two policy instruments are complementary: the higher t, the higher λ. Solving the first order conditions yields the following optimal values for the policy instruments and the corresponding level of domestic welfare:

$$t^* = \frac{1 + \theta}{2}$$

$$\lambda^* = \frac{5}{2} + \frac{1}{2\theta}$$

The introduction of emission taxes dramatically affects the optimal *ex post* tariff. In particular, the tariff is now increasing in θ so that a clean domestic firm receives less protection than a dirty one. However, this does not mean that freely managed trade necessarily *reduces* the firm's incentive to adopt the abatement technology. To assess this we must compare the firm's incentives to invest under free trade and freely managed trade. Using the expressions for the optimal policies and for the equilibrium quantities, we can derive the home firm's incentive to invest under freely managed trade:

$$B_{1a} = \frac{\theta(2-9\theta)}{4}$$

We now assume that the home country is committed to free trade so that $t = 0$. Maximizing W with respect to λ only, we get $\lambda^* = 2$. Substituting this into the equilibrium values of q_1 and q_2 and plugging these back into firm 1's profit function, we get the home firm's incentive to adopt:

$$B_{1f} = \frac{8\theta(1-2\theta)}{9}$$

Comparing the two expression, we find that $B_{1a} > B_{1f}$, i.e. incentives to invest in the abatement technology are still larger under freely managed trade. The intuition for this result is readily obtained by comparing the equilibrium value of the emission tax under the two trade regimes. Under free trade, the policy-maker has to weigh two opposing effects[13]: a higher tax reduces pollution, but it also reduces the home firm's competitiveness with respect to the foreign firm. On the other hand, when an active trade policy is feasible, the tariff can be used to preserve the competitive balance, allowing for more aggressive emission taxes. In other words, it is precisely because the two policy tools are complement that incentives to adopt the clean technology are once again higher under freely managed trade than under free trade.

4.4 Pollution Havens

Firms may migrate to countries with low pollution standards. This migration is facilitated by free trade since other markets can then be served more easily from a few 'pollution-tolerant' bases. Traditionally, countries are thought to be potential pollution havens if their policy-makers put a lower weight on the disutility from pollution than their trading partners. Such differences can in turn be linked to differences in tastes or income across countries or to differences in political institutions. If, for example, 'clean air' is a normal good, then countries with lower levels of income will have

[13] This is on top of the obvious negative effect that a tax has on consumer surplus.

lower pollution standards and attract more producers. Alternatively, a country is more likely to be a pollution haven if its government is heavily influenced by business interests.

Our analysis suggests that, *ceteris paribus*, countries that have stricter World Trade Organization (WTO) bounds on their tariffs are also more likely to emerge as pollution havens. This is because, for given preferences with respect to pollution, such countries are *less able* to enforce strict pollution standards than countries that retain more control over their trade policy.[14] Since countries with the tighter WTO commitments also tend to have higher income and – possibly – more representative political institutions, the implication of our model is perfectly compatible with the observation that countries with lesser WTO commitments seem, on average, to also have lower pollution standards.

5. Conclusion

The main point of this paper is very simple. In the presence of local pollution, a home government will protect its industry from foreign competition more if its domestic firms use a clean production technology. The anticipation of this *ex post* optimal trade policy gives domestic firms incentives to adopt clean(er) technologies in the first place. The ability to set an unconstrained tariff *ex post* is therefore valuable as an instrument to foster pollution-reducing investments. Countries that enter binding agreements limiting their ability to set tariffs might therefore lose a valuable tool to protect their environment. In that sense, a commitment to free trade can hurt the environment.

Whether the use of *ex post* tariffs provides adequate incentives for the adoption of cleaner technologies is less clear. With linear demand and Cournot duopoly, we show that the incentives provided are in fact insufficient. It follows that a commitment to trade liberalization never increases domestic (or world) welfare. In fact, trade liberalization that is drastic enough unambiguously decreases the welfare of the home country. In a more general setting, one cannot rule out the possibility that *ex post* tariffs provide domestic firms with excessive incentives to clean up their act. In such a case moderate trade liberalization increases the home country's welfare.

Appendix: Two Active Governments and Two Markets

We now briefly consider the case where both firms sell in both markets and both governments have an active trade policy. This does not affect the previous analysis much. The reason for this is that the trade policy of a government does not

[14] The mechanism analysed in the paper survives if some of the local firms are foreign-owned. The tariff levels chosen by the home government are of course lower when some local producers are foreign but it remains true that tariffs will be higher if local producers adopt the clean technology than if they rely on the dirty technology.

have any direct influence on the technology adoption decision of a foreign firm. In fact, what a firm produces at home in order to sell it in the foreign market does not affect the trade policy of *either* of the two governments. The foreign government does not care because the associated pollution does not occur in its own market. The firm's own government does not care either because its domestic trade policy can only affect how much the home firm sells in the home market. Production for sale abroad does generate local pollution, but there is nothing that the government can do about it when setting its ex post optimal tariff. Hence, as can readily be shown, each government's *ex post* optimal trade policy is exactly the same as in the one country case.

Turning to the first stage of the game, there is now a simultaneous adoption game between the domestic and foreign firms. The payoffs of the two firms for this game are shown in Table 1 below. While the payoffs that accrue to each firm depends on the adoption decisions of the other firm, the difference between a firm's payoff if it adopts and its payoff if it does not is independent of the other firm's technological choice. In fact, as shown in the last line of the table, a firm's incentives to adopt the new technology (B_I) are exactly the same as in our one country/one active government analysis. Hence a firm's own adoption decision does not depend on the action of its rival and the adoption decision is the same as in the one country case.

Table 1

	Adopt	**Not**
Adopt	$\frac{17}{81}$-*F*, $\frac{17}{81}$-*F*	$\frac{17+4\theta(1+\theta)}{81}$-*F*, $\frac{17+\theta(\theta-8)}{81}$
Not	$\frac{17+\theta(\theta-8)}{81}$, $\frac{17+4\theta(1+\theta)}{81}$-*F*	$\frac{17+\theta(4-5\theta)}{81}$, $\frac{17+\theta(4-5\theta)}{81}$
B_I	$\frac{\theta(\theta-8)}{81}$	$\frac{\theta(\theta-8)}{81}$

Still, two differences with respect to the 'one country' analysis might be worth mentioning:

1. *Scale Effect.* Private incentives to adopt are even more likely to be socially insufficient in a two-country framework. The private incentives to adopt (i.e. B_I) are still exactly the same since adoption by firm i only affects its profits in its home market. On the other hand, the social benefits of pollution are now greater since, if the firm sticks to the dirty technology, pollution is augmented by the domestic firm's sales in the foreign market.

2. *Adoption Externality.* From the point of view of country A, local adoption of clean technology is more desirable if the local output is large. The size of the local output depends on the adoption decision of the foreign firm. If the foreign firm adopts the clean technology then B's tariffs will be high so that the total local

output of A's own domestic firm will be low. In other words, adoption becomes less desirable for country A when adoption occurs in country B and vice-versa. Since trade policies and adoption decisions are the same as in the one country case, and the firms' incentives to adopt are still (even more) socially insufficient, this effect is of little consequence.

References

Barrett, S., 1994, "Strategic Environmental Policies and International Trade", *Journal of Public Economics*, 54(3), 325-338.

Brander, J. and B. Spencer, 1985, "Export Subsidies and Market Rivalry", *Journal of International Economics,* 18(1-2), 83-100.

Brander, J. and B. Spencer, 1987, "Foreign Direct Investment with Unemployment and Endogenous Taxes and Tariffs", *Journal of International Economics*, 22(3-4), 257-279.

Conrad, K., 1993, "Taxes and Subsidies for Pollution-Intensive Industries as a Trade Policy", *Journal of Environmental Economics and Management*, 25(2), 121-135.

Copeland, B.R. and M.S. Taylor, 1994, "North-South Trade and the Environment", *Quarterly Journal of Economics*, 109(3), 755-787.

Copeland, B.R. and M.S. Taylor, 2001, "International Trade and the Environment: A Framework for Analysis", *NBER Working Paper No. 8540.*

Eaton, J. and G. Grossman, 1986, "Optimal Trade and Industrial Policy under Oligopoly", *Quarterly Journal of Economics*, 101(2), 383-406.

Gallegos, A., 2004, *International Trade and Environmental Protection*, Ph.D Thesis, University of Essex, February.

Grossman, G.M. and A.B. Krueger, 1995, "Economic Growth and the Environment", *Quarterly Journal of Economics*, 110(2), 353-377.

Karp, L., S. Sacheti and J. Zhao, 2001, "Common Ground Between Free Traders and Environmentalists", *International Economic Review*, 42(3), 617-647.

Kennedy, P. W., 1994, "Equilibrium Pollution Taxes in Open Economies with Imperfect Competition", *Journal of Environmental Economics and Management*, 27(1), 49-63.

Maggi, G. and A. Rodriguez-Clare, 1998, "The Value of Trade Agreements in the Presence of Political Pressure", *Journal of Political Economy*, 106(3), 574-601.

Markusen, J.R., E.R. Morey and N.D. Olewiler, 1995, "Competition in Regional Environment Policies when Plant Locations are Endogenous", *Journal of Public Economics*, 56(1), 55-77.

Matsuyama, K., 1990, "Perfect Equilibria in a Trade Liberalization Game", *American Economic Review*, 80 (3), 480-492.

Staiger, R.W. and G. Tabellini, 1989, "Rules and Discretion in Trade Policy", *European Economic Review*, 33(6), 1267-1277.

Ulph, A., 1996, "Environmental Policy and International Trade when Governments and Producers Act Strategically", *Journal of Environmental Economics and Management*, 30(3), 265-281.

Vives, X., 1999, *Oligopoly Pricing :Old Ideas and New Tools*, MIT Press, Cambridge, Mass.

[4]

Pollution Havens and the Regulation of Multinationals with Asymmetric Information*

Xiaodong Wu

Abstract

This paper develops a common agency model to analyze the strategic interaction between governments in regulating polluting multinationals. We show that when a firm has private information about its production technology relating output to pollution that is difficult to monitor, the information rent extraction behavior of non-cooperative governments will work against the "pollution haven" hypothesis in a Nash equilibrium with or without pooling. The "pollution haven" result is more likely to be reversed in a separating equilibrium than in a pooling equilibrium as a firm's output is further away from the most efficient outcome. This result provides an explanation for why many empirical studies do not support the "pollution haven" hypothesis even after controlling for private non-environmental cost differentials.

KEYWORDS: multinational corporations, environmental regulation, common agency

*I would like to thank Gene Grossman, Avinash Dixit, Claudio Mezzetti, Eric Bond, and the editor Don Fullerton for their valuable comments and suggestions. I would also like to thank Gary Biglaiser, Patrick Conway, Robert Shimer, the conference participants at the Mid-West International Economics Meetings at University of Illinois and at the Econometric Society 8th World Congress at University of Washington, and the seminar participants at various universities and institutions. Financial support from the Ford Foundation is gratefully acknowledged. Address: Gardner Hall 300B, CB#3305, UNC-CH, Chapel Hill, NC 27599. Phone: (919) 966 5373. Email: wux@email.unc.edu.

One version of the "Pollution Haven Hypothesis" (PHH) is that less developed countries care more about economic growth than environmental protection and hence set lower environmental standards than developed countries.[1] Under freer trade, the hypothesis predicts that multinational firms will relocate to less developed countries to take advantage of their lax environmental standards. Over time, these countries will develop a comparative advantage in pollution-intensive industries and become the "havens" for the world's polluting industries.

In this paper, transboundary pollution can occur and decrease the welfare of the home country if a dirty industry is allowed to move to a foreign country with lax environmental controls so as to reduce its tax obligations. Hence, the creation of a "pollution haven" could reduce world welfare from the home country's point of view. However, the movement of the dirty industry to the foreign country could also raise world welfare because the residents of the foreign country are the ones more directly affected by the pollution but care more about jobs than about pollution (Oates and Schwab, 1988). Indeed, this paper finds that cooperation between governments increases welfare even as the dirty industry moves toward the foreign country with more pollution. Moreover, because of the information asymmetries, the concern that the dirty industry can and will over-pollute the foreign country is not justified.

Grossman and Krueger (1993) and Jaffe et al. (1995) demonstrate that there is little empirical evidence for the existence of "pollution havens" due to factor endowment and economies of scale considerations. Thus, freer trade may not harm the environment. To investigate this issue, Antweiler et al. (2001) and Grossman and Krueger (1993) divide trade's impact on pollution into a scale effect (i.e. aggregate demand for the polluting good), a technique effect (i.e. the pollution intensity of the dirty industry), and a composition effect (i.e. the share of the dirty industry in total output). They estimate the magnitude of these three effects and conclude that freer trade can be neutral or even beneficial to the environment.

This paper focuses on the technique effect under imperfect information and turns attention to the role of regulatory agencies (later referred to as governments). A key factor in regulating pollution is asymmetric information that arises when a government observes neither a multinational firm's pollution level nor its production technology to infer pollution from level of production. The paper demonstrates that the technique effect dominates, as in Antweiler et al., but due to informational reasons rather than a lower cost of capital. Although there have been many improvements in measuring pollution emissions directly and accurately with reasonable costs, incomplete information remains an important obstacle in regulating pollution.[2]

[1]Empirical studies on the relationship between pollution and income include Grossman and Krueger (1995), Komen et al. (1997), Selden and Song (1994), and the World Bank (1992).

[2]The World Bank (1992, p. 78) report also states that, "Ideally, regulators would attempt to change the behavior of resource users by means of direct policies — for instance, by taxing or regulating emissions. But these measures involve a heavy administrative burden because they target individual polluters or resource users."

Given that multinational firms have better information about the pollution intensity of their technology than do regulatory agencies, an efficiency-seeking government needs to design a regulatory regime that induces a firm to reveal its technology. Such a scheme will invariably generate informational rents to firms, which governments seek to minimize. In an open economy, what regulatory regime is optimal depends on the regulatory policies of other countries as well because a polluting multinational firm's output and pollution in one country affect both countries' welfare. By developing a common agency model with footloose multinationals as agents of separate governments, this paper demonstrates that the rent extraction behavior between governments can provide another rationale for why "pollution havens" may not exist and hence why freer trade may be good for the environment. The rent extraction incentive also provides an explanation to the following empirical findings.

Eskeland and Harrison (2003) and OECD (1995) find that, even after discounting factors such as transportation costs, factor costs, economies of scale, and country risks, there is still no significant production surge in the developing countries resulting from multinational firms' incentive to flee environmental costs at home. Javorcik and Wei (2004) extend this empirical pursuit by taking into account the domestic regulatory environment such as corruption. They find only weak evidence that supports the PHH. In testing the PHH in terms of revealed comparative advantages, the findings are mixed. Low and Yeats (1992) find that dirty industries account for a growing share of exports in some developing countries. Xu (1999) finds that export performance of "dirty" goods for most countries remained unchanged between the 1960s and 1990s.

Since the major concern of the PHH is the production reallocation from a developed country to a developing country, the following analysis will consider the case where all multinational firms originate in the home (developed) country that cares more about pollution. Also, the home government knows each firm's technology while the foreign government does not. This assumption is consistent with the finding in Caves (1996), the World Bank (1992) report, and UNCTAD (1993) that the developed countries in general have a lower administrative burden in regulating emissions directly than the developing countries. More generally, as discussed in Bond and Gresik (1997 and 2004), the case with only one uninformed government can capture the strategic interaction between governments just as well as the case where both governments are uninformed.

Without knowing a firm's technology, the foreign government either has to let all firms produce the same output (a pooling output scheme) or to internalize the externality of pollution by assigning different output levels to firms with different reported technologies (a separating output scheme). Thus, the home government can use its superior information to influence a firm's reported technology to the foreign government and hence to manipulate foreign regulatory policies so as to extract rents from the foreign government. On the other hand, since only the home government knows the firm's technology, the foreign government cannot do the same to the home government. The analysis in this paper finds two main conclusions.

First, if the foreign government chooses a separating output scheme, the home government has an incentive to encourage the firm to cheat the foreign government by increasing its production at home to compensate for its loss of production abroad due to cheating. This not only shifts information rent via tax collection from the foreign government to the home government, but also reduces the home government's welfare loss from transboundary pollution (if there is any). Foreseeing this, the foreign government reduces every firm's production abroad to induce truth telling. Alternatively, if the foreign government chooses a pooling output scheme, then a firm's output no longer depends on its reported technology and hence the home government can no longer induce cheating by increasing its domestic production. Therefore, the foreign government preempts any rent extraction from the home government. However, by doing this, the foreign government has to give up internalizing the externality of pollution.

Second, since the home (developed) country cares more about pollution, the above information rent extraction behavior between governments works against the PHH in both a pooling and a separating equilibrium. Moreover, the deviation from the cooperative most efficient outcome is bigger, and hence the "pollution haven" result is more likely to be reversed in a separating than in a pooling equilibrium.[3] At equilibrium, most firms face a separating output scheme in the foreign country. The paper also shows that this rent extraction effect is reinforced by profit repatriation and by transboundary pollution.

There are other cases where governments act strategically in regulating polluting multinationals, but under perfect information. In Dockner and Long (1993), Folmer et al. (1993), and Hoel (1997), this strategic consideration is to reduce transboundary pollution. Copeland (1996) analyzes how the home government affected by transboundary pollution can extract rents from foreign pollution quotas. Markusen et al. (1995) focus on the competition between governments in levying environmental taxes on multinational firms that generate local pollution. As in this paper, a government's major trade-off in regulating polluting multinationals is between the disutility of more pollution and the competition for additional tax revenues. For Markusen et al., the tax revenues come from monopoly profits. In this paper, tax revenues come from information rents.

Moreover, this paper focuses on the asymmetric information problem in Spulber (1988) and Dasgupta et al. (1980), and extends their analysis on regulating local firms in a single principal-agent problem to regulating multinational firms in a common agency problem. The analysis demonstrates that there can be over-production in the home country and under-production in the foreign country compared to the most efficient outcome. This asymmetry is not found in the previous studies and is due to the information rent extraction between the asymmetrically informed governments. This study also allows us to examine how firms with different technologies (private information) may respond differently to a policy change.

The literature on common agency models and regulation of multinationals with asymmetric information has been burgeoning since the early 1990s. Bernheim and Whinston

[3] Bond and Gresik (2004) also shows that, for firms with two discrete types, a pooling equilibrium can be more efficient than a separating equilibrium in a tax competition between governments over one final output.

(1985 and 1986), Laffont and Tirole (1991), Martimort (1992), and Stole (1991) were the first to introduce the common agency model with two uninformed principals and one common agent with either private information or unobservable action. In their models, the two principals have the same information set and the agent's type has the same effect on principals. Many studies have extended this homogeneity assumption. For examples, Bond and Gresik (1997 and 2004) study the case where there is one informed and one uninformed government rather than two uninformed governments, while Biglaiser and Mezzetti (1993) and Mezzetti (1997) extend the study of homogeneous principals to vertically and horizontally differentiated principals, respectively. Other extensions include Biglaiser and Mezzetti (2000), Bond and Gresik (1996), Gal-Or (1991), Ivaldi and Martimort (1994), Martimort (1996), and Olsen and Osmundsen (2001 and 2003).

To analyze the environmental regulation of multinationals by asymmetrically informed governments, this paper develops a common agency model that integrates Stole (1991) and Bond and Gresik (1997 and 2004). In the literature on common agency models, most studies focus on intrinsic common agencies only.[4] This paper shows that whether a firm has to produce in both or neither countries (an intrinsic common agency) or has the freedom to produce in only one country (a delegated common agency) affects only the tax payments at home and abroad, but leaves the equilibrium output levels unchanged. Thus, the rent extraction under asymmetric information works against the PHH regardless of the type of equilibrium. This analysis of the relationship between an intrinsic and a delegated common agency in a separating and a pooling equilibrium is applicable to other common agency problems as well.

The remainder of the paper is organized as follows. Section 1 describes the model. Section 2 shows that cooperation leads to the most efficient outcome, where the PHH always holds. Section 3 derives the non-cooperative Nash equilibrium with and without pooling and discusses the reversal of the "pollution haven" result under asymmetric information. Section 4 concludes and draws some empirical implications.

1. The Model

The model includes two countries, home and foreign, and a competitive industry with a continuum of firms. The production of this competitive industry generates a negative externality either by producing polluting by-products or by using polluting inputs. The product of the polluting industry is homogeneous and is sold at world price. All firms are price takers. Furthermore, the total output in each country is only a small proportion of world production so that each country's government cannot influence the world price by changing its domestic policy. Thus, we can single out distortions of asymmetric information in regulating polluting multinationals.

All firms are identical except for their production technologies indexed by θ. A higher θ means more pollution per unit of production. It is assumed that a firm's technology is

[4] Bernheim and Whinston (1986) provide a formal definition of an intrinsic and a delegated common agency.

determined by its technology know-how so that a firm is endowed with a given θ.[5] Also, each firm uses the same technology at home and abroad.[6] There is no strategic interaction between firms.

As discussed in the introduction, all multinational firms originate in the home (developed) country, whose residents have a higher marginal disutility of pollution. The home government also has superior information than the foreign government. In particular, the home government knows the exact value of each firm's technology parameter θ while the foreign government does not, but knows only the distribution of θ for the whole industry. Hence, θ is a random variable for the foreign government. The industry's technology θ ranges from $\underline{\theta}$ to $\overline{\theta}$ with $\underline{\theta} > 0$. Let $F(\theta)$ be the cumulative distribution function, and $f(\theta)$ be the density function of θ. $F(\theta)$ and $f(\theta)$ are continuous and differentiable with bounded derivatives. Since regulation does not change an existing firm's θ, regulation does not change $F(\theta)$ and $f(\theta)$. It is further assumed that $f(\theta) \neq 0$ for all $\theta \in \Theta$ so that the reciprocal of the hazard rate, $\frac{1-F(\theta)}{f(\theta)}$, is defined for all θ. This inverse hazard rate is assumed to be nonincreasing in θ, i.e. $\frac{d}{d\theta}(\frac{1-F(\theta)}{f(\theta)}) \leq 0$ for all θ.

Similar to the quadratic function used in Spulber (1988), a firm's cost of producing y units of total output in the home country is:

$$C(y, \theta) = \frac{1}{2}c_1 \cdot y^2 + (c_2 \cdot \theta + \xi)y + S, \tag{1}$$

where c_1 is the change of marginal cost with respect to a change in output, c_2 is the change of marginal cost with respect to a change in technology, ξ is a constant component of marginal cost, and $S > 0$ is a small sunk cost of setting up a plant.[7] It is assumed that there is a social welfare trade-off between pollution reduction and private cost reduction so that $c_2 < 0$.

---

[5]This is true, for example, if the pollution reduction technology is tied to human capital (e.g. R&D) or physical assets, which are either too costly to change after a firm sets up its plant or are not equally available to all firms. Alternatively, if pollution reduction is a matter of a firm's continuous effort or if a firm can choose its technology after knowing a country's environmental regulation, then it is more of a moral hazard problem than an adverse selection problem. In general, these two problems need to be analyzed separately. Under some circumstances, as discussed in Grossman and Hart (1983), a moral hazard problem can be transformed to an adverse selection problem discussed in this paper. The information cost of implementing a given effort under moral hazard is then equivalent to that of enforcing truth telling under adverse selection.

[6]Empirical studies suggest that firms usually use the same technology across countries (Bhagwati and Hudec, 1996; Caves, 1996; Ulph, 1998; UNCTAD, 1993). One reason is that it is more efficient for a firm to specialize in one technology than to operate under different technologies. The other reason is that, when a firm sets up a new plant in a developing country, the firm tends to make its production complying to the higher environmental standards in the developed countries just to avoid any sunk costs that may occur in the future as regulators in the developing countries tighten their requirements.

[7]If there are transportation costs, then the constant marginal cost should be ξ plus a per unit transportation cost if part or all of the output in one country is shipped to consumers in another country. For tractability, the following analysis assumes zero transportation cost. However, this assumption should not change the main results in this paper.

Also, it is assumed that production in the home country exhibits decreasing returns to scale so that $c_1 > 0$. If the sunk cost abroad S^* is not significant, then firms (originated in the home countries) have an incentive to open a new plant in the foreign country with the lower environmental standard.[8] However, firms will not stop producing in the home country because, as derived later, the tax scheme in the foreign country is non-linear and is increasing in a firm's total output to keep pollution under control. Finally, for the private marginal cost to be non-negative for all y and θ, $\xi \geq -c_2 \cdot \overline{\theta} > 0$.

Let $y \geq 0$ denote home production and $y^* \geq 0$ denote foreign production. With technology θ, a multinational firm's total cost of producing in both countries is:

$$TC(y, y^*) = C(y, \theta) + C(y^*, \theta) + cyy^*, \tag{2}$$

where $c \geq 0$ represents the additional operation cost of producing at two different locations rather than one, for example, coordination costs between plants to standardize their products.[9] However, $c_1 > c$ so that it pays firms to become multinationals. Equation (2) ensures that home and foreign production are perfect substitutes so that a firm's cost is minimized by choosing the same production in both countries if their environmental regulations were the same. This rules out the economies of scale effect already studied in Antweiler et al. (2001) and Grossman and Krueger (1993).

Usually, governments use only pure strategies to regulate pollution. Suppose the home government assigns each firm a pair of output and transfer, $\{y(\theta),\ t(\theta)\}$, according to a firm's **true** technology θ. Without knowing each firm's θ, the foreign government uses a direct mechanism to assign each firm a pair of output and transfer, $\{y^*(\hat{\theta}),\ t^*(\hat{\theta})\}$, according to a firm's **reported** technology $\hat{\theta}$.[10] If the foreign government chooses a pair of pooling schemes, then y^* and t^* are constants so that $y^{*\prime} = t^{*\prime} = 0$.[11]

It is assumed that the governments offer output and transfer schemes to firms and firms only have the choice of stay or exit. A firm receives a zero profit if it chooses exit. Thus, when its individual rationality constraint is satisfied so that it produces in both countries, a firm's objective function is its profit plus any transfers received from the governments, i.e.

$$u(\hat{\theta}, \theta) = \pi(y(\theta), y^*(\hat{\theta}), \theta) + t(\theta) + t^*(\hat{\theta}), \tag{3}$$

[8]A superscript * of a variable is hereafter used to denote its corresponding variable in the foreign country.

[9]The following discussion will focus on the case where $c > 0$ and the special case where $c = 0$ will be discussed in the appendix.

[10]Alternatively, as in Peters (2003), a direct mechanism can also specify transfer t^* as a function of y^* for each given technology. By the "taxation principle" developed by Guesnerie (1981 and 1995) and Rochet (1986) for a single agent problem and later extended to a common agency problem by Martimort and Stole (2002), we can implement $y^*(\hat{\theta})$ and $t^*(\hat{\theta})$ derived in this paper by specifying t^* as a function of y^* at each $\hat{\theta}$. Thus, the two definitions are compatible. Although direct mechanisms may not be the best mechanisms to model a common agency game as shown in Martimort and Stole (2002), this paper uses the direct mechanism approach in order to derive the equilibrium output levels across countries and hence compare non-cooperative output with cooperative output directly and avoid the complication in deriving explicitly the tax schemes to implement this equilibrium outcome, which is non-essential in this analysis.

[11]Hereafter, a prime denotes the first order derivative with respect to θ and any subscript to a variable denotes the partial derivative of the variable with respect to the subscript.

where

$$(4)\ \pi(y, y^*, \theta) = p\cdot(y+y^*) - \frac{1}{2}c_1(y^2+(y^*)^2) - (c_2\cdot\theta+\xi)(y+y^*) - cyy^* - S - S^*,$$

and p is the world price, which is assumed to be greater than or equal to the private average total cost of production so that all firms produce in both countries in the absence of regulation. Otherwise, regulation becomes redundant. Under a direct mechanism, all firms with the same θ have the same output and receive the same transfer.

Since consumers always buy at the world price p, following Copeland and Taylor (1995), it is assumed that pollution in each country only affects a representative consumer's utility through damage to health, but not his/her choice among goods. A firm's damages to the home and foreign consumers, respectively, are given by the following equations:

$$(5)\quad B(y(\theta), y^*(\theta), \theta) = \gamma b\theta(y(\theta)+\alpha y^*(\theta)),$$

$$(6)\quad B^*(y(\theta), y^*(\theta), \theta) = \gamma^* b\theta(y^*(\theta)+\alpha y(\theta)),$$

where γ is the home and γ^* is the foreign marginal disutility of pollution. $b\theta > 0$ is the rate of pollution emission per unit of output so that total emission increases not only as output rises but also as a more polluting technology is used. A more polluting firm causes more damage even with the same amount of output. Thus, damage depends on a firm's technology as well as its output. $\alpha \in [0, 1]$ is the rate at which pollution is transferred from one country to another. By assumption, $\gamma > \gamma^* > 0$. Thus, the PHH holds if all firms produce less at home than abroad.

Suppose each government's objective is to maximize social welfare, i.e. a weighted average of consumer surplus, producer surplus, and tax revenue. In order to compare with the most efficient outcome, the following discussion assigns the same weight to the part of profit that remains in the country (i.e. producer surplus), tax revenue, and the negative damage (i.e. consumer surplus as the rest of it stays the same). Shares (claims on after-tax profits) of multinational firms are assumed to be sold only to home and foreign residents.[12] Hence, the objective functions of the home and the foreign governments are

$$(7)\quad W = \int_{\underline{\theta}}^{\overline{\theta}} [\beta(\pi(s)+t(s)+t^*(s)) - t(s) - B(y(s), y^*(s), s)]f(s)ds$$

$$= \int_{\underline{\theta}}^{\overline{\theta}} [\beta\pi(s) - (1-\beta)t(s) + \beta t^*(s) - B(y(s), y^*(s), s)]f(s)ds,$$

and

$$(8)\ W^* = \int_{\underline{\theta}}^{\overline{\theta}} [(1-\beta)(\pi(s)+t(s)+t^*(s)) - t^*(s) - B^*(y(s), y^*(s), s)]f(s)ds$$

[12] This assumption can be relaxed if the home government has full information. The profit accrued to a third country works as if a third party levied a profit tax before home and foreign residents split the profit. However, if both governments do not know a firm's technology, then this profit tax to the third country causes the home and foreign governments to deviate from the most efficient outcome even under cooperation as discussed in the next section (see footnote 15).

$$= \int_{\underline{\theta}}^{\overline{\theta}} [(1-\beta)\pi(s) - \beta t^*(s) + (1-\beta)t(s) - B^*(y(s), y^*(s), s)] f(s) ds,$$

where $\beta \in (0, 1)$ is the percentage of shares purchased by home residents and hence is the share of a firm's retained profit contributing to the home government's welfare.[13]

The timing of the model is as follows. First, each firm is endowed with technology θ. Second, the home and foreign governments announce their joint output and transfer schemes if they cooperate. Otherwise, in an intrinsic common agency problem, the home and foreign governments announce their individual output and transfer schemes as functions of a firm's true and reported technology, respectively.[14] In a delegated common agency problem, each government, in addition, announces another set of output and transfer schemes if a firm chooses not to produce in the other country. After knowing the output and transfer schemes, firms decide whether to produce in a country and what technology to report to the foreign government. Finally, production and transfers occur.

2. Cooperative Governments

If the home and foreign governments set their environmental regulations cooperatively, they not only share information, but also choose production jointly. From equations (7) and (8), the home and foreign governments choose output y and y^* for each firm with technology θ to maximize their joint welfare, i.e.

$$\max_{y(\theta), y^*(\theta)} \quad W + W^* = \pi(y, y^*, \theta) - B(y, y^*, \theta) - B^*(y, y^*, \theta). \tag{9}$$

By the small country assumption, price is determined in the international market. The first order conditions are:

$$p = c_1 y + c_2\theta + \xi + cy^* + b\theta(\gamma + \alpha\gamma^*), \tag{10}$$

$$p = c_1 y^* + c_2\theta + \xi + cy + b\theta(\gamma^* + \alpha\gamma). \tag{11}$$

[13]Since this paper examines the competition between governments in extracting firms' profits, each government's welfare depends on the other's tax scheme as well as its own. This externality invalidates the revelation principal so that there may be other kind of equilibria as illustrated in Peters (2003) and also Martimort and Stole (2002), Epstein and Peters (1999) and Peters (2001). However, the rent extraction incentive studied in this paper should still play an important role in the other equilibria. Moreover, according to Theorem 2 in Peters (2003), the pure strategy equilibrium in direct mechanisms derived later in the paper is robust (stays an equilibrium) against the possibility that principals might offer far more complicated communication schemes. Hence, the analysis in this paper is still relevant as it captures the properties of at least one kind of equilibrium no matter what kind of communication mechanisms are feasible.

[14]Since the home government's schemes will not reveal to the foreign government any additional information about a firm's technology and the home government knows all the information, which government announces its schemes first does not change the information structure and hence is not crucial as long as the foreign government announces its schemes before each firm reports its technology. In any case, the home government always takes into consideration the effect of its schemes on the foreign government's schemes and chooses the best as to be discussed in section 3. On the other hand, the foreign government always tries to maximize its utility expecting any collusion between its informed opponents. As both the home government and firms have full information, and will make right expectations of the foreign government's schemes in equilibrium, the home government's optimal choice is renegotiation proof.

The second order condition is satisfied as $c_1 > c$ by assumption. From the first order conditions (10) and (11), the equilibrium output in each country achieves the most efficient outcome, where price p equals the private marginal cost of local production in that country plus its induced marginal damage of pollution to both home and foreign residents (the last term on the right hand side of each condition).[15]

For a given technology θ, solving y as a function of y^* from condition (10) gives the home government's output level that maximizes joint welfare for a given foreign output. To serve as a benchmark to compare with the non-cooperative case, this optimal choice of y at a given y^* is represented by the home government's cooperative first order condition, F_c, as shown in Figure 1. Similarly, solving y^* as a function of y from (11) gives the foreign output level that maximizes joint welfare at a given home output. This optimal choice of y^* at a given y is represented by the foreign government's first order condition, F_c^*. Since $c_1 > c$, F_c is steeper than F_c^*. Hence, the intersection of F_c and F_c^* defines a stable cooperative equilibrium, E_c.

Figure 1: Cooperative Equilibrium (for firms with a given θ)

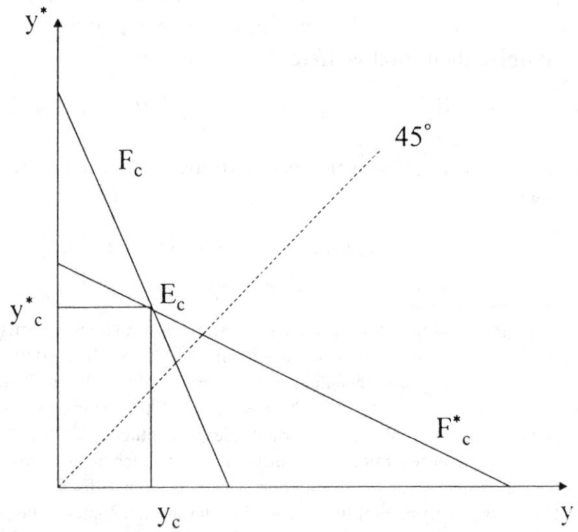

[15]This result hold even if neither government observes a firm's technology as long as all profits are distributed among home and foreign residents. Without profit outflow, it is easy to see from equation (9) that the joint welfare is independent of transfers. Thus, the governments can internalize the information cost and implement the most efficient outcome by letting each multinational firm be the residual claimant of all consumer surpluses and hence eliminating any incentive to cheat. This fails if part of a firm's profit goes to residents in a third country.

Subtracting condition (10) from (11) gives

$$(c_1 - c)(y - y^*) = b\theta(1 - \alpha)(\gamma^* - \gamma). \tag{12}$$

Since $\gamma > \gamma^*$, $0 \leq \alpha \leq 1$ and $c_1 > c$, $y \leq y^*$ for all θ. All firms produce less in the country with a higher marginal disutility of pollution, i.e. E_c lies above the 45° line, unless pollution is perfectly mobile ($\alpha = 1$) so that $y = y^*$. Moreover, although transboundary pollution reduces the production gap between countries compared with the case without transboundary pollution ($\alpha = 0$), transboundary pollution by itself cannot reverse the "pollution haven" result.

Proposition 1. *Cooperative governments implement the most efficient outcome that internalizes national as well as international externalities via information sharing. The PHH holds under cooperative governments regardless of transboundary pollution: all firms produce more in a country with a lower marginal disutility of pollution than in a country with a higher marginal disutility.*

3. Non-Cooperative Governments

In the non-cooperative case, governments no longer maximize joint welfare. Instead, each maximizes individual welfare. As a result, each cares how profits are distributed between them, which depends on both a country's profit share β and its tax revenue relative to the other government's. With asymmetric information, the home government has no incentive to reveal a firm's true θ to the foreign government. The foreign government has to design a pair of output and transfer schemes, $\{y^*(\hat{\theta}), t^*(\hat{\theta})\}$, based on a firm's reported technology $\hat{\theta}$. Thus, the crucial points of competition between governments in this common agency problem are the home government's attempt to extract information rent from the foreign government and the foreign government's effort to counteract such an attempt.

In a separating equilibrium, the fully informed home government has an incentive to encourage firms to cheat the foreign government so as to make the foreign government pay a higher information rent to firms, which eventually becomes the home government's tax revenue. As home and foreign production are substitutes in a firm's profit function and total production exhibits decreasing returns to scale, the home government wants to increase domestic production above a firm's most efficient level so that each firm has a bigger incentive to report a false θ to cut both its tax and production abroad. This forces the foreign government to pay a higher information rent in order to induce truth telling. Because this (first order) revenue gain outweighs the additional (second order) damage from more pollution, home welfare rises at the expense of foreign welfare. Meanwhile, to counteract the home government's rent extraction attempt, the foreign government further reduces each firm's production in the foreign country to discourage cheating.

By increasing its local output to extract information rents, the home government reduces production in the foreign country. Hence, if pollution is transboundary, asymmetric information provides the home government an opportunity to use its superior information not only to internalize the externality of home production, but also to reduce transboundary pollution from abroad.[16] Thus, the "pollution haven" result is more likely to be reversed if pollution crosses national boundaries.

To demonstrate the above arguments, we solve the common agency problem. A firm's net profit as a function of its true type (θ) and its reported type ($\hat{\theta}$) is

$$(13) \qquad U(\hat{\theta},\theta) \triangleq u(y(\theta), y^*(\hat{\theta}), t(\theta), t^*(\hat{\theta}), \theta).$$

In equilibrium, the foreign government's output and tax schemes, $y^*(\hat{\theta})$ and $t^*(\hat{\theta})$, must be implementable and feasible so that $U(\hat{\theta},\theta)$ is maximized at $\hat{\theta} = \theta$. Thus, the local first order necessary condition for implementability is

$$(14) \qquad t^{*'}(\theta) = -\pi_{y^*}(y(\theta), y^*(\theta), \theta) y^{*'}(\theta).$$

Applying Theorem 12 in Chapter 2 of Hurewicz (1958), there exists a transfer scheme $t^*(\theta)$, which satisfies the above condition at all points where y and y^* are differentiable.[17] From (A.3) in the appendix, the local second order necessary condition for implementability is

$$(15) \qquad (c_2 + cy^{'})y^{*'} \leq 0.$$

In an intrinsic common agency problem, the home government (with full information) extracts all retained profits from firms and leaves them with zero utility.[18] At a given pair of output and tax schemes $(y^*(\hat{\theta}), t^*(\hat{\theta}))$ set by the foreign government satisfying condition (14), the home government's maximization problem is:

$$(16) \qquad \max_{y(\theta), t(\theta)} \quad \beta\pi(y(\theta), y^*(\hat{\theta}), \theta) - (1-\beta)t(\theta) + \beta t^*(\hat{\theta}) - B(y(\theta), y^*(\hat{\theta}), \theta)$$

$$(17) \qquad s.t. \quad \pi(y(\theta), y^*(\hat{\theta}), \theta) + t(\theta) + t^*(\hat{\theta}) = 0,$$

where $\hat{\theta}$ is a function of y. Hence, when choosing y, the home government also considers how y affects $\hat{\theta}$ and thus y^* as discussed later in deriving the first order conditions. Substituting $t(\theta) = -(\pi(y(\theta), y^*(\hat{\theta}), \theta) + t^*(\hat{\theta}))$ from the constraint into the objective function gives the home government's simplified maximization problem:

$$(18) \qquad \max_{y(\theta)} \quad \Lambda = \pi(y(\theta), y^*(\hat{\theta}), \theta) + t^*(\hat{\theta}) - B(y(\theta), y^*(\hat{\theta}), \theta).$$

[16]In this case, although the world price stays the same, firms still have an incentive to reallocate their production across countries to take advantage of the differences in the tax schemes. Hence, as the governments change their environmental policies, they can affect the output and hence transboundary pollution across countries. This incentive is more apparent with the alternative direct mechanism specification discussed in footnote 10.

[17]In a pooling equilibrium, $y^{*'} = t^{*'} = 0$ so that (14) still holds. Firms will report their true θ to the foreign government as they cannot gain from cheating. Meanwhile, the foreign government chooses not to use this information to charge a more polluting firm a higher tax.

[18]The intuition of this policy is similar to that of capital export neutrality. Since the home government has full information so that taxation does not alter domestic production, extracting all retained profits minimizes the gain to foreign shareholders.

In a delegated common agency problem, a firm can choose to produce only in the foreign country. The above analysis stays the same except that $t(\theta) = -(\pi(y(\theta), y^*(\hat{\theta}), \theta) + t^*(\hat{\theta})) + \pi(0, y_0^*(\theta), \theta) + t_0^*(\theta)$, where $y_0^*(\theta)$ and $t_0^*(\theta)$ are the foreign government's output and tax schemes when $y = 0$.[19] Since $y_0^*(\theta)$ and $t_0^*(\theta)$ are independent of $y(\theta)$ at equilibrium, this will only change the home government's transfer scheme, but not output as the first order condition of the maximization problem (18) stays the same. Footnote 20 shows a similar property for the foreign government's maximization problem. Hence, whether a firm has to produce in both or neither countries (an intrinsic common agency) or can choose to produce in only one country or both (a delegated common agency) will only affect the home and foreign transfer schemes, but will have no effect on the output levels at home and abroad.

The foreign government's maximization problem is

$$\text{(19)} \quad \max_{y^*(\theta), t^*(\theta)} \int_{\underline{\theta}}^{\overline{\theta}} ((1-\beta)\pi - \beta t^* + (1-\beta)t - B^*) f(\theta) d\theta$$

$$\text{(20)} \quad s.t. \quad (IC)\ u(y(\theta), y^*(\theta), t(\theta), t^*(\theta), \theta) \geq u(y(\theta), y^*(\hat{\theta}), t(\theta), t^*(\hat{\theta}), \theta)$$

$$\text{(21)} \quad (IR)\ u(y(\theta), y^*(\theta), t(\theta), t^*(\theta), \theta) \geq 0.$$

Following Mirrlees (1971), let $U(\theta)$ be a firm's indirect utility function. Thus, (IC) implies that $U(\theta) = \max_{\hat{\theta}} u(\hat{\theta}, \theta) = \pi(y(\theta), y^*(\hat{\theta}), \theta) + t(\theta) + t^*(\hat{\theta})$, which yields equation (14). From the envelope theorem, $\frac{dU}{d\theta} = u_\theta = \pi_\theta + \pi_y y' + t'$ so that $U(\theta) = \int_{\underline{\theta}}^{\theta} u_\theta(y(s), y^*(s), s) ds + u(\underline{\theta})$. Since it is never profitable for the foreign government to give any information rent to the lowest type (the highest cost) firm, $u(\underline{\theta}) = 0$.[20] We can then write

$$\text{(22)} \quad U(\theta) = \int_{\underline{\theta}}^{\theta} [\pi_\theta(y(s), y^*(s), s) + \pi_y(y(s), y^*(s), s) y'(s) + t'(s)] ds.$$

Since $t^* = U - \pi - t$, the foreign government's objective function becomes

$$\text{(23)} \quad (1-\beta)\pi - \beta(U - \pi - t) + (1-\beta)t - B^* = \pi + t - B^* - \beta \int_{\underline{\theta}}^{\theta} [\pi_\theta + \pi_y y' + t'] ds.$$

[19] If $y = 0$, the problem degenerates to a single principal-agent problem with adverse selection. Following Mirrlees (1971) and Fudenberg and Tirole (1991), it is easy to prove that $y_0^*(\theta)$ satisfying $p = c_1 y^* + c_2\theta + \xi + \gamma^* b\theta - \beta \frac{1-F(\theta)}{f(\theta)} c_2$ and $t_0^*(\theta)$ satisfying (14) with $t_0^*(\underline{\theta}) = -\pi(0, y_0^*(\underline{\theta}), \underline{\theta})$ are the solution to the single principal-agent problem. Clearly, the monotonicity condition, $y^{*'} > 0$, is satisfied.

[20] This holds in both an intrinsic and a delegated common agency problem. For an intrinsic common agency, the home and foreign tax schemes for a firm with $\underline{\theta}$ are jointly determined by the condition, $t(\underline{\theta}) + t^*(\underline{\theta}) = -\pi(y(\underline{\theta}), y^*(\underline{\theta}), \underline{\theta})$ so that $u(\underline{\theta}) = 0$. For a delegated common agency, $u(\underline{\theta}) = \pi(0, y_0^*(\underline{\theta}), \underline{\theta}) + t_0^*(\underline{\theta}) + \pi(y_0(\underline{\theta}), 0, \underline{\theta}) + t_0(\underline{\theta})$, where $y_0(\underline{\theta}) = \frac{p - (c_2\underline{\theta} + \xi + \gamma b\underline{\theta})}{c_1}$ and $t_0(\underline{\theta}) = -\pi(y_0(\underline{\theta}), 0, \underline{\theta})$ if $\beta < 1$. If $\beta > 0$, then at $y = 0$, $t_0^*(\underline{\theta}) = -\pi(0, y_0^*(\underline{\theta}), \underline{\theta})$. Hence, for $0 < \beta < 1$, $u(\underline{\theta}) = 0$. In fact, from equations (14) and (17), $U_\theta = 0$ so that all firms get zero utility at equilibrium as the fully informed home government taxes away all firms' residual surpluses.

Integration by parts gives

$$\int_{\underline{\theta}}^{\overline{\theta}}\int_{\underline{\theta}}^{\theta}[\pi_\theta(y(s),y^*(s),s)+\pi_y(y(s),y^*(s),s)y'(s)+t'(s)]f(\theta)dsd\theta$$
$$=\int_{\underline{\theta}}^{\overline{\theta}}\frac{1-F(\theta)}{f(\theta)}[\pi_\theta(y(\theta),y^*(\theta),\theta)+\pi_y(y(\theta),y^*(\theta),\theta)y'(\theta)+t'(\theta)]f(\theta)d\theta.$$

Under the monotonicity condition, $y^{*'}\geq 0$, proved in the appendix, the foreign government's maximization problem at a given pair of home output and tax schemes, $(y(\theta), t(\theta))$, reduces to the following:

$$(24)\quad \max_{y^*(\theta)}\quad \Lambda^* = \int_{\underline{\theta}}^{\overline{\theta}}[\pi(y,y^*,\theta)-\beta\frac{1-F}{f}(\pi_\theta(y,y^*,\theta)+\pi_y(y,y^*,\theta)y')]f(\theta)d\theta$$
$$+\int_{\underline{\theta}}^{\overline{\theta}}\left(t-\beta\frac{1-F}{f}t'-B^*(y,y^*)\right)f(\theta)d\theta.$$

It is easy to see that the transfer schemes are differentiable and the appendix proves that Λ and Λ^* are strictly concave in y and y^*, respectively, and that both have an interior maximum. Given the foreign government's tax scheme derived from (14), the first order conditions of (18) and (24) by piecewise differentiation are

$$(25)\qquad p = c_1y+c_2\theta+\xi+\gamma b\theta+cy^*+\alpha\gamma b\theta y^{*'}\frac{d\hat{\theta}}{dy},$$

$$(26)\qquad p = c_1y^*+c_2\theta+\xi+\gamma^*b\theta+cy-\beta\frac{1-F(\theta)}{f(\theta)}(c_2+cy'),$$

where $y^{*'}\frac{d\hat{\theta}}{dy}$ is 0 if $y^{*'}=0$ and is $\frac{cy^{*'}}{c_2+cy'}$ if $y^{*'}\neq 0$.[21] This term, calculated in the appendix, reflects a strategic interaction between governments. It gives how much the home government can affect the foreign government's output by changing a firm's production at home to encourage the firm to report a false θ to the foreign government. As to be explained in the following subsections, this strategic effort of the home government becomes fruitless ($y^{*'}\frac{d\hat{\theta}}{dy}=0$) if the foreign government chooses a pooling output scheme ($y^{*'}=0$) or if there is no transboundary pollution ($\alpha=0$).[22]

[21] There is no loss of generality here in restricting analysis to piecewise differentiation as the home government has full information and all the terms in the foreign government's objective function have second order derivatives.

[22] Obviously, the last term also becomes zero if the production at home and abroad are unrelated in a firm's profit function ($c=0$). This case is discussed in the appendix. Clearly, if $\alpha=\beta=c=0$, then the non-cooperative outcome collapses to the most efficient cooperative outcome. In this extreme case, there is no incentive for the home government not to convey its information to the foreign government.

Since all functions are continuous and have bounded derivatives to the third order, the Lipschitz condition is satisfied uniformly. From Theorem 11 and 12 in Chapter 2 of Hurewicz (1958), there exists a unique solution satisfying the above conditions from any initial point (y_0, y_0^*, θ). Thus, although different initial conditions lead to different equilibrium output schemes, (25) and (26) characterize the common property of all output schemes that maximize the non-cooperative individual welfare of home and foreign governments.[23]

These schemes are locally and globally implementable if and only if the monotonicity condition $(y^{*'} \geq 0)$ and the local second order necessary condition for implementation $((c_2 + cy')y^{*'} \leq 0)$ are satisfied. The appendix proves that the monotonicity condition holds if $-c_2 \geq \gamma b$, which ensures that the production cost advantage of a more polluting firm is big enough for both home and foreign governments to let this firm produce more than a cleaner firm.[24]

3.1 Separating Equilibrium

From the monotonicity and local implementability conditions, the appendix proves that the home and foreign outputs are strategic substitutes. If the foreign government chooses a separating output scheme so that $y^{*'} \neq 0$, then $y^{*'} \frac{d\hat{\theta}}{dy} \leq 0$ and $c_2 + cy' < 0$. Since the last term in condition (25) is negative unless $y^{*'} = 0$ or $\alpha = 0$ or $c = 0$, as discussed at the beginning of this section, the home government's rent extraction incentive raises the home non-cooperative output above the output that maximizes joint welfare at any given foreign output. Hence, the home government's non-cooperative first order condition, F_{nc}, shifts to the right of F_c as shown in Figure 2.

If there is no transboundary pollution $(\alpha = 0)$, then the last term in (25) becomes zero so that F_{nc} stays at F_c. This is because the informed home government collects a firm's

[23] A common agency problem with asymmetrically informed principals usually results in multiple equilibria. This is because the uninformed principal's incentive compatible constraint depends on the informed principal's schemes (as in this case, U_θ depends on $y(\theta)$ and $t(\theta)$), which in turn, depend on how the uninformed principal responds. Bond and Gresik (2004) illustrate this possibility of multiple equilibria as the informed principal chooses different transfer schemes. This multiplicity usually disappears as we move to a common agency with symmetrically uninformed principals. As studied in Martimort and Stole (2003), less information imposes more restrictions on the set of reasonable conjectures and hence can lead to a unique robust equilibrium. In this paper, since the relation of output levels between countries is derived from the first order necessary conditions, which must hold in all equilibria, multiplicity does not matter too much in deriving the effect of asymmetric information on the "pollution haven" hypothesis.

[24] Otherwise, the optimal output should be decreasing in θ. Any output scheme decreasing in θ will not be implementable when the marginal cost of production decreases in θ, i.e. $c_2 < 0$. This is because $y^{*'} \geq 0$ is necessary and sufficient for implementability under the single crossing condition $\pi_{y\theta} = -c_2 > 0$ (Chapter 7, Fudenberg and Tirole, 1991). However, if $c_2 > 0$, then there will be a second-best implementable output scheme which is decreasing in θ. In this case, a cleaner firm with a lower θ also has a lower marginal cost. There will be no social trade-off between cost reduction and pollution reduction. Hence, this case is less interesting.

residual profit in equilibrium. Without transboundary pollution, the home government's preferences over output and tax abroad coincide with a firm's private preferences.

The foreign government's incentive compatibility constraint ensures that the last term in condition (26), $-\beta\frac{1-F(\theta)}{f(\theta)}(c_2 + cy')$, is positive for all firms with $\theta < \bar{\theta}$. This demonstrates that, at each given home output, the foreign government chooses a lower output than the output that maximizes joint welfare to induce truth telling. cy' in the last term reflects the impact of the home government's rent extraction attempt on the foreign government's equilibrium output scheme as discussed at the beginning of this section. Asymmetric information shifts the foreign government's non-cooperative first order condition, F^*_{nc}, to the left of F^*_c. Transboundary pollution has no effect on F^*_{nc} as the foreign government does not have the information advantage to affect the home government's output decision.

Figure 2: Non-Cooperative Equilibrium (for firms with a given $\theta < \bar{\theta}$)

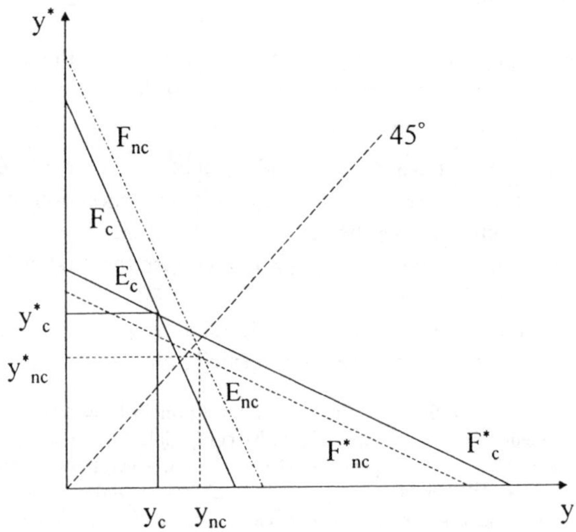

In sum, asymmetric information in a common agency model has two effects. One is to **decrease** the output in a country with inferior information further below its output in a single principal-agent problem with asymmetric information. The other is to **increase** the output in a country with superior information. This latter rent extraction effect is absent in a single principal-agent problem. In this paper, both of these two effects weaken the "pollution haven" effect and can even reverse it as shown in Figure 2 for firms with $\theta < \bar{\theta}$.[25]

[25]For firms with $\theta = \bar{\theta}$, both governments choose their cooperative best responses if $\alpha = 0$. This is consistent with the standard result that there is no distortion "at the top", i.e. for those firms with the most

Moving from cooperation to non-cooperation, the governments not only stop sharing information as discussed above, but also exclude the pollution damage imposed on consumers in another country from their individual welfare functions. Compared with the cooperative case, transboundary pollution further shifts F_c out by $\gamma^* \alpha b\theta$ and F_c^* up by $\gamma \alpha b\theta$. This reinforces the previous effect on the home government's first order condition, but makes the overall effect on the foreign government's first order condition ambiguous, depending on the relative size of $-\beta\frac{1-F(\theta)}{f(\theta)}(c_2 + cy')$ and $\gamma \alpha b\theta$.

We can address this ambiguity mathematically and analyze the effects of asymmetric information more rigorously by subtracting (25) from (26). If $y^{*'} \neq 0$,

$$(c_1 - c)(y - y^*) = b\theta(\gamma^* - \gamma) - \beta\frac{1-F}{f}(c_2 + cy') - \frac{\alpha\gamma b\theta c y^{*'}}{c_2 + cy'}. \tag{27}$$

From condition (27), a country's relative output depends on the country's relative marginal disutility of pollution (the first term on the right hand side), the rent extraction effect (the second term), and the transboundary pollution effect resulting from the home government's attempt to use its superior information to reduce transboundary pollution (the third term) by adjusting its transfer scheme. Obviously, the third term becomes zero if pollution has local effects only.

Since $c_2 + cy' < 0$, the rent extraction effect works against the PHH as long as some profits accrue to home residents, i.e. $\beta > 0$, so that the foreign government does lose part of its information rent payments to firms. This rent extraction effect increases with the profit share accruing to home residents as the second term is increasing in β. The transboundary pollution effect reinforces the rent extraction effect as $y^{*'} > 0$ and $\frac{\alpha\gamma b\theta c y^{*'}}{c_2+cy'} < 0$ in a separating equilibrium. This is consistent with the empirical result in Bui (1998) that the gain from cooperation is large when there is severe transboundary pollution. Compared with the cooperative outcome, asymmetric information under non-cooperation always decreases foreign output and increases home output. This distorts the "pollution haven" result and can completely reverse it if $-\beta\frac{1-F}{f}(c_2 + cy') - \frac{\alpha\gamma b\theta c y^{*'}}{c_2+cy'} > b\theta(\gamma - \gamma^*)$.[26]

3.2 Pooling Equilibrium

If the foreign government chooses a pooling output scheme, then the last term in (25) and that in (27) become zero. F_{nc} stays at F_c for all firms and the foreign government pays no information rent. The "pollution haven" result is reversed only if $-\beta\frac{1-F}{f}(c_2 + cy') >$

---

polluting technology.

[26]If the home and foreign governments are both uninformed of a firm's production technology, then the rent extraction incentive will shift each government's non-cooperative first order condition to the right while the usual incentive compatibility constraint will shift each government's non-cooperative first order condition to the left. Thus, the overall effect of asymmetric information on a country's output and hence the "pollution haven" hypothesis in a common agency problem depends on the relative effectiveness of the rent extraction behavior between governments as analyzed in Wu (2000).

$b\theta(\gamma - \gamma^*)$. As studied in Bond and Gresik (1997 and 2004) and Mezzetti (1997), the appendix shows that such a Nash equilibrium with pooling can exist. Also, as in Bond and Gresik (2004), a non-cooperative Nash equilibrium with pooling is closer to the co-operative equilibrium and hence the most efficient outcome than a separating equilibrium because $-\frac{\alpha\gamma b\theta c y^{*'}}{c_2+cy'} > 0$.

However, in a pooling equilibrium, the foreign government has to give up using information on a firm's production technology to internalize the externalities of pollution. The appendix proves that it is only optimal for the foreign government to choose a pooling equilibrium over some narrow intervals of θ and only if $\gamma/\gamma^* < c_1/c$. If the interval is wide, then the externality of pollution will be too high for the foreign government to choose pooling. If the marginal disutility in the home country is much bigger than that in the foreign country (i.e. γ/γ^* is large), but the degree of decreasing returns to scale is not much bigger than the degree of diseconomies of scope (i.e. c_1/c is small), then it is much more profitable for firms to concentrate their production in the foreign country regardless of the home output level so that the home government has limited influence over a firm's production abroad and hence its ability to extract information rent is limited. The benefit of a separating over a pooling equilibrium is bigger than the cost. Thus, the foreign government chooses a separating output scheme for all firms with $\theta < \bar{\theta}$ if $\gamma/\gamma^* > c_1/c$.

Proposition 2. *The rent extraction effect between non-cooperative governments with asymmetric information can by itself reverse the "pollution haven" result: polluting firms can produce more in a developed country than in a developing country. This rent extraction effect is reinforced by profit repatriation and transboundary pollution. These results hold for firms both as intrinsic common agents and as delegated common agents. The results also do not depend on whether pooling exists for some firms or not. Although the outcome with pooling is closer to the most efficient outcome than without, pooling is possible only if $\gamma/\gamma^* < c_1/c$ and can only exist for firms with very similar technologies.*

4. Conclusions and Extensions

This paper develops a common agency model to analyze the strategic interaction between two governments in regulating polluting multinationals. When the foreign government (in a developing country) cannot observe a multinational firm's level of pollution and the cleanliness of its production technology, the strategic rent extraction behavior between governments weakens the "pollution haven" effect. The output levels in a country not only depend on that country's marginal disutility of pollution (γ/γ^*), but also depend on its profit share (β), the degree of transboundary pollution (α), and most importantly, the increase in marginal cost for a firm with a more polluting technology (c_2). The rent extraction effect is stronger if a larger share of multinational firms' profits accrues to home residents or if pollution is transboundary. Thus, "pollution havens" may not exist and freer trade may be good for the environment.

The above results apply regardless of whether the foreign government chooses a pooling or a separating output and tax schemes and whether firms have to choose to produce in both or neither countries (intrinsic agents) or they have the freedom to produce in just one country (delegated agents). These results can also be extended to cases where both governments are uniformed about a firm's technology and where governments assign different weights on consumer's and producer's surpluses and tax revenues. The restrictions on the cost function can also be relaxed.[27] The analysis of an intrinsic and a delegated common agency in a separating and a pooling equilibrium in this paper can also be applied to other common agency models on the regulation of multinational firms or the coordination problem between state governments. For example, we can extend the study on the relationship between trade policies and environmental cooperation between governments from the case with local firms to one with multinational firms.

The results in the paper also have some empirical implications. First, for a particular country, we will observe a negative relationship between pollution and that country's marginal disutility of pollution (or income as its instrumental variable) even with asymmetric information. However, for two or more countries, with asymmetric information, we may fail to observe that production is lower in a developed country with a higher marginal disutility of pollution or that revealed comparative advantage concentrates in relatively more polluting industries in the developing countries. Moreover, as firms use technologies with different pollution intensities, the PHH can hold for some most polluting firms, but fail for some relatively cleaner ones. Javorcik and Wei (2004) also emphasize the value of using a firm-level data set in searching for empirical support for the PHH. Finally, the PHH is more likely to hold for industries with relatively standard technology so that there exists a Nash equilibrium with pooling.

Appendix

1. Strategic revelation effects

Rewrite equation (13) gives:

$$\text{(A.1)}\, U(\hat{\theta},\theta) \triangleq p(y(\theta)+y^*(\hat{\theta})) - \frac{1}{2}c_1(y(\theta)^2+y^*(\hat{\theta})^2) - (c_2\theta+\xi)(y(\theta)+y^*(\hat{\theta})) \\ -cy(\theta)y^*(\hat{\theta}) + t(\theta) + t^*(\hat{\theta}) - S - S^*.$$

The first order condition is:

$$\text{(A.2)}\ U_{\hat{\theta}} = py^{*'}(\hat{\theta}) - c_1y^*(\hat{\theta})y^{*'}(\hat{\theta}) - (c_2\theta+\xi)y^{*'}(\hat{\theta}) - cy(\theta)y^{*'}(\hat{\theta}) + t^{*'}(\hat{\theta}) = 0.$$

[27] Although these extensions will make no inherent differences on the strategic interaction between governments, the overall effect can be ambiguous (Wu, 2000). As explained in footnote 26, the home government's non-cooperative output scheme with asymmetric information can be either above or below its output scheme with perfect information if the home government also does not observe each firm's θ. A more general cost function can bring in additional effects, such as economies of scale, which will complicate a firm's production decision.

Totally differentiating the above condition at $\hat{\theta} = \theta$ gives $U_{\hat{\theta}\theta} + U_{\hat{\theta}\hat{\theta}} = 0$. Hence,

(A.3) $$U_{\hat{\theta}\hat{\theta}} = -U_{\hat{\theta}\theta} = (c_2 + cy^{'})y^{*'}.$$

If the local second order necessary condition for implementation, i.e. $(c_2 + cy^{'})y^{*'} \leq 0$, is satisfied, then $U_{\hat{\theta}\hat{\theta}} < 0$. Hence, $U(\hat{\theta}, \theta)$ is strictly concave in $\hat{\theta}$ so that there is a unique $\hat{\theta}$ that maximizes $U(\hat{\theta}, \theta)$.

Following the derivations in Stole (1991) and Mezzetti (1997), to evaluate the strategic revelation (interaction) effect in a pure-strategy differentiable Nash equilibrium, represented by $\frac{\partial \hat{\theta}^*}{\partial y}$ and $\frac{\partial \hat{\theta}^*}{\partial t}$, totally differentiating the first order condition (A.2) with respect to y, t, and $\hat{\theta}$ gives:

(A.4) $$(py^{*''} - c_1(y^{*'})^2 - c_1 y^* y^{*''} - (c_2\theta + \xi)y^{*''} - cyy^{*''} + t^{*''})d\hat{\theta} - cy^{*'}dy + 0dt = 0.$$

Clearly, $\frac{\partial \hat{\theta}}{\partial t} = 0$ and when $y^{*'} = 0$, $\frac{\partial \hat{\theta}}{\partial y} = 0$. In equilibrium, the first order condition (A.2) must hold for $\hat{\theta} = \theta$. Totally differentiating (A.2) with respect to $\hat{\theta}$ at $\hat{\theta} = \theta$ yields

(A.5) $$(py^{*''} - c_1(y^{*'})^2 - c_1 y^* y^{*''} - (c_2\theta + \xi)y^{*''} - cyy^{*''} + t^{*''} - c_2 y^{*'} - cy^{'}y^{*'})d\hat{\theta} = 0.$$

Subtracting (A.4) from (A.5) gives:

$$cy^{*'}dy = (c_2 + cy^{'})y^{*'}d\hat{\theta}.$$

Thus,

(A.6) $$\frac{d\hat{\theta}}{dy} = \begin{cases} \frac{c}{c_2 + cy^{'}} & \text{if } y^{*'} \neq 0 \\ 0 & \text{if } y^{*'} = 0 \end{cases}.$$

2. Strict concavity of the objective functions

Given the incentive compatibility constraints and the assumption that all the cross-partial derivatives are constant, the second order derivatives of Λ defined by equation (18) and Λ^* defined by equation (24) with respect to y and y^*, respectively, are the following:

(A.7) $$\Lambda_{yy} = -c_1 - \gamma \alpha b \theta \frac{d}{dy}(y^{*'} \frac{\partial \hat{\theta}}{\partial y}),$$

(A.8) $$\Lambda_{y^* y^*} = -c_1 < 0.$$

Since the optimal output scheme (y^*) is only a function of the true technology θ, $y^{*'}$, representing the rate of changes of the optimal output with respect to changes of θ, is independent of the variations of y in the process of maximizing the integrand (Λ) at any given θ. From (A.6),

$$\frac{d}{dy}(y^{*'} \frac{\partial \hat{\theta}^*}{\partial y}) = \frac{d}{dy}(\frac{y^{*'}c}{c_2 + cy^{'}}) = 0.$$

Substituting the above expression into (A.7) gives

$$\Lambda_{yy} = -c_1 < 0. \tag{A.9}$$

Hence, Λ is strictly concave in y and Λ^* is strictly concave in y^*.

3. Monotonicity and global implementability of the foreign output scheme

In this part, we will prove the monotonicity of the foreign government's output scheme so that a local implementable (incentive compatible) output scheme is also globally implementable. We will first prove this for a special case where foreign and domestic outputs are independent ($c = 0$), and then for the general case where $c > 0$.

If $c = 0$, from conditions (25) and (26), $p = c_1 y + c_2\theta + \xi + \gamma b\theta + cy^*$ so that the home government does not deviate from its non-cooperative perfect information best responses. However, $p = c_1 y^* + c_2\theta + \xi + \gamma^* b\theta + cy - \beta\frac{1-F}{f}c_2$ so that the foreign government still chooses a lower output for all firms below its non-cooperative perfect information best responses. Thus, asymmetric information works against the PHH.

Totally differentiating conditions (25) and (26) yields:

$$c_1 y' + c_2 + \gamma b = 0, \tag{A.10}$$

$$c_1 y^{*'} + c_2 + \gamma^* b - \beta\frac{d}{d\theta}\left(\frac{1-F}{f}\right)c_2 = 0. \tag{A.11}$$

Hence, if $-c_2 \geq \gamma b$, then since $\gamma > \gamma^*$ and $\frac{d}{d\theta}\left(\frac{1-F}{f}\right) \leq 0$,

$$y' = \frac{-c_2 - \gamma b}{c_1} \geq 0, \tag{A.12}$$

$$y^{*'} = \frac{-c_2 - \gamma^* b + \beta\frac{d}{d\theta}\left(\frac{1-F}{f}\right)c_2}{c_1} > 0. \tag{A.13}$$

Also, $c_2 + cy' = c_2 - \frac{c}{c_1}(c_2 + \gamma b) = \frac{c_1 - c}{c_1}c_2 - \frac{c}{c_1}\gamma b < 0$.

To prove the monotonicity for the general case of $c > 0$, let $MC_y = c_1 y + (c_2\theta + \xi) + cy^*$ and $MC_{y^*} = c_1 y^* + (c_2\theta + \xi) + cy$. Rearranging conditions (25) and (26) yields:

$$c(p - MC_y - \gamma b\theta)y' - \alpha\gamma b\theta c y^{*'} = -c_2(p - MC_y - \gamma b\theta), \tag{A.14}$$

$$\beta\frac{1-F}{f}cy' = -(p - MC_{y^*} - \gamma^* b\theta) - \beta\frac{1-F}{f}c_2. \tag{A.15}$$

Applying the Cramer's rule, for $\theta < \bar{\theta}$,

$$y' = -\frac{f}{c\beta(1-F)}(p - MC_{y^*} - \gamma^* b\theta + \beta\frac{1-F}{f}c_2), \tag{A.16}$$

$$y^{*'} = -\frac{f}{\alpha\gamma b\theta c\beta(1-F)}(p - MC_y - \gamma b\theta)(p - MC_{y^*} - \gamma^* b\theta). \tag{A.17}$$

Also, $c_2+cy' = -\frac{f}{\beta(1-F)}(p-MC_{y^*}-\gamma^*b\theta)$ so that $(c_2+cy')y^{*'} = \frac{1}{\alpha\gamma b\theta c}(\frac{f}{\beta(1-F)})^2(p - MC_{y^*} - \gamma^*b\theta)^2(p - MC_y - \gamma b\theta)$. Since $(c_2 + cy')y^{*'} \leq 0$ from the local second order necessary condition for implementability, $p - MC_y - \gamma b\theta \leq 0$ so that the home production is either at or above the home government's non-cooperative perfect information best responses. If $p - MC_y - \gamma b\theta = 0$ at $\theta_1 < \bar{\theta}$, then $y^{*'}(\theta_1) = 0$. From (A.10), $y'(\theta_1) = -\frac{c_2+\gamma b}{c_1}$ so that $c_2 + c_1y' = \frac{(c_1-c)c_2-c\gamma b}{c_1}$. Substituting this into (26) gives $p - MC_{y^*} - \gamma^*b\theta = -\beta\frac{1-F}{f}\frac{(c_1-c)c_2-c\gamma b}{c_1}$. Totally differentiating this gives $\beta\frac{d}{d\theta}(\frac{1-F}{f}) = 1 + \frac{\gamma^*bc_1}{(c_1-c)c_2-c\gamma b}$ at $\theta = \theta_1$. Since $\frac{d}{d\theta}(\frac{1-F}{f}) \leq 0$, $c_1 > c$, and $c_2 < 0$, $p - MC_y - \gamma b\theta = 0$ and hence $y^{*'}(\theta) = 0$ can hold at $\theta_1 < \bar{\theta}$ only if $c\gamma < c_1\gamma^*$ or $\gamma/\gamma^* < c_1/c$ so that $(c_1 - c)c_2 - c\gamma b < 0$.

From (26), $p - MC_{y^*} - \gamma^*b\theta = 0$ at $\theta = \bar{\theta}$. If $p - MC_{y^*} - \gamma^*b\theta = 0$ at $\theta_2 < \bar{\theta}$, then (26) gives $c_2 + cy' = 0$ so that $y' = -c_2/c$. Substituting $y^{*'} = 0$ into (25) gives $p - MC_y - \gamma b\theta = 0$. Hence, $y' = (-c_2 - \gamma b)/c_1 \neq -c_2/c$. Thus, $p - MC_{y^*} - \gamma^*b\theta \neq 0$ for all $\theta < \bar{\theta}$. From (A.16), $y'(\bar{\theta}) = 0$. Hence, totally differentiating (26) gives $\frac{d}{d\theta}(p - MC_{y^*} - \gamma^*b\theta) = -\beta\frac{d}{d\theta}(\frac{1-F(\theta)}{f(\theta)})c_2 < 0$. Thus, $p - MC_{y^*} - \gamma^*b\theta > 0$ so that $c_2 + cy' < 0$ for all $\theta < \bar{\theta}$. $p - MC_{y^*} - \gamma^*b\theta = 0$ and $c_2 + cy' = \frac{c_1-c}{c_1}c_2 - \frac{c}{c_1}\gamma b < 0$ at $\theta = \bar{\theta}$, which is consistent with the standard result that there is no distortion "at the top".[28]

From (A.17), $y^{*'} \geq 0$. From (14), $t^{*'} = -(p-MC_{y^*}-\gamma^*b\theta)y^{*'} \leq 0$. Since $t(\underline{\theta}) < 0$, all firms pay a tax ($t < 0$) that increases with θ. According to Chapter 7 in Fudenberg and Tirole (1991), Martimort (1992) and Stole (1991), $y^{*'} \geq 0$ is necessary and sufficient for global implementability under the single crossing condition $\pi_{y\theta} = -c_2 > 0$. Thus, the first order conditions (25) and (26) yield a pair of implementable output schemes for any given initial condition, provided $-c_2 \geq \gamma b$.

It is easy to check that, in a Nash equilibrium, the foreign government never chooses pooling for all firms, otherwise, the two first order conditions would give different values of y'. Moreover, if $y^{*'} = 0$ over an interval $\theta' < \theta < \theta''$, then $p - MC_y - \gamma b\theta = 0$ on (θ', θ''). Following the calculation in the case where $c = 0$, $y'' = 0$ on (θ', θ'') so that $y^{*'} = \frac{-c_2-\gamma^*b+\beta\frac{d}{d\theta}(\frac{1-F}{f})(c_2+cy')}{c_1} > 0$ at $\theta = \theta'$ and $\theta = \theta''$. Hence, the equilibrium foreign output increases for firms with a θ bigger than θ'' and decreases for firms with a θ less than θ'. Thus, the foreign government chooses a pooling equilibrium, i.e. $y^{*'} = 0$, only in a narrow interval of θ in Θ and only if $\gamma/\gamma^* < c_1/c$. Otherwise, $y^{*'} > 0$.

References

Antweiler, Werner, Brian R. Copeland, and Scott M. Taylor, "Is Free Trade Good for the Environment?," *American Economic Review*, 91(4):877-908, September 2001.

[28] The H-type separating equilibrium in Bond and Gresik (2004), where $p - MC_{y^*} - \gamma^*b\theta = 0$ for firms with $\underline{\theta}$, does not exist here because the marginal social cost is lower for a firm with a higher θ by assumption so that, at its perfect information best responses, the foreign government always sees a higher extractable profit from firms with a higher θ.

Bernheim, Douglas B. and Michael Whinston, "Common Marketing Agency as a Device for Facilitating Collusion," *RAND Journal of Economics*, 16:269-81, 1985.

– – and – –, "Common Agency," *Econometrica*, 54:923-43, 1986.

Bhagwati, Jagdish and Robert E. Hudec (eds.), *Fair Trade and Harmonization: Prerequisites for Free Trade? Volume 1. Economic Analysis*, Cambridge and London: MIT Press, 1996.

Biglaiser, Gary and Claudio Mezzetti, "Principals Competing for an Agent in the Presence of Adverse Selection and Moral Hazard," *Journal of Economic Theory*, 61:302-30, 1993.

– – and – –, "Incentive Auctions and Information Revelation," *RAND Journal of Economics*, 31(1):145-64, Spring 2000.

Bond, Eric W. and Thomas A. Gresik, "Regulation of Multinational Firms with Two Active Governments: A Common Agency Approach," *Journal of Public Economics*, 59:33-53, 1996.

– – and – –, "Competition between Asymmetrically Informed Principals," *Economic Theory*, 10:227-40, 1997.

– – and – –, "Information Sharing among Asymmetrically Informed Governments," *Working paper*, Vanderbilt University, 2004.

Bui, Linda T. M., "Gains from Trade and Strategic Interaction: Equilibrium Acid Rain Abatement in the Eastern United States and Canada," *American Economic Review*, 88(4):984-1001, 1998.

Caves, Richard, E. *Multinational Enterprise and Economic Analysis*, 2nd edition, New York and Melbourne: Cambridge University Press, 1996.

Copeland, Brian R., "Pollution Content Tariffs, Environmental Rent Shifting, and the Control of Cross-Border Pollution," *Journal of International Economics*, 40(3-4):459-76, May 1996.

– – and Scott M. Taylor, "Trade and Transboundary Pollution," *American Economic Review*, 85:716-37, 1995.

Dasgupta, Partha, Peter Hammond and Eric Maskin, "On Imperfect Information and Optimal Pollution Control," *Review of Economic Studies*, 47:857-60, 1980.

Dockner, Engelbert J. and Ngo Van Long, "International Pollution Control: Cooperative versus Non-Cooperative Strategies," *Journal of Environmental Economics and Management*, 25(1):13-29, July 1993.

Epstein, Larry G. and Michael Peters, "A Revelation Principle for Competing Mechanisms," *Journal of Economic Theory*, 88(1):119-60, 1999.

Eskeland, Gunnar S. and Ann E. Harrison, "Moving to Green Pastures? Multinationals and the Pollution Haven Hypothesis," *Journal of Development Economics*, 70(1):1-23, February 2003.

Folmer, Henk, Pierre V. Mouche, and Shannon Ragland, "Interconnected Games and International Environmental Problems," *Environmental and Resource Economics*, 3(4):313-35, 1993.

Fudenberg, Drew and Jean Tirole, *Game Theory*, Cambridge: MIT Press, 1991.

Gal-Or, Esther, "A Common Agency with Incomplete Information," *RAND Journal of Economics*, 22:274-86, 1991.

Grossman, Gene M. and Alan B. Krueger, "Environmental Impacts of a North American Free Trade Agreement," in Peter M. Garber (ed.), *The U.S.-Mexico Free Trade Agreement*, Cambridge: MIT Press, 13-56, 1993.

– – and – –, "Economic Growth and the Environment," *Quarterly Journal of Economics*, 110(2):353-77, May 1995.

Grossman, Sanford and Oliver Hart, "An Analysis of the Principal-Agent Problem," *Econometrica*, 51:7-45, 1983.

Guesnerie, Roger, "On Taxation and Incentives: Further Remarks on the Limits to Redistribution," *Bonn Discussion Paper*, 89, 1981.

– –, *A Contribution to the Pure Theory of Taxation*, Cambridge, New York and Melbourne: Cambridge University Press, 1995.

Hoel, Michael, "Coordination of Environmental Policy for Transboundary Environmental Problems?," *Journal of Public Economics*, 66(2):199-224, November 1997.

Hurewicz, Witold, *Lectures on Ordinary Differential Equations*, Cambridge: MIT Press, John Wiley & Sons Inc., 1958.

Ivaldi, M. and David Martimort, "Competition under Nonlinear Pricing," *Annales d'Economie et de Statistique*, 34:72-114, 1994.

Jaffe, Adam B., Steven R. Peterson, Paul R. Portney, and Robert N. Stavins, "Environmental Regulation and the Competitiveness of U.S. Manufacturing: What does the Evidence Tell Us?," *Journal of Economic Literature*, 33(1):132-63, March 1995.

Javorcik, Beata K. Smarzynska and Shang-Jin Wei, "Pollution Havens and Foreign Direct Investment: Dirty Secret or Popular Myth?," forthcoming in *Contributions to Economic Analysis & Policy*, 2004.

Komen, Marinus H. C., Shelby Gerking and Henk Folmer, "Income and Environmental R&D: Empirical Evidence from OECD Countries," *Environment and Development Economics*, 2(4):505-15, October 1997.

Laffont, Jean-Jacques and Jean Tirole, "Privatization and Incentives," *Journal of Law, Economics, and Organization*, 7:84-105, Special Issue, 1991.

Low, Patrick and Alexander Yeats, "Do 'Dirty' Industries Migrate?," in Patrick Low (ed.), *International Trade and the Environment*, World Bank Discussion Papers, No. 159: 89-103, Washington DC.: World Bank, 1992.

Markusen, James, R., Edward R. Morey and Nancy Olewiler, "Competition in Regional Environmental Policies when Plant Locations are Endogenous," *Journal of Public Economics*, 56:55-77, 1995.

Martimort, David, "Multi-Principaux avec Anti-Selection," *Annales D'Écono-mie et de Statistique*, 28:1-37, 1992.

– –, "Exclusive Dealing, Common Agency, and Multiprincipals Incentive Theory," *RAND Journal of Economics*, 27:1-31, 1996.

– – and Lars Stole, "The Revelation and Delegation Principles in Common Agency Games," *Econometrica*, 70(4):1659-73, July 2002.

– – and – –, "Contractual Externalities and Common Agency Equilibria," *Working Paper*, University of Chicago, 2003.

Mezzetti, Claudio, "Common Agency with Horizontally Differentiated Principals," *RAND Journal of Economics*, 28:323-45, 1997.

Mirrlees, James, A., "An Exploration in the Theory of Optimum Income Taxation," *Review of Economic Studies*, 38:175-208, 1971.

Oates Wallace E. and Robert M. Schwab, "Economic Competition among Jurisdictions: Efficiency Enhancing or Distortion Inducing?," *Journal of Public Economics*, 35(3): 333-54, April 1988.

OECD, "Report on Trade and Environment to the OECD Council at Ministerial Level," *OECD Working Paper*, 3(47), 1995.

Olsen, Trond E. and Peter Osmundsen, "Strategic Tax Competition; Implications of National Ownership," *Journal of Public Economics*, 81:253-77, 2001.

– – and – –, "Spillovers and International Competition for Investments," *Journal of International Economics*, 59(1):211-38, January 2003.

Peters, Michael, "Common Agency and the Revelation Principle," *Econometrica*, 69:1349-72, September 2001.

– –, "Negotiation and Take it or Leave it in Common Agency," *Journal of Economic Theory*, 111(1):88-109, July 2003.

Rochet, Jean C., "Le Controle des Equations aux Derivees Partielles Issues de la Theorie des Incitations," *Thesis Universite Paris*, IX, 1986.

Selden, Thomas M. and Daqing Song, "Environment Quality and Development: Is There a Kuznets Curve for Air Pollution Emissions?," *Journal of Environmental Economics and Management*, 27(2):147-62, September 1994.

Spulber, Daniel F., "Optimal Environmental Regulation under Asymmetric Information," *Journal of Public Economics*, 35:163-81, 1988.

Stole, Lars A., "Mechanism Design under Common Agency," *Working Paper*, MIT, 1991.

Ulph, Alistair, "The International Trade and the Environment: A Survey of Recent Economic Analysis," in Henk Folmer and Tom Tietenberg (eds.), *The International Yearbook of Environmental and Resource Economics 1997/1998: A Survey of Current Issues*, Cheltenham: Edward Elgar, NH: UK and Lyme, 1998.

UNCTAD, *Environmental Management in Transnational Corporations*, United Nations: New York, 1993.

World Bank, *World Development Report 1992: Development and the Environment*, New York: Oxford University Press, 1992.

Wu, Xiaodong, *Three Essays on Multinational Firms*, Princeton University, 1-151, April 2000.

Xu, Xinpeng, "Do Stringent Environmental Regulations Reduce the International Competitiveness of Environmentally Sensitive Goods? A Global Perspective," *World Development*, 27(7):1215-26, 1999.

[5]

Trade Integration and Political Turbulence: Environmental Policy Consequences*

Per G. Fredriksson and Muthukumara Mani

Abstract

This paper contributes to the unresolved issue regarding the effect of economic integration on environmental policymaking. In particular, we discuss the joint impact of trade openness and political stability on environmental policymaking. Our theory predicts that the effect of trade integration on environmental policy is conditional on the degree of political stability. Trade integration affects the stringency of environmental policies due to changes in industry bribery behavior, and the effect is conditional on the degree of political stability. The empirical findings support the theory and are robust to alternative specifications. The stringency enhancing effect on environmental policy of trade integration is greater in politically stable countries.

KEYWORDS: Trade liberalization, environmental regulations, political instability, pollution haven, trade policy

*Per G. Fredriksson is at the Department of Economics, Southern Methodist University, Box 0496, Dallas, TX 75275-0496 (pfredrik@mail.smu.edu), and Muthukumara Mani is at the Environment Department, The World Bank, MC 5-511, 1818 H St., NW, Washington, DC 20433 (mmani@worldbank.org). We would like to thank the helpful referees, Rimjhim Aggerwal, Andrea Bigano, Richard Damania, Tom Fomby, Mads Greaker, Eckhard Janeba, Michael Keen, Daniel Millimet, Muhammet Pakdil, Arvind Subramanian, Olaf Unteroberdoerster, Jim Wollscheid, participants at presentations at Göteborg University, SMU, the Industrial Research Institute, the ASSA Meetings in Atlanta, the EAERE meetings in Bilbao, and in particular the editor, Don Fullerton, for many helpful comments and suggestions. Parts of this paper were written while the first author visited the Beijer Institute at the Royal Academy of Sciences, Stockholm, and the Department of Economics, Göteborg University, Sweden. He thanks these institutions for their hospitality. Travel support from SMU is also gratefully acknowledged. The views expressed in this paper are those of the authors and not those of the World Bank. The usual disclaimers apply.

This paper examines how increased economic integration influences the determination of environmental policies in countries with unstable political regimes.

The relationship between trade policy and environmental policy and quality is an unresolved issue in the scholarly and public debate. For example, Antweiler et al. (2001) and Dean (2002) find that trade liberalization, on the whole, is likely to improve environmental quality. Others argue that trade openness threatens environmental goals (see Esty and Geradin, 1998). The current discussion on the effects of economic integration on environmental policymaking has risen to the forefront of the public debate in part due to the well-publicized protests at recent major trade negotiation meetings (for example, at the WTO meeting in Seattle in 1999).[1]

Nonetheless, economic integration is progressing rapidly in several regions of the world, for example in the southern cone of Latin America (MERCOSUR), Asia (AFTA), North America (NAFTA), Europe (EU), and negotiations are under way for a Free Trade Area of the Americas (FTAA). However, some fundamental questions are still being asked, both in developed and developing economies, on the implications of further economic integration. In particular, does trade liberalization affect governments' ability to manage economic activity, i.e. does it limit governments' freedom of action on regulatory policies such as on the environment? If this would be the case, trade liberalization would contribute to creating "pollution havens".

When analyzing the effect of trade liberalization on environmental policymaking, we believe it beneficial to take domestic political conditions into account. It is often argued that political instability has large adverse economic and social effects, especially in countries where the government is driven by short-term motives, supporting narrow interest groups, and ignoring long-term costs (see, e.g., Rodrik, 1991).

While the effects of economic integration and political instability have been studied in separate strands of the literature, no attempt has been made to study the *interaction* between them (on economic integration see, for example, Fredriksson, 1999; Antweiler et al., 2001; and Dean, 2002; on political instability see, for example, Bohn and Deacon, 2000; Brett and Keen, 2000; and Fredriksson and Svensson, 2003).[2] We

---

[1] In the context of increasing globalization of capital flows, some observers have argued that gaps in national environmental standards draw the most polluting industries to developing countries, creating "pollution havens" and propelling a global "race to the bottom" in environmental standards (see Wilson, 1996, for a survey). Esty and Geradin (1998, p.7) argue that: "considerable evidence suggests that government officials, facing the prospect of reduced sales, lost jobs and diminished investment in domestic industries caused by competition with foreign companies whose costs are lower due to more lax environmental requirements, often choose not to elevate environmental standards and sometimes even relax enforcement of current standards."

[2] Antweiler et al. (2001) study the effect of trade openness on pollution levels, and show theoretically and empirically the scale, composition, and technique effects of trade liberalization. In their study, all countries will set stricter environmental policies due to real income increases following trade liberalization, but the total effect depends on factor endowments. On average, the pollution levels decline, however. Copeland (1994, 1996) investigates the welfare effects of trade and environmental policy reforms, Copeland and Taylor (1994, 1995) discuss the linkages between income, pollution, and trade, López (1997) analyzes the effect of trade liberalization on environmental resources (biomass) and deforestation in Ghana, Rauscher (1994), Bommer and Schulze (1999), Fredriksson (1999), and Damania

aim to close this gap in the literature. We seek to shed light on whether the effects of trade openness and political stability reinforce each other, or whether their impacts instead are opposed.

Our focus is on the impact of trade liberalization on the political forces forming environmental policy, in particular in the presence of domestic political instability. Our theoretical model builds on Bernheim and Whinston's (1986) and Grossman and Helpman's (1994) common agency model, which recently has been extended to environmental policymaking by, for example, Damania (2001).[3] In our model, a lobby group representing the polluting industry aims to influence the incumbent government's environmental policy choice by promising a bribe (or campaign contribution) in return for a favorable policy decision. The incumbent government values both the bribe and aggregate social welfare, and the relative weight on welfare is viewed as a useful measure of government corruptibility.[4] We focus the discussion on relatively dishonest (corrupt) regimes, which put a great weight on bribes relative to social welfare.

In a three-stage small open economy model, we analyze the effects of economic integration on the behavior of the incumbent government and a producer lobby group. In the first stage, the government receives a bribe schedule by the lobby group, which relates the size of the bribe to the attractiveness of the government's environmental policy choice. When formulating its bribe offer, the lobby takes the probability of eventual policy implementation into account, i.e. the level of political stability. For policy implementation to occur, the incumbent government must remain in power throughout the (third) policy implementation stage. In the second stage, the government chooses its optimal environmental policy, taking into account its own turnover probability. It receives in return a bribe corresponding to its policy choice. In the third stage, the environmental policy previously selected by the government is implemented, given that it remains in power. The political crisis causing the departure of the government could take the form of, for example, a coup d'etat or a vote of no

et al. (2003) study the effect of trade liberalization on environmental policymaking, the latter in the presence of corruption. Hillman and Ursprung (1994) study the effect of environmental lobby groups on trade and environmental policies. These papers ignore political stability, however. Bohn and Deacon (2000) study the effect of ownership security (a function of political violence and instability) on deforestation and oil extraction, Brett and Keen (2000) explore politicians' incentive to earmark pollution tax revenues in the presence of political uncertainty, and Fredriksson and Svensson (2003) study the interaction between corruption and political instability on agricultural sector environmental policies. This literature ignores the issue of trade liberalization. Thus, no previous study exists on the joint effects of trade openness and political stability on environmental policymaking.

[3] While several authors in the literature use the basic structure of Grossman and Helpman's (1994) model to study environmental and trade policies, the joint effects of trade integration and political instability have not been analyzed using this common agency model.

[4] While in some countries the transfer of funds to politicians is legal (e.g., campaign contributions from political action committees in the U.S.), in this paper we view all monetary gifts to the government corruption. Corruption here implies that the policy maker is willing to deviate from a welfare maximizing policy in return for monetary gifts from a lobby group.

confidence. If the incumbent government's policy is not implemented, the lobby group does not benefit from the bribe paid in the second stage.[5]

In this paper, we show that trade integration affects both the producer lobby's and the government's incentives in the policy formation process, and that these incentives are conditional on the degree of political stability.[6] We find that trade integration in a polluting sector has two effects on the pollution tax. First, as output falls, the lobby has less at stake and its bribery effort declines accordingly. This "bribery reduction effect" of trade integration causes the pollution tax to rise. Second, note that the government has a welfare maximizing incentive to tax excess output and pollution created by protection. This incentive declines as a result of trade integration, a "welfare effect". When the government is sufficiently dishonest (corrupt), the bribery reduction effect dominates. In this case, trade integration raises the pollution tax. The model also predicts that the effect of trade integration is conditional on the degree of political stability.

Since political instability reduces both the lobby's incentive to seek influence, and the government's incentive to deliver welfare, the changes in the intensities of these motives resulting from trade integration are moderated by political instability. In the limit when political stability is extremely low, trade integration has close to no effect on environmental policy, because both the bribery and welfare effects of trade integration are negligible. This is a novel finding in the literature, to our knowledge.

The prediction generated by the model is evaluated using cross-country data from 92 non-OECD and 26 OECD countries. Our empirical work supports our theory, since we find that trade openness is associated with more stringent environmental policy. This suggests that the "bribery reduction effect" dominates the "welfare effect". We also find that the marginal effect of openness is conditional on the degree of political stability. In particular, the stringency enhancing effect of trade integration on environmental policy is *greater* in more stable countries. Our empirical results are robust to several alternative measures of the degree of economic integration and political stability. We believe they may serve to inform the policy debate of the likely effects on environmental policymaking of further trade integration.

The paper is organized as follows. Section 1 sets up the theoretical model and Section 2 discusses the effects of trade integration and political stability on environmental policy. Section 3 presents our empirical model and data, and Section 4 discusses the empirical results. Section 5 offers a brief conclusion.

---

[5] In other words, bribe payments are not conditional on the policy being implemented and there is no legal recourse forcing the return of given bribes. Another implicit assumption is that a government dislodged from power will not seek reelection, and hence has no incentive to either placate the lobby or address welfare questions (see also Coate and Morris, 1999).

[6] Rose (2004) reports that GATT/WTO members do not appear to be more open to international trade, and we therefore disregard the possibility that reductions in trade barriers are reciprocal.

1. The Model

A small open economy has two sectors. The "clean" sector produces a numeraire good *z*, and the polluting sector produces a good *x*. The economy is populated by two types of individuals, consumers (denoted by *S*) and factor owners (*F*). The two population groups are of size *s* and *f*, respectively, and the population is normalized to 1. We assume that the factor owners are a highly concentrated group such that their share of the population is approximately zero, i.e. f=0 and s =1. All individuals have labor income, factor owners in addition have factor income from ownership of a sector-specific factor. The consumers derive disutility from the pollution associated with the local production. An individual *k*, k=*S,F,* has a utility given by[7]

$$U^k = c^{zk} + u^k(c^{xk}) - \delta^S \theta X, \tag{1}$$

where c^{zk} and c^{xk} are consumption of the numeraire good *z* and good *x* by a type *k*, with world market prices equal to 1 and p^w, respectively. The world market price p^w is exogenously given as the country is a price taker. $u(c^{xk})$ is a strictly concave and differentiable sub-utility function. δ^S is an indicator variable which takes a value of one if the individual is a consumer, and zero otherwise.[8] While *s* is defined as the share of consumers in the total population, it is also the marginal disutility of pollution, since the share *s* has disutility 1 and the others have disutility 0. Production of *x* is given by *X*, and θ is the per-unit damage function. The government regulates pollution by levying a pollution tax $t \in T$, $T \subset \Re$, on each unit of damage from polluting production activities. If the polluting sector is import-competing (exporting), it benefits from an ad-valorem import tariff (export subsidy), τ.[9] Whereas the pollution tax is set by the government (relatively few pollutants are covered by international agreements), the tariff is assumed determined entirely by multilateral trade negotiations which this small country government is unable to influence. All individuals thus take the tariff rate as given.[10] Let subscripts denote partial derivatives. The per-unit damage is a function of the amount of abatement, *A*, carried out per unit of output, such that $\theta = \theta(A)$, where $\theta_A < 0$, and $\theta_{AA} > 0$.

An individual *k* spending Y^k consumes $c^{xk}=d(p^w(1+\tau))=u_c^{-1}$ and $c^{zk}=Y^k-p^w(1+\tau)d(p^w(1+\tau))$, where *d*(.) is the demand function for good *x*. Thus, the indirect

---

[7] Corner solutions may result with quasi-linear preferences. We assume interior solutions, however.

[8] Our results would be unaltered if we assumed that factor owners also suffered disutility from pollution. as long as they remain in favor of a lower pollution tax. We opt to keep the model as simple as possible.

[9] From now on we discuss only the import-competition (tariff) case, since import competing sectors generally receive more protection. Identical results hold in the case of an exporting sector with an export subsidy. Note that export taxes are illegal in some countries, for example the U.S. We therefore ignore the issue of export taxes, although we recognize that they are used in some developing countries.

[10] The exogeneity assumption is supported by our empirical findings, see Section 3 below. Moreover, as pointed out by Ederington (2000), the success of GATT negotiations in reducing worldwide tariff barriers has resulted in a shift of attention to domestic policy instruments as secondary trade barriers.

utility function of a consumer is expressed as $V^S(p^w,\tau,Y^S)=Y^S+C(p^w(1+\tau))-\theta X$, where $C(p^w(1+\tau))= u[d(p^w(1+\tau))]-p^w(1+\tau)d(p^w(1+\tau))$ is the consumer surplus derived from consumption of good x. There is no consumer surplus from consumption of good z.

Each individual has a unit of labor and the total labor endowment equals l. Good z is produced by labor alone with a constant returns to scale technology, and an input-output coefficient equal to one. The labor supply is sufficiently large for the supply of this good to be positive which implies a wage rate equal to one. The inputs into production of good x are labor and a sector-specific factor. The technology is constant returns to scale. Ignoring labor costs, producers of good x face a net price given by $p = p^w(1+\tau) - t\theta(A) - A$, and the specific factor reward depends entirely on the producer's net price, p, i.e. $\pi(p)$. The first-order condition with respect to abatement equals $\pi_A = -X(t\theta_A + 1) = 0$, which in turn is used to find $A_t = -\theta_A/(t\theta_{AA}) > 0$. The supply curve for good x is given by Hotelling's Lemma, i.e. $X(p) = \pi_p(p)$, where $X_p > 0$, and $X_{pp} = 0$.[11] Imports of the polluting good are given by $M(p) = p^w[d(p^w(1+\tau)) - X(p)]$. Net aggregate tax and tariff revenues are given by

$$R(t,\tau) = t\theta X(p) + \tau M(p), \tag{2}$$

and are assumed distributed equally to all individuals as lump-sum.

The income obtained by the owners of the sector-specific factor depends on the environmental policy (as well as on trade policy). Factor owners are assumed able to organize into a lobby group that coordinates a prospective bribe offer to the incumbent government. The consumers are assumed to face sufficiently severe free-riding problems to be unable to organize political action (see Olson, 1965). The model defines a three-stage game between the government and the lobby. Both players are risk neutral. The timing assumptions are as follows:

Stage 1. In stage one the lobby group offers the incumbent government a bribe schedule $\Lambda(t)$. The lobby's strategy hence consists of a continuous function $\Lambda(t): T \rightarrow \Re$, i.e., it offers a specific bribe for selecting a policy t. The lobby faces uncertainty on whether the incumbent government will remain in power long enough for the lobby to reap a reward to its bribe (policy implementation occurs only in stage three). In stage one, the lobby assigns a probability $0 < \gamma < 1$ that the government will remain in office, and a probability $(1-\gamma)$ that it will thrown out. Thus, γ is a measure of the degree of political stability. All players also assign a probability $0 < \lambda < 1$ that the chosen policy is implemented by the *new* government, in the event the incumbent leaves office early.[12] We make a simplifying assumption about this parameter. **Assumption 1:** The probability λ is small.[13]

---

[11] This assumption simplifies the presentation of our theoretical results, but does not change them.

[12] Fredriksson and Svensson (2003) make an identical assumption. For example, the new government may have other policy priorities before the next election, the process of hiring new officials in charge of

Stage 2. In the second stage the incumbent government proceeds to set its optimal environmental policy, given the lobby group's strategy. The government also collects the associated bribe from the producer lobby.[14] Bribes are used for the incumbent politicians' personal consumption during this stage.

Stage 3. In the third stage, the selected policy is implemented, given that the incumbent government remains in power, or in the event that the successor keeps the policy selected by its predecessor. Producers determine output and abatement levels. Turnover could occur, for example, because of a vote of no confidence, or a coup attempt. From the lobby's perspective in stage one, policy implementation occurs with a probability $\gamma+(1-\gamma)\lambda$. This is the probability that all policy-favors "purchased" by the lobby will be delivered. With a probability $(1-\gamma)(1-\lambda)$ the incumbent is removed from office and the policy is not implemented.[15] For simplicity, the game between the lobby and the incumbent simply ends here in this event. Since the challenger did not receive a bribe within the period, it is not committed to its predecessor's policy promises, although it may choose to implement its predecessor's policy choice. In the event the challenger decides to change the predecessor's environmental policy in stage three, the new government is assumed to set an exogenous tax t^c, until a new lobbying game starts between itself and the lobby.

The lobby takes the political stability level into account in its formulation of its bribe schedule. The gross (indirect) utility of the lobby group is therefore given by the expected value of specific factor income,

$$E[\Omega^F(t,\tau)]\equiv\pi(p)[\gamma+(1-\gamma)\lambda]+\pi(p^c)(1-\gamma)(1-\lambda), \tag{3}$$

where $E[.]$ is the expectations operator and $\pi(p)$ is aggregate factor income. This factor income is received in the event that the policy determined by the incumbent materializes. Since $f=0$, the lobby ignores tax and tariff revenues, as well as consumer surplus (it receives a negligible share). If the incumbent government is

---

carrying out environmental policy and the process of agenda setting may involve delays, or the government may want to avoid that environmental policy becomes an election issue. Thus, we view the new government as having an objective function identical to the incumbent, but due to some exogenous circumstance it may delay the taking of bribes in return for environmental policy choices.

[13] This assumption implicitly implies that the new government must care sufficiently about receiving a bribe in return for setting environmental policy.

[14] Neither the lobby group, nor the government, is assumed to renege on their promises in the second or third stages. Since corrupt government politicians likely have a monopoly on selling policy favors, it appears plausible that they have a relatively great bargaining power, and are able to demand that a bribe must be paid before any industry friendly policy action is taken. Our timing assumption also sharpens the focus on the considerate resources that industry groups invest in political connections and bribery, which is supported by empirical evidence.

[15] We abstract from possible strategic choices by the lobby, and we do not model the bribery game between the lobby and the new government. While it would be preferable to include the new government's actions into the model, we opt to keep it as simple as possible and thus focus on the incumbent government policy choice only.

removed and the successor chooses a different policy, the lobby's exogenous factor income equals $\pi(p^c)$, where $p^c = p^w(1+\tau) - t^c\theta(A) - A$.

The incumbent government values bribes and aggregate social welfare. Bribes are used for personal consumption, and social welfare is of relevance because the incumbent is more likely to win future re-election, the greater is average welfare. However, aggregate welfare is assumed of value to the incumbent government only if it stays in office. The incumbent government's policy choice influences its own welfare only if it stays in power, in which case aggregate social welfare is given by

(4) $$\Omega^A(t,\tau) \equiv \pi(p) + l + C(p^w(1+\tau)) + R(t,\tau) - s\theta X(p),$$

which expresses the sum of all individuals' aggregate factor rewards, labor income, consumer surplus, tax and tariff revenues, take the consumers' aggregate disutility from pollution. Note that taking the first-order condition of the expression for aggregate social welfare, (4), yields the second-best tax, $t^* = s + \tau p^w X_p \theta / (\theta^2 X_p - \theta_A A_t X) > s$, where s is the first-best pigouvian tax (i.e., under free trade). Since the presence of a tariff stimulates output, the optimal tax exceeds the marginal disutility of pollution, s. If the incumbent government loses power, it has no interest in aggregate social welfare, and from the incumbent government's point of view it equals zero.

The incumbent government thus has an objective function equal to

(5) $$E[I(t,\tau)] \equiv \Lambda(t) + h\gamma\Omega^A(t,\tau),$$

a weighed sum of the bribe and the expected aggregate social welfare. The exogenous parameter h is the government's weight on welfare relative to bribes, which in our view reflects the degree of government honesty (absence of corruption). In our model, the bribe aims to influence government policy and not elections (see also Schulze and Ursprung, 2001; Fredriksson and Svensson, 2003). López and Mitra (2000) employ a similar formulation in their investigation of the effect of corruption on the relationship between income and environmental quality. The government trades off the size of the bribe (which it enjoys with probability 1) with the expected value of aggregate social welfare.

The subgame perfect Nash equilibrium in the well-known model by Grossman and Helpman (1994) can be found using two necessary conditions:

(i) $$t^\circ = \arg\max_t \Lambda^\circ(t) + h\gamma\Omega^A(t,\tau) \text{ on } \mathrm{T};$$

(ii) $$t^\circ = \arg\max_t [\Omega^F(t,\tau) - \Lambda^\circ(t)] + [\Lambda^\circ(t) + h\gamma\Omega^A(t,\tau)] \text{ on } \mathrm{T}.$$

The equilibrium pollution tax policy t° simultaneously maximizes the government's utility function [condition (i)] and the joint utility of the lobby and the incumbent [condition (ii)], given the turnover probability. The equilibrium characterization is found by taking the first-order conditions of (i) and (ii), which yields

(6) $$\Lambda_t^{\circ}(t^{\circ}) + h\gamma\Omega_t^A(t^{\circ},\tau) = 0,$$

and

(7) $$[\Omega_t^F(t^{\circ},\tau) - \Lambda_t^{\circ}(t^{\circ})] + [\Lambda_t^{\circ}(t^{\circ}) + h\gamma\Omega_t^A(t^{\circ},\tau)] = 0.$$

Substituting (6) into (7) yields $\Omega_t^F(t^{\circ},\tau) = \Lambda_t^{\circ}(t^{\circ})$, which reflects the fact that the bribe schedule is locally truthful, as discussed by Grossman and Helpman (1994). The characterization of the equilibrium pollution tax is found by substituting this condition into (6), which yields

(8) $$\Omega_t^F(t^{\circ},\tau) + h\gamma\Omega_t^A(t^{\circ},\tau) = 0.$$

Differentiation of equations (3) and (4) with respect to the pollution tax yields

(9) $$\Omega_t^F(t,\tau) = -\theta X(\gamma + (1-\gamma)\lambda),$$

and

(10) $$\Omega_t^A(t,\tau) = (t-s)(\theta_A A_t X - \theta^2 X_p) + \tau p'' X_p \theta.$$

Substituting expressions (9) and (10) into equation (8), and rearranging, we find an explicit expression for the equilibrium characterization given by

(11) $$\underbrace{-\theta X[\gamma + (1-\gamma)\lambda]}_{A} + \underbrace{h\gamma[(t^{\circ} - s)(\theta_A A_t X - \theta^2 X_p) + \tau p'' X_p \theta]}_{B} = 0.$$

Note that the second term in (11) is adjusted by γ, rather than $(\gamma + (1-\gamma)\lambda)$, since the incumbent government does not benefit from its policy choice in the event it leaves office. Note also that the equilibrium tax rate t° is smaller than the second-best tax, which the government would choose in the absence of instability and lobbying. Since term A in (11) is negative, term B is positive, which holds only if $t^{\circ} < s + \tau p'' X_p \theta / (\theta^2 X_p - \theta_A A_t X)$.

2. Trade Integration and Political Stability

We now analyze the effects of trade integration on environmental policymaking, accounting for political stability. Note that "trade integration" in our analysis is identical to a cut in the tariff rate. The aim is to derive testable hypotheses for our empirical work carried out in the subsequent sections. We find the following proposition.

Proposition 1: (*i*) *In equilibrium, trade integration unambiguously causes the pollution tax rate to rise when (a) the degree of government honesty is low, or (b)* $(s-t)\theta_A A_t > \theta$; (*ii*) *The effect is conditional on the degree of political stability. For sufficiently low government honesty, the effect is increasing in political stability.*

Proof: (*i*) Differentiation of Eqn. (11) with respect to the tariff equals

$$\frac{\partial t}{\partial \tau} = \frac{\{\theta\overbrace{[\gamma+(1-\gamma)\lambda]}^{A} - h\gamma\overbrace{[(t-s)\theta_A A_t + \theta]}^{B}\}X_p p^w}{|D|}, \tag{12}$$

where $|D| = \theta^2 X_p[\gamma+(1-\gamma)\lambda] + h\gamma\psi$, and where $\psi \equiv \{(\theta_A A_t X - \theta^2 X_p) + (t-s)[X(\theta_{AA}A_t^2 + \theta_A A_{tt}) - 3\theta\theta_A A_t X_p] + \tau p^w X_p \theta_A A_t)\}$. $|D|$ is the second-order condition of the government's maximization (8), which is required to be negative for a maximum. We assume this to be the case, which requires $\psi < 0$. The denominator of (12) is consequently negative. In the numerator, term A is positive, whereas term B is ambiguous in sign. We have two cases: (a) Term B is positive. For sufficiently low values of honesty, h, (equivalent to sufficiently *high* corruption), such that $h < \theta[\gamma+(1-\gamma)\lambda]/\{\gamma[(t-s)\theta_A A_t + \theta]\}$, the numerator is positive and thus expression (12) is negative. Trade integration leads to an increase in the pollution tax if government honesty is sufficiently low. (b) Term B is negative. In this case, (12) is unambiguously negative for all values of h. Thus, trade integration leads to an increase in the pollution tax.

(*ii*) Differentiation of (12) yields

(13)
$$\frac{\partial^2 t}{\partial \tau \partial \gamma} = \underbrace{\frac{\overbrace{\{\theta(1-\lambda) - h\overbrace{[(t-s)\theta_A A_t + \theta]}^{A}\}}^{B} X_p p^w}{|D|}}_{I} - \underbrace{\frac{\overbrace{\overbrace{(\theta^2 X_p(1-\lambda) + h\psi)}^{C}\overbrace{\{\theta[\gamma + (1-\gamma)\lambda] - h\gamma[(t-s)\theta_A A_t + \theta]\}}^{E}}^{F} X_p p^w}{|D|^2}}_{II}.$$

The sign of (13) is indeterminate. We seek to evaluate the sign of (13), focusing on parts *I* and *II* separately (one at a time). The focus is on the role of *h* and cases where (13) is negative. ***Part I***: (a) If term *A* in the numerator is negative, term *B* is unambiguously positive for all values of *h*. (b) If term *A* is positive, term *B* in the numerator is positive for sufficiently low honesty, such that $h < \theta(1-\lambda)/[(t-s)\theta_A A_t + \theta]$. **Part *II***: Term *C* is positive (negative) for $h < (>)\theta^2 X_p(\lambda - 1)/\psi$. Term *E* is positive (negative) if $h < (>)\theta[\gamma + (1-\gamma)\lambda]/\gamma[(t-s)\theta_A A_t + \theta]$, when $[(t-s)\theta_A A_t + \theta] > 0$. Term *E* is unambiguously positive when $[(t-s)\theta_A A_t + \theta] < 0$. Term *F* is positive if *sign* (term *C*) = *sign* (term *E*).

Expression (13) shows how the effect of trade openness on the pollution tax $(\partial t / \partial \tau)$ is affected by a change in political stability, γ. For sufficiently low honesty *h*, terms *B*, *C*, and *E* are positive, yielding a negative sign of (13). In this case (see part (*i*) of the proposition), greater political stability raises the effect of trade integration on the pollution tax. However, (13) is negative also with a high *h* and $[(t-s)\theta_A A_t + \theta] > 0$, such that terms *C* and *E* are negative and thus *F* is positive. In this case, if part *II* dominates part *I* in absolute value, or when part *I* is negative, (13) is negative.

In sum, for sufficiently low *h*, $\partial^2 t / (\partial \tau \partial \gamma) < 0$, such that the effect of trade integration on the pollution tax is increasing with political stability. For other values of *h*, we know only that $\partial^2 t / (\partial \tau \partial \gamma) \neq 0$, i.e. the effect of trade integration on the pollution tax is conditional on political stability. ■

The intuition is the following.[16] (*i*) In this model, trade integration (a cut in the tariff rate) has two effects on environmental policy, and their size both depend on the degree of political stability. Term *A* in the numerator of (12) represents the reduction in the producers' incentive to seek a lower pollution tax as a result of trade integration

[16] The result is independent of environmental policy instrument used. With an emissions standard, trade integration would still result in a "bribery reduction effect" and a "welfare effect". Revenues do not drive our result. Note that an identical result is obtained if the trade policy used is an export subsidy.

(note from (9) that term A is the second derivative $\Omega_{t\tau}^{F}$). When output shrinks, less is at stake for the lobby in the political process. Thus, the bribe offer is reduced, and the pollution tax rises through this channel. We denote this the "bribery reduction effect" of trade integration.

Term B in the numerator of (12) reflects the change in the government's incentive to tax pollution for welfare reasons (note from (10) that term B is the second derivative $\Omega_{t\tau}^{A}$). As discussed above, the tariff introduces distortions in production and consumption, and increases the aggregate level of pollution damage. In this small open economy, only one policy instrument is available to address these distortions: the pollution tax. As discussed above, the second-best pollution tax therefore exceeds marginal damage, i.e. $t^{*} > s$. Trade integration reduces the marginal incentive to tax pollution for this reason. However, the net effect on the pollution tax is ambiguous, because of taxpayers' interest in tax revenues and the effect on profits. We denote this the "welfare effect" of trade integration.[17] If term B in (12) positive, the relative importance of the two effects (represented by terms A and B in (12), respectively) depends on the degree of honesty, h. In particular, when the degree of government honesty is low (corruption is high) such that $h < \{\theta[\gamma + (1-\gamma)\lambda]\} / \{\gamma[(t-s)\theta_A A_t + \theta]\}$, the bribery reduction effect dominates the welfare effect and trade integration raises the pollution tax. If term B is negative, trade integration unambiguously raises the pollution tax.

(*ii*) The two effects discussed above are adjusted for the probability of successful eventual policy implementation, reflected by the level of political stability, γ. Bribery and social welfare are important (to the lobby and the government, respectively) only as long as the government stays in power long enough to deliver and reap the benefit of its policy choice. In addition, the lobby also benefits from bribery to the extent that a new government would implement the policy paid for (without extra cost), which occurs with probability $(1-\gamma)\lambda$.

Changes in γ affect $\partial t / \partial \tau$ both via its impact on the "bribery reduction effect" and on the "welfare effect". For low government honesty, h, terms B, C, and D in (13) are all positive, and thus (13) is negative. In this case, the effect of γ occurs via the "bribery reduction effect", since welfare in this case is of minor importance. The main effect is here that political stability raises the lobby's incentive to offer a sizeable bribe. Changes in this incentive due to trade integration are therefore more pronounced with greater political stability.

For high levels of honesty, all terms in (13) have ambiguous sign, and the relative impact of political stability on the (i) bribery reduction effect and (ii) the welfare effect determines the sign of $\partial^2 t / (\partial \tau \partial \gamma)$. Greater political stability implies

---

[17] Note that since trade integration shrinks the pollution intensive sector, environmental damage falls, given the pollution tax. A falling pollution tax would therefore not necessarily be welfare-reducing. Our focus is on the determination of the pollution tax, however, and we abstract from welfare issues in this paper.

that social welfare is of greater concern to the incumbent government. As discussed above, trade integration lowers the incentive to tax pollution for second-best welfare reasons, and this incentive-reducing effect is relatively stronger where political stability is high.

3. Empirical Specification and Data

The theoretical model developed in the previous sections yields testable implications of the relationships between trade openness (trade integration), political stability, and environmental policy formation, expressed in Proposition 1. Our objective is to test these implications using cross-country data on environmental policy. Proposition 1 shows that for low levels of government honesty, trade integration raises environmental policy stringency, and this effect is stronger the greater the degree of political stability. However, the same outcome may occur also for high levels of honesty, and our empirical work will also seek to determine the sensitivity of our results to the level of government honesty.

The empirical estimation can be formulated as follows,

$$t_i = x'_i\beta^x + \beta^{\gamma}\gamma_i + \beta^{\tau}\tau_i^{-1} + \beta^{\gamma\tau}\gamma_i\tau_i^{-1} + \beta^h h_i + \varepsilon_i, \quad (14)$$

where t_i is the stringency of environmental policy in country i, x_i is a vector of controls, γ_i is the degree of political stability, τ_i^{-1} is the degree of trade openness (the inverted tariff rate), h_i is government honesty (absence of corruption), and ε_i is a zero mean error term. Whereas β^{γ}, β^{τ}, $\beta^{\gamma\tau}$ and β^h are coefficient scalars, β^x is a coefficient vector.

We now describe the variables used to test the prediction generated by our theory. Table 1 summarizes the descriptive statistics of variables used in the regression. Table A.1 in the Appendix contains a further description of the data and sources.

Our measure of the stringency of environmental regulations is a sub-index from the *2001 Environmental Sustainability Index* (ESI) developed by CIESIN (2001). We employ one of the core indicators as our measure of the stringency of environmental policies for 2000, which we denote **Environmental Stringency**. This is an aggregation of various indicators consisting of regulation and management, private sector responsiveness, science and technology, capacity for debate, environmental information, and eco-efficiency. It captures the extent to which the country has in place institutions and policies that result in effective responses to environmental problems (i.e., laws on the book, implementation, monitoring, and enforcement). It takes values between 0 and 100, where a higher value implies greater environmental policy stringency.[18]

[18] To our knowledge, our dependent variable is the best measure of environmental policy stringency available for a large number of countries during the time period studied.

Table 1.A. Descriptive Statistics: Full Sample

Variable	Number of Observ.	Mean	Standard Deviation	Minimum	Maximum
Environmental Stringency	118	46.94	18.38	18.1	92.3
Log (GDP)	114	8.37	1.09	6.2	10.3
Openness	113	2.78	1.19	1.0	5.0
Dishonesty	77	5.32	2.35	0.0	8.3
Honesty	113	0.02	0.91	-1.6	2.1
Political Stability	113	0.06	0.82	-2.4	1.7
% Non-Ag Labor	117	63.06	27.57	5.9	98.8
Economic Freedom	99	6.60	1.41	3.7	9.1
Democracy	86	0.23	0.42	0.0	1.0
Federal	86	0.20	0.40	0.0	1.0
War Dummy	98	0.30	0.46	0.0	1.0
Civil War Dummy	98	0.27	0.44	0.0	1.0
Common Law	86	0.29	0.46	0.0	1.0
Racial Tension	87	3.72	1.62	0.0	6.0
Ethno-religious Fractionalization	97	0.44	0.23	0.0	0.9
Ethno-linguistic Fractionalization	98	0.33	0.30	0.0	0.9
Civic Freedom	118	3.51	1.64	1.0	7.0
Dismag	77	0.55	0.39	0.0	1.0

Table 1.B. Descriptive Statistics: Developing Country Sample

Variable	Number of Observ.	Mean	Standard Deviation	Minimum	Maximum
Environmental Stringency	92	39.63	10.98	18.1	72.9
Log (GDP)	88	7.96	0.85	6.2	9.8
Openness	87	2.43	1.13	1.0	5.0
Dishonesty	51	6.53	1.27	2.7	8.3
Honesty	87	-0.35	0.55	-1.6	1.3
Political Stability	87	-0.21	0.67	-2.4	1.1
% Non-Ag Labor	92	56.38	26.91	5.9	98.8
Economic Freedom	73	6.10	1.25	3.7	8.4
Democracy	60	0.05	0.22	0.0	1.0
Federal	60	0.13	0.34	0.0	1.0
War Dummy	72	0.38	0.49	0.0	1.0
Civil War Dummy	72	0.33	0.47	0.0	1.0
Common Law	60	0.33	0.48	0.0	1.0
Racial Tension	61	3.25	1.51	0.0	6.0
Ethno-religious Fractionalization	71	0.49	0.22	0.0	0.9
Ethno-linguistic Fractionalization	72	0.40	0.31	0.0	0.9
Civic Freedom	92	4.01	1.43	2.0	7.0
Dismag	51	0.53	0.40	0.0	1.0

Political stability is not directly observable. However, a measure of political stability has been developed by Kaufmann *et al.* (1999a, 1999b) for the years 1997-98. The **Political Stability** index combines several indicators seeking to measure perceptions of the likelihood that the government in power will be destabilized or overthrown. It takes values from –2.5 to 2.5, where a higher value represents greater political stability. We also use **Racial Tension** as an alternate measure of political (in-)stability (Knack and Keefer, 1995). It is an index for countries experiencing racial tension and takes values from 1 (high tension) to 6 (low tension).

Our trade openness measure is an index developed by the Heritage Foundation and the *Wall Street Journal* (O'Driscoll *et al.*, 2000) (**Openness**). An economy earns a 5 if it has an average tariff rate of ≤4% and/or has very few non-tariff barriers, and a 1 if the average tariff rate is >19% and/or there are very high non-tariff barriers that virtually prohibit imports. Thus, we expect that **Openness** will take a positive sign. We also use a more comprehensive (general) measure of economic liberalization and openness, the **Economic Freedom** index, compiled by the Fraser Institute (Gwartney *et al.*, 2000). The **Economic Freedom** index is a composite index of ten individual indices for 1997. It takes values between 0 and 10, where a higher value for the index indicates a greater degree of economic liberalization and freedom.

A main implication of our theory is that the effect of trade integration is conditional on the degree of political stability. We therefore include the relevant interaction variables **Openness×Political Stability** and **Economic Freedom×Political Stability**.

We have two proxies for government honesty. One is an index developed by Kaufmann *et al.* (1999a, 1999b), **Honesty**.[19] It measures perceptions of honesty in government in a country, or more precisely, the absence of the use of public power for private gain. The index takes values from –2.5 to 2.5, where a higher value implies more honesty, and thus we expect a positive sign. Our second government honesty measure is the Corruption Perceptions Index (**Dishonesty**) developed by Transparency International, which measures the "perceptions of the degree of corruption as seen by business people, risk analysts, and the general public." The index is computed as the sample average of a number of different surveys assessing each country's performance. **Dishonesty** ranges between 0 (perfectly clean) and 10 (highly corrupt).[20] We expect a negative sign.

Proposition 1 suggests that the level of government honesty may play a role for the effect of openness on environmental policy, and the interaction effect with -political stability. To investigate the impact of government honesty on the relationships of interest, we therefore experiment with different cut-off values for **Honesty**.

---

[19] The indicator reflects the statistical compilation of perceptions of the quality of governance of a large number of survey respondents in industrial and developing countries, as well as non-governmental organizations, commercial risk rating agencies, and think tanks during 1997 and 1998.

[20] The index is inverted in the scale from the original data by subtracting values from 10 to make results more intuitive. A number of recent empirical studies of corruption have employed this index, including Persson et al. (2000), Fisman and Gatti (2002), and Fredriksson et al. (2004).

Given the possible endogeneity of the **Honesty**, **Political Stability**, and **Openness** variables, OLS is expected to lead to biased results. We therefore test for the necessity of an instrumental variable approach (whether the set of estimates obtained by least squares are consistent or not) by using an augmented regression test (the Durbin-Wu-Hausman test) (Davidson and MacKinnon, 1993). This is carried out by including the residuals of each endogenous right-hand side variable as a function of all exogenous variables in a regression of the original model.[21] A smaller p-value for **Honesty** and **Political Stability** in our model indicates that they are endogenous and that OLS is inconsistent. However, trade openness was found to be *exogenous* and hence can be treated as predetermined. We correct for the endogeneity of corruption and political stability by using Two-Stage Least Squares (2SLS). We also test for possible heteroskedasticity and correct it using White's correction.[22]

Our theoretical model, as well as the previous empirical literature (see, for example, Fredriksson and Svensson, 2003), is informative concerning control variables to include in regression (14). Two sets of controls capture demand factors and structural features of an economy that may influence environmental policy in alternative ways that are not the focus of the present paper. We thus include the log of purchasing power adjusted per capita GDP, **Log GDP**.

Industrial environmental policies are also influenced by the workers employed in polluting sectors. The greater their stake in the policy outcome, the greater their political pressure and success. However, Olson's (1965) theory of free-riding implies that political influence may decline as the size of an interest group increases. We use the proportion of the total labor force working in non-agricultural sectors (**% Non-Ag. Labor**) as our measure of worker political pressure on industry environmental policies. The labor force here comprises all individuals who meet the International Labor Organization's definition of the economically active population. The expected sign is indeterminate. The remaining control variable seeks to adjust for structural differences between countries.[23]

Given the endogeneity of **Honesty**, **Dishonesty** and **Political Stability**, we use a number of instrumental variables to test and correct for the bias. For example, La Porta et al. (1999), Fisman and Gatti (2002), and Persson et al. (2003) discuss factors determining corruption. We model the determinants of government honesty into two main categories, namely standard economic controls, and political and legal history. Based on the literature, we selected the following instruments: a dummy for common law system (**Legal Origin**) from La Porta et al. (1999) (also used by Fredriksson and Svensson, 2003), an indicator of the federal nature of government (**Federation**) (see Fisman and Gatti, 2002), and a variable measuring electoral district magnitudes

[21] This is an augmented form of the Hausman test for contemporaneous correlation between the error term and the regressors, used to test exogeneity of variables (Hausman, 1983).

[22] See White (1980) and Greene (1997) for White's correction to standard errors for an unknown form of heteroskedasticity.

[23] We do not include a measure of the marginal damage from pollution since this will be determined by environmental policy.

(**Dismag**). **Dismag** comes from Persson et al. (2003) who argue that electoral competition becomes stiffer in countries with single-member districts than in countries with a single nation-wide district. As a result, candidates become more focused and disciplined and hence demonstrate less rent seeking behavior and more honesty.

We hypothesize that political stability is to a large extent a function of the prevailing economic, political and social factors. We use **Racial Tension, Ethno-religious Fractionalization**, dummies for countries with recent history of **War** and **Civil War** as our instruments for **Political Stability**. The test of over-identifying restrictions was applied to the various sets of instruments. It indicates that our instruments are valid.[24]

4. Empirical Results

We empirically test our model for a sample of 118 countries for which the **Environmental Stringency** variable is available, of which 92 are non-OECD countries. Table 2 presents the first-stage regressions for **Dishonesty**, **Honesty**, and **Political Stability**, identifying various instruments. They are able to explain a large portion of the variation in **Dishonesty**, **Honesty**, and **Political Stability**. A number of instruments are significant at conventional levels, in addition to the level of income (**Log GDP**). **Dishonesty** is driven by the degree of decentralized structure of government (**Federal Dummy**), consistent with the literature suggesting that a federal structural of government adds further layers of bureaucracy and increases rent seeking opportunities. The existence of common law system (**Legal Origin**) appears to reduce **Dishonesty** levels, consistent with earlier research findings which suggest that 'civil law' countries with higher expected duration of judicial proceedings, less consistency, less fairness in judicial decisions and more corruption compared with 'common law' countries (see Djankov et al., 2002).[25] The **Honesty** equation suggests that curtailing basic civic rights may reduce government honesty by vesting power in a few hands.

Political Stability appears to be associated primarily with low levels of ethno-religious fractionalization and racial tension, and higher levels of economic development. This is consistent with the literature which argues that that ethnic and religious fractionalization proxies for the degree of conflict in society, which fuels political instability (see Annet, 2001).

The estimation results for the **Environmental Stringency** equation for the full sample are presented in Table 3. The models contain estimates from both OLS and 2SLS regressions using different controls. The OLS results are presented for comparison. The results suggest that the point estimates for our key variables such as **Openness** and **Political Stability** are more robust and significant at higher levels when

---

[24] The order condition necessary for identification is satisfied since the **Dishonesty**, **Honesty** and **Political Stability** equations are over-identified.

[25] La Porta et al. (1999) and Fredriksson and Svensson (2003) have previously found that legal origin is a good instrument for the degree of corruption.

Table 2: First Stage Regressions

Variable	Dishonesty	Honesty	Political Stability
Log GDP	-2.04 (6.1)***	0.91 (7.2)***	0.44 (3.4)***
War Dummy			-0.17 (1.3)
Civil War Dummy			-0.21 (1.1)
Ethno-religious Fractionalization			0.74 (2.8)***
% Non-Ag. Labor	0.01 (0.4)	-0.007 (1.2)	-0.002 (0.4)
Openness	-0.09 (0.6)	0.06 (1.0)	0.04 (0.6)
Federal Dummy	0.69 (1.8)*	-0.17 (1.18)	
Racial Tension			0.14 (2.6)***
Dismag	-0.10 (0.2)	-0.13 (0.7)	
Legal Origin	-1.44 (3.8)***	0.46 (3.1)***	
Civic Freedom		0.32 (2.0)*	
Constant	23.3 (11.0)***	-7.38 (9.61)***	-4.34 (5.3)***
R^2	0.716	0.750	0.585
Observations	77	79	87

Notes: t-statistics in parenthesis. *(**) [***] statistically significant at the 10 (5) [1] percent level.

the endogeneity is taken into account.[26] The results support our theory and the estimates appear robust under alternate specifications. Moreover, all variables that are statistically significant have the expected sign.

The **Openness** and **Economic Freedom** variables are significant across most models, indicating that countries with more liberal trade policies, which are more economically integrated with the rest of the world, tend to set more stringent environmental policies. First, protection tends to worsen pollution because it stimulates output. Tariff reduction (**Openness**) by itself reduces output and pollution, which makes

[26] As Table 2 reveals, we are able to explain a large portion of the variation in **Dishonesty** and **Political Stability** in our first stage regressions.

it less profitable to bribe the government to reduce stringency. Thus, the "bribery reduction effect" of trade openness is to make environmental policy more stringent. In the opposite direction, because openness reduces the need for domestic production of the polluting good, the "welfare effect" means a lower need for stringent environmental policy. The empirical finding is that liberal trade policies lead to more stringent environmental policies. This indicates that the bribery reduction effect of trade integration dominates the welfare effect.

To gain an understanding of the full effect of trade openness on environmental policy, we need to consider the interaction effect identified by our theory. The interaction variables **Openness×Political Stability** and **Economic Freedom×Political**

Table 3: Environmental Stringency Regressions (OLS and 2SLS)

Variable	OLS				2SLS		
Log GDP	5.39 (2.3)**	4.42 (2.6)**	4.07 (2.1)**	3.15 (1.5)	3.18 (0.8)	3.80 (1.0)	1.91 (0.4)
Dishonesty	-4.52 (7.3)***		-4.05 (6.6)***	-4.99 (9.3)***	-3.42 (2.4)**	-2.10 (1.1)	
Honesty		12.75 (7.8)***					8.86 (2.2)**
% Non-Ag. Labor	-0.11 (1.7)*	-0.11 (2.3)**	-0.06 (1.0)	-0.01 (0.2)	-0.05 (0.5)	-0.02 (0.3)	-0.03 (0.0)
Openness	2.26 (2.1)**	2.73 (3.5)***		-0.98 (0.6)	3.74 (2.3)**		3.71 (2.3)**
Economic Freedom			2.95 (3.2)***			3.86 (3.6)***	
Political Stability	-10.72 (2.3)**	-6.10 (7.8)***	-17.62 (2.3)**		-20.77 (3.9)***	-51.33 (4.6)***	-20.3 (3.8)***
Racial Tension				-3.83 (2.7)***			
Openness× Polit. Stability	3.82 (2.9)***	2.34 (2.2)**			8.38 (4.8)***		8.17 (4.6)***
Ec. Freedom× Polit. Stability			2.68 (2.5)**			7.99 (5.2)***	
Openness× Racial Tension				1.23 (2.8)***			
Constant	29.70 (1.7)*	8.47 (2.2)**	21.77 (1.6)	56.18 (3.5)***	31.7 (0.9)	1.82 (0.0)	19.0 (0.6)
R^2	0.870	0.853	0.879	0.884	0.818	0.856	0.816
Observations	76	107	74	69	68	68	68

Notes: t-statistics in parenthesis. *(**) [***] statistically significant at the 10 (5) [1] percent level.

Stability are positive and significant in all models. Thus, the effect of **Openness** is *conditional* on the level of **Political Stability**. As the degree of trade integration rises, so does the stringency of environmental policy, and this is *particularly* true in politically stable countries. The reduction of industry bribery as a result of trade integration is particularly strong when influence-seeking and bribery takes place under predictable (stable) political conditions. Using **Racial Tension** as a proxy for political (in-)stability in the fourth column, our results largely continue to hold. **Openness×Racial Tension** is significant, although **Openness** becomes insignificant when this interaction term is included.

To see the complete effect of **Political Stability**, we also need to consider its interactions. For example, using the estimates from the 5th model (5th column) of Table 3 (2SLS), we find that ∂**Environmental Stringency**/∂**Political Stability** = 2.53 (= -20.77 + 8.38×2.78) (evaluated at the mean of **Openness**). That is, political stability is

Table 4: Environmental Stringency Regressions: Developing Countries

Variable	OLS			2SLS	
Log GDP	4.82 (1.9)*	2.73 (1.2)	2.91 (1.4)	3.52 (0.8)	2.58 (0.5)
Dishonesty	-4.16 (4.4)***	-4.3 (4.8)***	-4.14 (5.6)***	-2.74 (1.6)	-2.85 (1.3)
% Non-Ag. Labor	-0.13 (1.9)*	-0.07 (1.3)	-0.03 (0.5)	-0.05 (0.5)	-0.04 (0.3)
Openness	2.98 (2.7)***		-0.27 (0.2)	5.14 (3.2)***	
Economic Freedom		3.52 (3.2)***			4.08 (2.9)***
Political Stability	-8.63 (1.7)*	-17.89 (2.1)**		-17.90 (3.0)***	-48.48 (3.4)***
Racial Tension			-3.94 (2.9)***		
Openness× Political Stability	3.05 (1.8)*			6.73 (2.5)**	
Economic Freedom× Political Stability		2.49 (2.0)*			7.44 (3.5)***
Openness× Racial Tension			1.31 (2.7)***		
Constant	31.4 (1.5)	30.58 (2.0)*	51.97 (3.0)***	22.34 (0.6)	15.78 (0.3)
R^2	0.761	0.781	0.817	0.738	0.733
Observations	52	50	46	45	45

Notes: t-statistics in parenthesis. *(**) [***] statistically significant at the 10 (5) [1] percent level.

associated with greater environmental policy stringency, except in relatively closed economies (where the total marginal effect turns negative).

Turning to our control variables, **Dishonesty** and **Honesty** are significant across the models with the expected signs, suggesting that lower levels of corruption are associated with stricter environmental policies. **Log GDP** is significant in only some of the OLS regressions, as is **% Non-Ag. Labor**.[27] This may suggest that the variables of main interest are relatively important determinants of environmental policy.[28]

Table 4 replicates Table 3 using only the developing country observations. Policy advice may have particular relevance for this group of countries. In addition, this group of countries has a lower average of Honesty, and these regressions may consequently shed some light on the role of the honesty variable for the direction of the interaction in focus (see the discussion in connection with Proposition 1). We find that **Openness** and **Economic Freedom** are associated with more stringent environmental policies in developing countries, and again this is particularly the case where political

Table 5: Environmental Stringency Regressions: Varying Corruption Levels

	OLS			**2SLS**		
Sample Range Dishonesty Index	**5-10**	**4-10**	**2-10**	**5-10**	**4-10**	**2-10**
Log GDP	5.30 (2.36)**	5.20 (2.4)**	5.83 (2.6)**	1.51 (0.3)	0.75 (0.2)	2.22 (0.5)
Dishonesty	-1.91 (1.3)	-1.89 (1.5)	-3.92 (4.2)*	-2.10 (1.0)	-2.3 (1.5)	-3.1 (2.0)**
% Non-Ag. Labor	-0.12 (1.7)*	-0.10 (1.6)	-0.12 (1.8)*	-0.02 (0.3)	0.01 (0.1)	-0.01 (0.1)
Openness	2.90 (2.9)***	2.60 (2.7)**	2.36 (2.3)**	3.46 (3.1)***	3.70 (3.3)***	4.01 (3.2)***
Political Stability	-8.51 (1.6)	-8.35 (1.8)**	-9.92 (2.0)**	-12.1 (3.0)***	-12.0 (3.0)***	-17.5 (3.9)***
Openness× Political Stability	3.60 (2.3)**	3.53 (2.60)**	3.57 (2.60)**	3.90 (3.0)***	3.96 (3.1)***	5.98 (3.8)***
Constant	12.0 (0.7)	12.7 (0.7)	22.3 (1.2)	35.0 (0.8)	41.2 (1.1)	34.2 (0.9)
R^2	0.526	0.591	0.742	0.561	0.585	0.721
Observations	50	58	65	42	49	56

Notes: t-statistics in parenthesis. *(**) [***] statistically significant at the 10 (5) [1] percent level.

[27] To adjust for structural differences between countries, we also tried including an OECD country dummy. This was insignificant in all regressions.

[28] This can also be interpreted that the traditional Kuznets type results may not hold when structural variables such as dishonesty and political stability are controlled for. This is a topic for future research.

stability is high. The only difference between Tables 3 and 4 appear to be that **Dishonesty** is insignificant in both 2SLS models, possibly due to insufficient variation.

To further explore the impact of honesty for our main results, we performed a sensitivity analysis for our regressions of main interest. The samples used in Table 5 are determined by the countries' **Dishonesty** scores. We used three different cut-off levels of **Dishonesty** (i) 5-10; (ii) 4-10; (iii) 2-10, where the first sample contains only the most dishonest countries. Table 5 reveals that our main results appear quite insensitive to the **Dishonesty** scores. **Openness** and its interaction with **Political Stability** remain significant and positive in all models. **Dishonesty** is here significant only in the largest samples (including countries with **Dishonesty** index 2-10).

5. Conclusion

This paper develops a theory of how environmental policy formation is affected by the degree of economic integration and the level of political turbulence. The prediction that emerges is that trade integration is associated with stricter environmental policies in countries with relatively corrupt governments, and this effect may occur also in less corrupt countries. Moreover, the theory predicts an interaction effect between the degree of trade integration and the level of political stability.

These predictions are supported by our empirical findings from developed and developing countries. Trade integration leads to more stringent environmental policies, and the effect is greater in politically stable countries. Our findings suggest that predictions of environmental policy outcomes in connection to trade reform programs should take the degree of political stability into account.

References

Alvarez, M., J.A. Cheibub, F. Limongi, and A. Przeworski, "Classifying political regimes," *Studies in Comparative International Development* **31**, 2-36 (1996).

Antweiler, W., B.R. Copeland, and M.S. Taylor, "Is free trade good for the environment?" *American Economic Review* **91**, 877-908 (2001).

Annet, A., "Social fractionalization, political instability, and the size of government," *IMF Staff Papers* **48**, 561-592 (2001).

Bernheim, B.D. and M.D. Whinston, "Menu auctions, resource allocation, and economic influence," *Quarterly Journal of Economics* **101**, 1-31 (1986).

Bohn, H. and R.T. Deacon, "Ownership risk, investment, and the use of natural resources," *American Economic Review* **90**, 526-549 (2000).

Bommer, R. and G.G. Schulze, "Environmental improvement with trade liberalization," *European Journal of Political Economy* **15**, 639-661 (1999).

Brett, C. and M. Keen, "Political uncertainty and earmarking of environmental taxes," *Journal of Public Economics* **75**, 315-340 (2000).

Center for International Earth Science Information Network (CIESIN), Global Leaders for Tomorrow, and Yale Center for Environmental Law and Policy, Columbia University, World Economic Forum, and Yale University, www.ciesin.org (2001).

Coate, S. and S. Morris, "Policy persistence," *American Economic Review* **89**, 1327-1336 (1999).

Copeland, B.R., "International trade and the environment: policy reform in a polluted small open economy," *Journal of Environmental Economics and Management* **26**, 44-65 (1994).

--, "Pollution content tariffs, environmental rent shifting, and the control of cross-border pollution," *Journal of International Economics* **40**, 459-476 (1996).

-- and M.S. Taylor, "North-South trade and the environment," *Quarterly Journal of Economics* **109**, 755-787 (1994).

-- and --, "Trade and transboundary pollution," *American Economic Review* **85**, 716-737 (1995).

Damania, R., "When the weak win: the role of investment in environmental lobbying," *Journal of Environmental Economics and Management* **42**, 1-22 (2001).

--, P.G. Fredriksson, and J.A. List, "Trade liberalization, corruption, and environmental policy formation: theory and evidence," *Journal of Environmental Economics and Management* **46**, 490-512 (2003).

Davidson, R. and J. G. MacKinnon, *Estimation and Inference in Econometrics*, New York: Oxford University Press (1993).

Dean, J.M., "Does trade liberalization harm the environment: a new test," *Canadian Journal of Economics* **35**, 819-842 (2002).

Djankov, S.D., R. La Porta, F.Lopez de Silanes and A. Shleifer, "Courts: the Lex Mundi project," NBER Working Paper No. 8890, Cambridge, MA (2002).

Ederington, J., "International coordination of trade and domestic policies," *American Economic Review* **91**, 1580-1593 (2000).

Esty, D.C. and Geradin, D., "Environmental protection and international competitiveness," *Journal of World Trade* **32**, 5-46 (1998).

Fisman, R. and Gatti, R., "Decentralization and corruption: evidence across countries," *Journal of Public Economics* **83**, 325-345 (2002).

Fredriksson, P.G., "The political economy of trade liberalization and environmental policy," *Southern Economic Journal* **65**, 513-525 (1999).

-- and J. Svensson, "Political instability, corruption and policy formation: the case of environmental policy," *Journal of Public Economics* **87**, 1383-1405 (2003).

--, H.R.J. Vollebergh and E. Dijkgraaf "Corruption and energy efficiency in OECD countries: theory and evidence," *Journal of Environmental Economics and Management* (2004), forthcoming.

Greene, W.H., *Econometric Analysis*, 3rd ed., Upper Saddle River, N.J.: Prentice Hall (2003).

Grossman, G.M. and E. Helpman, "Protection for sale," *American Economic Review* **84**, 833-850 (1994).

Gwartney, J., R. Lawson, and D. Samida, *Economic Freedom of the World 2000 Annual Report*, Vancouver: The Fraser Institute (2000).

Hausman, J., “Specification and estimation of simultaneous equations models,” in Z. Griliches and M. Intriligator, eds., *Handbook of Econometrics*, Amsterdam: North-Holland (1983).

Hillman, A.L. and H.W. Ursprung, “Greens, supergreens, and international trade policy: environmental concerns and protectionism,” in C. Carraro, ed., *The International Dimension of Environmental Policy*, Dordrecht: Kluwer (1994).

Kaufmann, D., A. Kraay, and P. Zoido-Lobaton, “Aggregating Governance Indicators,” Policy Research Working Paper No. 2195, Washington, DC: The World Bank (1999a).

--, -- and --, "Governance Matters," Policy Research Department Working Paper No. 2196, Washington, DC: The World Bank (1999b).

Knack, S. and P. Keefer, “Institutions and economic performance: cross-country tests using alternative institutional measures,” *Economics and Politics* **7**, 207-227 (1995).

La Porta, R., F. Lopez-de-Silanes, A. Shleifer, and R. Vishny, “The quality of government,” *Journal of Law, Economics, and Organization* **15**, 222-279 (1999).

López, R., “Environmental externalities in traditional agriculture and the impact of trade liberalization: the case of Ghana,” *Journal of Development Economics* **53**, 17-39 (1997).

-- and S. Mitra, “Corruption, pollution and the Kuznets environment curve,” *Journal of Environmental Economics and Management* **40**, 137-150 (2000).

O'Driscoll, Jr., G.P., K.R. Holmes, and M. Kirkpatrick, *2000 Index of Economic Freedom*, The Heritage Foundation and *The Wall Street Journal* (2000).

Olson, M., *The Logic of Collective Action*, Cambridge: Harvard University Press (1965).

Persson, T., G. Tabellini, and F. Trebbi, “Electoral rules and corruption,” *Journal of the European Economic Association* **1**, 958-989 (2003).

Rauscher, M., “On ecological dumping,” *Oxford Economic Papers* **46**, 822-840 (1994).

Rodrik, D., “Political uncertainty and private investment in developing countries,” *Journal of Development Economics* **36**, 229-42 (1991).

Rose, A.K., "Do WTO members have more liberal trade policy?" *Journal of International Economics* (2004), forthcoming.

Schulze, G. and H. Ursprung, "The political economy of international trade and the environment," in G. Schulze and H. Ursprung, eds., *International Environmental Economics: A Survey of the Issues*, Oxford: Oxford University Press (2001).

Transparency International, www.transparency.de (1999).

Wilson, J.D., "Capital mobility and environmental standards: is there a theoretical basis for a race to the bottom," in J. Bhagwati and R.P Hudec, eds., *Fair Trade and Harmonization: Prerequisites for Free Trade*, Vol. 1, Cambridge and London: MIT Press (1996).

White, H., "A heteroscedasticity-consistent covariance matrix estimator and a direct test for heteroscedasticity," *Econometrica* **48**, 817-838 (1980).

World Development Indicators, Washington DC: The World Bank (2000).

Appendix

Table A.1. Variable Definitions and Data Sources

Variable	Definition and Source
Environ-mental Stringency	Captures the extent to which the country has in place institutions and policies that result in effective responses to environmental problems. It takes values between 0 and 100, where a higher value implies greater environmental policy stringency. Source: Center for International Earth Science Information Network (2001), www.ciesin.org.
GDP	GDP Per Capita (PPP) or Purchasing power adjusted GDP is obtained when GDP is converted to international dollars using purchasing power parity rates. An international dollar thus has the same purchasing power over GDP as the U.S. dollar in the United States. Source: World Development Indicators (2000).
Dishonesty	Corruption Perceptions Index published by Transparency International, describes the level of perceived corruption in the public sector using a poll of political risk indexes. Original scores range from 0 (completely corrupt) to 10 (clean). Average of CPI indexes for years 1997, 1998, and 1999. The index is inverted in scale by subtracting values from 10 to make the results more intuitive. Available at: www.transparency.de.
Honesty	Measures perceptions of honesty (absence of corruption) corruption in a country, or more precisely, the absence of the use of public power for private gain. The index takes values from –2.5 to 2.5, where a higher value implies greater control over corruption. Source: Kaufmann et al. (1999a, 1999b).
Political Stability	Measures perceptions of the likelihood that the government in power will be destabilized or overthrown. Takes values from –2.5 to 2.5, where a higher value represents greater political stability. Source: Kaufmann et al. (1999a, 1999b).
Openness	Index of trade openness developed by the Heritage Foundation and the *Wall Street Journal*. Takes values from 1 to 5. An economy earns a "5" if it has average tariff rate of less than or equal to 4 percentage points and/or has very few non-tariff barriers, and "1" if the average tariff rate is greater than 19% and/or there are very high non-tariff barriers that virtually prohibits imports. Source: O'Driscoll et al. (2000).
% Non-Ag. Labor	Proportion of the total labor force recorded as working in non-agricultural sectors. Source: World Development Indicators (2000).
Economic Freedom	Measure of economic liberalization and openness. A composite index of ten individual indices for 1997. Takes values between 0 and 10, where a higher value indicates a greater degree of economic liberalization and freedom. Source: Gwartney et al. (2000).
Racial Tension	Index for countries experiencing racial tension. It takes values from 1 (high tension) to 6 (low tension). Source: Knack and Keefer (1995).
Dismag	A measure of the average number of representatives elected in each district taking values between 0 and 1. 0 represents a system with only single-member districts, 1 a system with a single electoral district. Persson et al. (2003).
Federal Dummy	Federal structure dummy variable. It takes the value 1 for federation, and 0 otherwise.
Legal Origin	Legal Origin Dummy taking the value 1 for countries with history of Anglo-Saxon Common Law, and 0 otherwise. Source: La Porta et al. (1999).
Civic Freedom	Index that indicates the freedom enjoyed by the civil society. Take a value from 1 (most free) to 7 (least free). Source: Gwartney et al. (2000).
War Dummy	Dummy variable which takes the value 1 if the country was involved in at least one incidence of war in the last 30 years, 0 otherwise. Source: Knack and Keefer (1995).
Civil War Dummy	Civil war dummy variable taking the value 1 if the country experienced civil war in the last 30 years, 0 otherwise. Source: Knack and Keefer (1995).

[6]

Pollution Abatement Expenditure by U.S. Manufacturing Plants: Do Community Characteristics Matter?*

Randy A. Becker

Abstract

This paper examines whether community characteristics impact the air pollution abatement (APA) activity at nearby manufacturing plants, using establishment-level data from the U.S. Census Bureau's Pollution Abatement Costs and Expenditures (PACE) survey. Controlling for facility characteristics and various forms of *formal* environmental regulation, certain community characteristics are found to have additional effects on local APA expenditures. In particular, for the most pollution-intensive industries, larger per capita income is found to increase plant-level APA activity, as is a higher degree of homeownership, a greater concentration of Democratic voters, and being located in a metropolitan area. Meanwhile, a greater presence of manufacturing employees is found to decrease APA expenditures, suggesting a constituency that is more resistant to additional regulation. If local populations can indeed affect pollution abatement activity, they can impact the spatial distribution of pollution, thereby creating pollution haven effects.

KEYWORDS: environmental regulation, costs, air pollution abatement, community characteristics, manufacturing

*This paper has benefited from the extremely helpful comments of Don Fullerton and three anonymous referees as well as from valuable discussions at the Second World Congress of Environmental and Resource Economists, the 2003 Allied Social Science Association conference, and the Center for Economic Studies. The opinions and conclusions expressed herein are those of the author and do not necessarily represent those of the U.S. Bureau of the Census. This paper has been screened to ensure that it does not disclose any confidential information. All errors are my own. Contact: Randy A. Becker, Center for Economic Studies, U.S. Census Bureau, 4700 Silver Hill Road, Washington, DC 20233-6300.

A growing literature examines the ways community characteristics affect local environmental outcomes. Recent studies have shown the importance of various economic and political factors in explaining the level of toxic releases or local ambient pollution (e.g., Kriesel et al. 1996; Arora and Cason 1999), the siting of polluting facilities (e.g., Hamilton 1993, 1995; Wolverton 2002), local regulatory enforcement (e.g., Gray and Deily 1996; Helland 1998a, 1998b), voting patterns on environment-related ballot initiatives (Kahn and Matsusaka 1997), and general attitudes towards the environment (e.g., Elliott et al. 1997). Explanatory variables common to many of these studies include household incomes, house values, population density, educational attainment, poverty rate, racial composition, voting behavior, and other measures of local demographics, economic conditions, and political opposition.

In this paper, I examine whether local factors help explain the level of pollution abatement expenditures by manufacturing plants. If certain populations are indeed exposed to lower levels of pollution, as previous studies have suggested, it may be that (a) a smaller number of polluting facilities are located in those communities and/or (b) facilities in those areas invest much more in abatement than they would have otherwise. The latter is the subject of this study.

Here, plant-level air pollution abatement (henceforth, "APA") expenditure per dollar of output is modeled as a function of *formal* air quality regulation at the federal, state, and county levels, as well as county characteristics potentially affecting such expenditures, either because facilities behave differently in such areas and/or local regulators regulate more (or less) stringently in such areas. In certain contexts, this might be called "informal regulation" (see Pargal and Wheeler 1996); in the United States, however, community influence may still be transmitted through established regulatory channels. "Less formal" regulation, therefore, may be a better description of the focus of this paper.

The principle motivation here is to provide additional evidence on which community characteristics impact local environmental outcomes. The literature to date simply has not reached any consensus on this front.[1] This paper's chief innovation in this realm is that it is the first to look at the effects on the environmental expenditures of firms. In principle, this should yield the same expected results as studies based on other outcomes since they are all obviously interrelated: community preferences manifest themselves in voting and lobbying, which in turn may affect local regulatory enforcement, which in turn affects abatement expenditures, which in turn affect facility emissions and/or plant location, which in turn affect local ambient pollution. However, given potentially serious concerns with the measurement of some of these environmental outcomes, the examination of several different ones—including my focus on pollution abatement expenditure—is certainly a worthwhile endeavor, in order to gauge

[1]Part of the issue may certainly be that these studies have chosen to examine—not simply different outcomes, but perhaps more importantly—different pollutants, industries, geographies, and/or time periods.

the robustness of previous and future findings in this area.[2]

Another desirable feature of this paper's approach over several others' is that it peers deeper inside the "black box". Besides the effect of community characteristics, firms' decisions regarding pollution abatement expenditure may obviously also depend on the industry a plant is in, the *formal* regulation it faces, and its age, size, corporate ownership, productivity, profitability, and so forth—factors that cannot be adequately explored (or controlled for) using anything but establishment-level data. Here, I use establishment-level data on APA expenditures from the Census Bureau's Pollution Abatement Costs and Expenditures (PACE) survey, with additional establishment-level controls coming from the Annual Survey of Manufactures (ASM).

Results show that, after controlling for establishment characteristics and various forms of formal regulation, some community characteristics have an additional effect on APA expenditures, others do not. Some of the most noteworthy results are for a sub-sample of establishments that are particularly high air polluters. In this group, some of the political/self-interest factors are found to be statistically significant determinants of APA activity. As in some other studies, the percent of homes in an area that are owner-occupied is found to improve local environmental outcomes—in this case, APA investment. Homeowners care about maintaining their properties' value and may lobby for more stringent regulation than normally required by their state or federal government. Others have argued that a higher proportion of renters implies lower sunk costs and higher mobility by local residents—and therefore less compensation is required in the event of damages.

The proportion of the local population that is engaged in manufacturing or is unemployed has the opposite effect, at least among the most polluting industries. This is also consistent with some previous findings and suggests that those that feel their livelihoods threatened by increased regulation will oppose it. Political ideology—as measured by the percent who vote Democrat—also appears to have an impact on the local environment, as in the work of Kahn, but voter turnout—a measure of a populace's potential for political action, introduced into this literature by Hamilton (1993)—generally does not exhibit the hypothesized effect.

Larger per capita income is also found to have a positive effect on the APA expenditure of plants, consistent with the notion that the environment is a normal good and/or that plants abate more in areas where potential economic damages are higher. Few of the other variables related to would-be compensation, however, such as population density, education, and home value, are found to have the expected effects.

---

[2]Take, for instance, pollution-based measures. At the plant-level, "emissions" are very often just engineering estimates based on input usage, emissions factors, and various other assumptions, rather than actual readings from monitoring devices. Meanwhile, data on ambient air pollution levels are also quite problematic. Monitor placement is almost certainly endogenous; that is, monitors are placed in areas that are suspected to have high pollution concentrations and/or their placement may be determined community characteristics. Furthermore, most areas simply have no pollution monitors at all, which obviously limits analyses—e.g., by eliminating important control groups. Another obvious question is which pollution statistics are most relevant—e.g., peak readings or averages—and how one aggregates the readings from several nearby monitors.

And in terms of "environmental injustice," I find, *at best*, mixed evidence that plants in areas with large nonwhite and/or foreign-born populations have significantly different APA expenditures.

Overall, these results suggest that some communities can and do adopt stricter environmental standards, over and above any *formal* regulatory requirements. This can lead to a shift in the relative spatial distribution of pollution, resulting in pollution haven effects. Furthermore, if local populations can indeed affect the expenditures of polluting facilities, they can (at the margin) potentially impact the location of such facilities, further reinforcing pollution haven effects. Previous research has already shown that *formal* air quality regulation—particularly the Clean Air Act and its amendments—has led to (perhaps unintended) shifts in the location of air polluting activity (e.g., Henderson 1996; Kahn 1997a; Becker and Henderson 2000; Greenstone 2002), presumably due to the spatial variation inherent in these regulatory costs (Becker and Henderson 2001; Becker 2005). This current study is the first to directly examine the effects of *"less formal"* regulation on the costs of air polluters, using the establishment-level data developed by Becker (2005) to study the impact of formal air quality regulation.

In the next section, I review some of the previous literature in this area. This is followed by Section 2, which discusses the community characteristics under study here, as well as the data and methodology that are used. Empirical results are presented in Section 3, followed by a section of concluding remarks.

1. Previous Literature

Much of the previous literature in this area has explored the apparent correlation between minority populations and local pollution levels and the siting of polluting facilities. Several recent studies have begun debunking the notion of "environmental racism," finding that other factors—some correlated with race—may play a more dominant role.

Hamilton (1993), for example, examines the expansion plans of hazardous waste processing facilities and finds that a county's potential for collective action—as measured by voter turnout—is the most influential community characteristic in determining where noxious activities were (and were not) sited—demand, supply, and factor costs being equal. "Coasian" variables related to the would-be compensation from environmental damages—for example, median household income, house value, college education, population density, and urban population—are generally found not to have an effect, with the exception perhaps on sitings prior to the mid-1980s. A county's racial (nonwhite) composition does not matter in the principle specifications, but may have played a role with respect to which sites had planned contractions.

In Hamilton (1995), he performs similar analyses at the zip code level and expands the set of community characteristics examined. Again, the primary explanation is the threat of collective action—i.e., political opposition. The only economic characteristics that ever matter are total population—i.e., the number of potential "victims"—and the percent of the population that rents—i.e., a proxy for attachment to a locale and sunk costs. As before, race is not a statistically significant factor in the regression analyses even though nonwhite population and hazardous waste capacity expansion are positively correlated.[3] "Pure discrimination" therefore is *not* a likely explanation.

Kriesel et al. (1996) continue this line of investigation by exploring toxic emissions inside Census block groups in Georgia and Ohio. The authors show that statistical evidence of environmental racism disappears as explanatory variables are added to the analysis. In the end, only poverty rate, educational attainment, and the presence of an interstate highway—an industrial location factor—helped explain local exposure to toxic pollutants. Voter turnout had an effect that was opposite of what Hamilton found, and house value, population density, and manufacturing employment had no statistically significant effect. The authors also explore the toxic releases of *new* hazardous waste facilities and find that race and income do not matter, while voter turnout continues to have a positive effect, as does interstate highways. Population density has a negative effect on toxic emissions, and some evidence is found that average house value also has a negative effect.

More recently, Arora and Cason (1999) examined toxic emissions at the zip code level, employing a sample selection model and a long list of local economic, demographic, and political characteristics. No strong, consistent results emerge across areas of the country. Race matters, but only under very specific conditions: "non-urban" areas of the South. A number of economic and "political" variables also matter in the South, such as household income, housing vacancy, unemployment, and the percent who rent homes. Many fewer factors are significant outside the South, limited at times to percent foreign-born, educational attainment, and total population. The political/ collective action variables are generally either statistically insignificant or wrong-signed.

Like the previous two papers, Kahn (1997b) also looks at toxic emissions using data from EPA's Toxics Release Inventory. He also examines county-level ambient air quality with respect to two pollutants: particulates and sulfur dioxide. In this study, the effects of local economic and political characteristics—in particular, per capita income, manufacturing's share of local employment, percent who voted Democrat, college graduates, and population density—are only of secondary interest, meant to control for differential regulatory enforcement across counties as well as notions of Coasian compensation (see Hamilton 1993). Nonetheless, some evidence is found that all these factors matter, in the expected ways.

Other strands of literature have focused on quite different environmental outcomes. Gray and Deily (1996), for example, look at air pollution inspections and enforcement actions taken against U.S. steel mills. They hypothesize that regulators may be more reluctant to act against plants when the political costs are high—for example, when the

[3]Wolverton (2002) also finds that race and the presence of TRI facilities (in Texas) are *currently* correlated, but finds no such correlation *at the time of siting*. This suggests some sort of *ex post* sorting. Perhaps the presence of such facilities suppresses local property values, subsequently attracting populations with lower incomes.

firm accounts for a large share of the locale's total employment and/or unemployment is high in the county. They find strong evidence that at least the former is true. Eric Helland, in a series of papers, also explores the determinants of environmental inspections and discovers that local characteristics often play a role. In Helland (1998a), for example, he finds that pulp and paper mills experience more inspections in counties with higher per capita income and fewer inspections in states with larger numbers of manufacturing facilities. The latter is thought to measure the caseload burden faced by regulators.

Meanwhile, Kahn and Matsusaka (1997) examine cross-county voting outcomes in the State of California with respect to 16 environment-related ballot initiatives, including ones on toxic disclosure, environmental protection & conservation, expanding parklands, pesticide bans, etc. Here, the environment is treated as—and found to behave much like—any other good: its demand is driven by both income and "prices". In particular, the authors find that many of these "goods" are normal at mean income levels and inferior (perhaps) at higher incomes. Prices are proxied by measures of "self-interest"—namely, the share of local employment and income derived from affected sectors (e.g., manufacturing, construction, farming, and/or forestry) as well as the educational attainment of the populace (i.e., a measure of potential worker displacement). These price variables had their expected effect in the majority of cases. They also find an added effect from political ideology—or "preferences"—as measured by the percent who register or vote Democrat.

Elliott et al. (1997), in their examination of survey data on public attitudes toward environmental spending, also find that (self-identified) Democrats and liberals are more likely to support environmental causes. In addition, they find that environmentalism is higher among women, those with some college education, wealthier households, nonwhite populations, and residents of major metropolitan areas. They hypothesize that the views of the latter two groups are perhaps influenced by their exposure to pollution and other environmental hazards. Meanwhile, full-time workers are *less* likely to support additional environmental spending, as are older individuals. The authors hypothesize that the former may reflect less leisure time to enjoy the environment, and the latter may suggest cohort effects and/or the smaller present value of a cleaner environment for older people.

2. Data and Methodology

2.1. Community Characteristics

In this study, I examine expenditures on air pollution abatement by manufacturing establishments and the effect community characteristics have on the decisions of these plants and/or the regulators who regulate them. I draw upon the previous literature for local factors that may influence this outcome as well as contribute a few new variables to the usual mix. They can be broadly—and somewhat arbitrarily—classified as belonging to one of four groups: (1) political/self-interest variables, (2) income and preferences, (3) Coasian compensation variables, and (4) other factors driving regulators and/or facilities. These variables are described briefly in Table 1 and **highlighted** and further discussed below. As in Hamilton (1993), Gray and Deily (1996), Kahn and Matsusaka (1997), Kahn (1997b), and Helland (1998a, 1998b), my geographic unit of analysis is the county.

First, in terms of political/self-interest variables, some local constituencies may demand that their regulators adopt policies more stringent than required by their state or federal government—e.g., under the Clean Air Act. A higher **density of population** surrounding a plant, for example, may increase the pressure placed on it since its emissions (left unabated) adversely affect that many more people. And **homeowners** in particular may demand greater pollution abatement efforts from their nearby plants; numerous studies have demonstrated a relationship between ambient air quality and property values (e.g., Chay and Greenstone 2004). Other groups, however, may be more hostile to additional environmental protection. The **unemployed** and those engaged in **manufacturing**, in particular, may feel their very livelihoods threatened by increased regulation. And, as in several of the cited studies, I also include **voter turnout** as a measure of a populace's ability to take collective action.

Second, apart from these potential political factors, the demand for environmental quality, like that for any other good, may simply derive from income and preferences. If environmental protection is indeed a normal good, wealthier counties—as measured by per capita **income**—might be expected to press for stricter requirements on their polluting plants. Demand might also be expected to depend on one's vulnerability to air pollution, and the EPA has identified children and the elderly among those particularly at risk (**sensitive population**).[4] I further proxy for a county's "taste for health" by observing the number of active **doctors** per capita. Finally, I also allow political ideology to shape a county's preference for environmental protection. Here, as in Kahn and Matsusaka (1997), this is measured by the percent of (major party) votes cast for the **Democratic** presidential candidate.

Third, an establishment may consider would-be compensation for environmental damages posed by its operation in deciding where to locate and how much to invest in pollution abatement. Areas with high rates of **poverty** and/or low levels of **educational attainment** may result in lower damages being awarded in the event of injuries and/or deaths from accidental exposure. Similarly, litigation may result in compensation for reduced property values. Here, I include the median **value of houses** in the county. **Population density** and per capita **income**—mentioned previously—are also likely to be considered Coasian-type variables.

Finally, other local characteristics may also explain the "supply" of environmental protection. As noted, much has been written on the issue of environmental injustice (or environmental inequity, racism, discrimination, etc.)—the notion that polluting facilities are more likely to be situated near minority communities and/or toxic emissions are higher in such areas. Here, I allow for the possibility of such discrimination on the part

[4] Gray and Shadbegian (2004) indeed show that pulp and paper mills that are surrounded by larger proportions of kids and elders have lower sulfur dioxide (SO_2) and particulate (PM-10) emissions.

Table 1: County Characteristics

	Mean	Std. dev.	Description†
Population density	1,611	5,446	Persons per square mile.
Homeownership	65.8	10.7	Percent of occupied housing units that are owner-occupied.
Unemployment	6.66	2.39	Percent of civilian labor force (16 years and older) that is unemployed.
Manufacturing	26.2	9.3	Percent of employed civilian labor force (16 years and older) employed in manufacturing.
Voter turnout	53.6	8.4	Percent of voting-age population that cast a vote for president.
Income	7,271	1,301	Per capita money income (in dollars).
Sensitive population	18.4	2.6	Percent of population under 5 years of age or 65 years and over.
Doctors	195	139	Active, non-federal physicians per 100,000 population (1985).
Democrats	40.5	10.2	Percent of (major party) votes cast for Democratic presidential candidate (1984).
Poverty	9.15	4.10	Percent of families below the poverty level.
Education	15.6	6.0	Percent of population (25 years and older) with at least a bachelor's degree.
Home value	48,638	18,929	Median value of owner-occupied non-condominium housing units (in dollars).
Nonwhite	15.8	13.5	Percent of population that is nonwhite.
Foreign-born	5.3	5.8	Percent of population that is foreign-born.
Air polluters	93	38	Air polluting manufacturing establishments per 100,000 population (1982).
Metropolitan area	.766	–	County is in an MSA (June 1998 definition).

†Unless noted otherwise, all variables are for reference year 1980.

of firms—and obviously, by extension, regulators—by including the **nonwhite** proportion of a county's population and the percent that is **foreign-born**. Unequal regulatory treatment and dissimilar pollution abatement expenditure, however, may also be due to disparate caseloads among local regulators. One might expect, for example, all else being equal, that a plant in a region with more polluters may face a lower probability of being inspected and perceive a greater opportunity to shirk environmental obligations. A measure of **air polluters per capita** is meant, in part, to capture this effect. Whether a plant is in a **metropolitan statistical area**—where regulators are likely based—is also a potential factor.

With a few exceptions, these county-level variables are based on data from the 1980 Census of Population and Housing and extracted from the Census Bureau's *USA Counties 1998* (CD-ROM).[5] The underlying voting data were compiled by the Elections Research Center, while the number of physicians is based on data assembled by the American Medical Association from its Physician Masterfile and extracted from the Census Bureau's *County and City Data Book 1988*. The air polluters variable was constructed using 1982 Census of Manufactures microdata and the definition of high air pollution emitting industries from Becker (2005).

2.2. The Establishment-level Data

The establishment-level data for this study come from the Census Bureau's Pollution Abatement Costs and Expenditures (PACE) surveys of 1979-1982, 1984-1986, and 1988. My particular variable of interest is **air pollution abatement operating costs**, which includes salaries & wages, parts & materials, fuel & electricity, capital depreciation, contract work, equipment leasing, and other operating costs associated with a plant's abatement of air pollution in that calendar year.[6,7] To this I merge data that these plants may have reported in the contemporaneous Annual Survey of Manufactures (ASM). The ASM data provide me with some important, basic information on these plants, including establishment employment, value of output, *total*

---

[5]The local characteristics these various variables attempt to capture are, for the most part, ones that are unlikely to have changed very much, or in important ways, during the time period under study here. They are therefore treated as time-invariant in the specification below. To mitigate endogeneity concerns, the explanatory variables generally predate the expenditure being analyzed. The fact that APA expenditures, as collected by the PACE survey, is not observable to the public by locality further weakens potential endogeneity.

[6]See U.S. Bureau of the Census (various years) for additional details.

[7]It has been suggested that a focus on only air pollution abatement may yield misleading conclusions if establishments merely substitute away from other types of environmental expenditure—e.g., water pollution abatement—to address a concern that is perhaps more noticeable to the general public. Greenstone (2003), however, finds that reductions in facilities' air pollution emissions, in response to air quality regulation, did not result in increased releases into other media, namely water and ground. Concerns about cross-media substitution, therefore, appear unfounded.

operating costs, location, industry, age, and ownership.[8] After restricting my attention to cases that were in both the PACE and ASM samples in a given year, and after eliminating survey non-respondents, certain item non-respondents, non-manufacturing establishments, and inactive and otherwise out-of-scope plants, I am left with 89,889 plant-years of observations for my empirical analyses. Additional detail regarding data sources, variable construction, and the research sample can be found in Becker (2005) and in a separate working paper available from the author.[9]

Besides community characteristics, establishments' air pollution abatement (APA) expenditures may be driven by a host of other factors. Outlays, for example, will obviously depend on the size of the establishment. I therefore define my dependent variable as:

$$APA\ intensity_i = (APA_i \,/\, Q_i) \qquad (1)$$

where i indexes an establishment; *APA* is dollars of air pollution abatement operating costs; and Q is dollars of output.[10,11] This too may vary by the size of the plant however, if returns to scale in pollution abatement are present, for example. Establishment employment and employment squared, therefore, will be used as control variables in the empirical analyses. Other establishment characteristics possibly affecting the intensity of APA activity are the age of the plant and whether or not it belongs to a multi-plant firm. The former will be represented by a series of dummy variables indicating the time elapsed since the establishment's first appearance in the Census of Manufactures (conducted every five years); the latter is captured by a simple dummy variable. I also control for plant productivity, as measured by real value added per worker, using data from the ASM.

2.3. Measures of Formal Air Quality Regulation

Formal regulation obviously also plays a role in determining how much APA is done by plants. At the national level, certain industries are inherently more polluting and are therefore regulated more intensely. Here, I control for the stringency of this regulation with an industry-level air pollution intensity index constructed using the IPPS air emissions coefficients described by Hettige et al. (1994). In particular, for each 4-digit SIC industry, I sum the estimates of pounds of emissions per dollar of output for sulfur dioxide (SO_2), nitrogen dioxide (NO_2), carbon monoxide (CO), total suspended

---

[8]The establishment-level survey data in both the PACE and ASM are confidential, collected and protected under Title 13 of the U.S. Code. Restricted access to these data can be arranged through the Census Bureau's Center for Economic Studies. See http://www.ces.census.gov/ for details.

[9]Becker (2005) also addresses the quality of the PACE data—a concern that has been raised by a number of observers.

[10]Output in the ASM is measured by total value of shipments adjusted for net changes to inventory and the value of products resold as originally purchased.

[11]The mean APA intensity in this sample is 0.00242 (i.e., APA operating costs are 0.242% of total output) with a standard deviation of 0.06356. In all, the plants in this sample incurred about $28.5 billion worth of APA operating costs (in 1988 dollars).

particulates (TSP), and volatile organic compounds (VOC)—all pollutants covered under the Clean Air Act.[12] Beyond this "federal" regulation, state-level variation in environmental regulations also exists, and a number of indexes measuring the relative stringency of such regulation have been developed, some of which even employ data from the PACE survey (see Levinson 2001 for a review). Here, however, I will simply use state dummy variables to control for interstate differences in regulation and APA costs.

Below the state level, the Clean Air Act requires that all locales achieve the National Ambient Air Quality Standards (NAAQS) set for the six "criteria" air pollutants. A growing literature is finding that county-level non-attainment of these NAAQS significantly impacts heavy air polluters, presumably through more stringent regulation (e.g., Henderson 1996; Kahn 1997a; Becker and Henderson 2000, 2001; List and McHone 2000; Greenstone 2002; Becker 2005). Here, as in Becker (2005), I capture this regulation through a series of indicator variables, indicating whether the establishment is a "high emitter" of any of the criteria air pollutants in a county that is non-attainment of the NAAQS for the respective pollutant(s).[13] It is important to note that this county-level NAAQS-related regulation is *not* determined by the community characteristics under study here.

2.4. Empirical Approach

Table 2 shows the correlation coefficients between plant-level air pollution abatement activity and the community characteristics. Seven of the sixteen correlations are statistically significant, with homeownership and unemployment positively related to air pollution abatement expenditure, and income, doctors, education, home value, and metropolitan area status negatively correlated. With the exception of homeownership, the direction of these statistically significant correlations are the opposite of what might be hypothesized; however, multiple regression analysis is obviously the much more

---

[12]Note that equal weight is given to each of these pollutants though their actual toxicities may differ. Also, the IPPS data come on the 1987 SIC basis while the establishments in my sample are classified according to 1972 SIC definitions. The IPPS data are converted to the 1972 SIC basis using a 1987–1972 SIC concordance table with shipment weights. Data are missing for about a dozen of the 450 or so industries. These industries are treated with dummy variables instead. The empirical results that follow validate this measure of national regulation, with strong positive coefficients in all specifications, suggesting that the most air pollution intensive industries face the most regulation and, hence, spend the most on APA.

[13]I classify establishments as (potential) heavy emitters of the criteria air pollutants according to their 4-digit Standard Industrial Classification (SIC) code. In turn, I categorize industries as high emitters based on data in the EPA's Aerometric Information Retrieval System (AIRS). Among other things, the AIRS database maintains an inventory of facilities in the United States (classified by industry) that emit a threshold amount of any of the six criteria pollutants. I label an industry—and all its plants—a "high emitter" of a particular air pollutant if it had a minimum number of establishments emitting above that threshold. See Becker (2005) for additional details, including the list of industries defined as high air pollution emitters.

Table 2: Correlations between APA Operating Costs Per Dollar of Output and County Characteristics

Population density	–0.0026
Homeownership	+0.0064*
Unemployment	+0.0092**
Manufacturing	–0.0026
Voter turnout	–0.0008
Income	–0.0088**
Sensitive population	+0.0042
Doctors	–0.0055*
Democrats	–0.0000
Poverty	+0.0047
Education	–0.0071**
Home value	–0.0075**
Nonwhite	–0.0022
Foreign-born	–0.0043
Air polluters	–0.0040
Metropolitan area	–0.0094**

** Significant at the 5% level. * Significant at the 10% level.

appropriate way to explore these various relationships.[14]

To summarize my empirical approach, plant-level air pollution abatement operating costs, as specified in equation (1), are modeled as a function of the community characteristics that are of interest in this study (i.e., "less formal" regulation), establishment characteristics, and *formal* regulation from various sources.[15] Year effects are also incorporated. A double log functional form is assumed here because it is found to fit the data much better than alternative forms (e.g., linear, semi-log). Since the value of the dependent variable is bounded from below, Tobit specifications are employed.[16] The next section will present results, first using only community characteristics as explanatory variables, then adding plant characteristics, and then state and NAAQS-related regulation. Results are presented for both the full sample as well as for a

[14]It should be noted that—while some of these community characteristics may be well correlated with each other (never more than ρ=0.8 however)—multicollinearity is not found to be an issue in the analyses and results that follow.

[15]While this study attempts to control for all the relevant determinants of APA, we cannot rule out the possibility of some remaining omitted variable bias.

[16]In particular, establishments are asked to report their APA expenditures in *thousands* of dollars. Therefore, with rounding, an entry of zero reflects expenditures of less than \$500. The *cnreg* (censored normal regression) command in Stata is a generalization of the standard Tobit procedure that allows the censoring point to vary by observation. In this case, left-censoring occurs at $ln(0.5/Q_i)$ for about 53.5% of the plants in this sample.

subsample of establishments that are particularly heavy air polluters. In particular, (approximately) the top quartile of plants—in terms of their air pollution intensity index—are analyzed separately. These plants should be the most sensitive to the factors under examination here. Some 100 unique 4-digit SIC industries are represented in this subsample, compared to 435 in the full sample.[17]

3. Results

Table 3 contains results for the full sample of establishments; Table 4 contains results for the subsample of high air polluters. In each table, the first column presents model coefficients controlling only for industry effects—via the air pollution intensity index described above—and year effects. The model in the second column adds establishment-level controls: employment and employment squared, age variables, productivity, and a dummy variable indicating "multi-unit" plants. The model in the third column additionally controls for state-level effects—including differences in air quality regulation between states—via state dummy variables, as well as county-level regulation resulting from the Clean Air Act—i.e., indicators of county NAAQS non-attainment of the respective criteria air pollutants and interactions between county non-attainment statuses and indicators of whether the establishment is a high emitter of the respective air pollutants. Given the double log functional form, reported coefficients are elasticities. Statistical significance at the 10% and 5% level—using robust Huber-White standard errors—is indicated by single and double asterisks, respectively.

In Table 3, most of the coefficients on the community characteristics are statistically significant. Most of these effects, however, are opposite of what might be expected. Among the *expected* effects, higher rates of homeownership are found to increase APA expenditure, as are higher incomes, higher propensities to vote Democrat, and being in a metropolitan statistical area. And once the fuller set of controls is in place, in column 3, APA activity is also found to be lower in counties with larger nonwhite populations, suggesting environmental injustice. The percent of the population that is foreign-born, however, is found to have the opposite effect on environmental expenditures. Also having effects opposite of what is expected are population density, unemployment, manufacturing, sensitive populations, poverty, and educational attainment.[18] The coefficient on voter turnout, meanwhile, has the expected sign—only after state and NAAQS-related regulation are controlled for—but does not quite attain statistical significance. And the number of other local air polluters per capita has the right sign and is statistically significant *until* those final controls are put in place.

This mix of expected and unexpected results using the full sample can be somewhat puzzling. It is important to note, however, that most of the plants in this sample are *not* large air polluters. In fact, I estimate that the median plant in this sample had about 22

[17]The list of 100 industries is available from the author upon request.

[18]The result on manufacturing would suggest that affected employees actually lobby for *additional* APA, perhaps in lieu of possible plant downsizing or closure. However, the inclusion of a plant-specific measure of employment expansion/contraction in these regressions does not alter this result.

Table 3: APA Operating Costs Per Dollar of Output: *Full Sample*†

	(1)	(2)	(3)
Population density	–0.120**	–0.095**	–0.139**
	(0.019)	(0.019)	(0.025)
Homeownership	+1.312**	+1.373**	+0.835**
	(0.136)	(0.145)	(0.164)
Unemployment	+0.278**	+0.210**	+0.173**
	(0.052)	(0.053)	(0.080)
Manufacturing	+0.405**	+0.054	+0.235**
	(0.051)	(0.052)	(0.064)
Voter turnout	–0.509**	–0.629**	+0.250
	(0.128)	(0.129)	(0.191)
Income	+2.350**	+1.668**	+0.668**
	(0.224)	(0.228)	(0.290)
Sensitive population	–0.832**	–0.724**	–1.044**
	(0.162)	(0.164)	(0.186)
Doctors	+0.140**	+0.053	+0.050
	(0.037)	(0.037)	(0.039)
Democrats	+0.142*	+0.141*	+0.363**
	(0.072)	(0.073)	(0.088)
Poverty	+0.457**	+0.389**	+0.174*
	(0.085)	(0.086)	(0.104)
Education	–0.447**	–0.588**	–0.302**
	(0.089)	(0.089)	(0.098)
Home value	–0.189**	+0.019	–0.139
	(0.086)	(0.087)	(0.145)
Nonwhite	–0.017	–0.026	–0.063**
	(0.021)	(0.021)	(0.026)
Foreign-born	+0.106**	+0.127**	+0.089**
	(0.025)	(0.025)	(0.038)
Air polluters	–0.368**	–0.095**	–0.005
	(0.043)	(0.044)	(0.048)
Metropolitan area	+0.211**	+0.194**	+0.140**
	(0.047)	(0.048)	(0.049)
Industry: Emissions per employee	+0.727**	+0.727**	+0.626**
	(0.007)	(0.007)	(0.007)
Establishment characteristics	–	*yes*	*yes*
State effects	–	–	*yes*
NAAQS-related regulation	–	–	*yes*
Pseudo R^2	0.0635	0.0737	0.0804

†All samples consist of 89,889 observations. All models include year effects. Coefficients are elasticities. Robust Huber-White standard errors are in parentheses. Statistical significance at the 10% and 5% level are indicated by single and double asterisks, respectively.

tons of annual emissions of *all* criteria air pollutants.[19] As a point of reference, this is far from being a facility that the EPA would list as a "major" source, which generally requires at least 100 tons of emissions of *one* of these air pollutants. Many of these plants are likely to be exempt from most (if not all) forms of formal air quality regulation. And indeed, nearly two-thirds of the establishments that are minimal air polluters—i.e., they are not in my subsample of high emitters—have no APA operating costs at all.

Why do the rest of these minimal polluters have non-zero APA expenditures, and why should they be related to community characteristics? It is not absolutely clear. It may suggest some form of reporting error. In particular, if one believes that these plants should have little or no APA expenditure, the results here would suggest potential over-reporting of expenditure in areas that are less densely settled, have higher unemployment, have more manufacturing, are less college educated, etc. Perhaps an unmeasured sentiment exists in such areas regarding government regulation (and/or government surveys) that lead respondents to intentionally misstate their costs.[20] This issue is certainly worth further study—especially if reporting error is suggested—but is well beyond the scope of this current study. The remainder of the paper will focus on just the highest emitting industries, as is typical in the environmental literature.[21]

Table 4 contains the results for this subsample of high air polluters.[22] We see in column 1 that most of these sixteen community characteristics have a statistically significant effect on plant-level APA operating costs, including eight that have their hypothesized effect. Of those that don't have their expected effect: Increased voter turnout is found to *decrease* APA expenditure. Its statistical significance diminishes however once plant characteristics and state and NAAQS-related regulation are controlled for (column 3) though it still has a *t*-statistic of 1.44. Only Hamilton (1993, 1995) has really ever found this factor to have a positive effect on the local environment. Educational attainment is also found to have a statistically significant negative effect on APA activity in a community, but this too becomes statistically zero once state dummies and NAAQS-related regulation are added to the specification. Meanwhile, higher nonwhite populations are found to *increase* local APA expenditure,

[19]Plant-level emissions of the five criteria air pollutants (SO_2, NO_2, CO, TSP, and VOC) are estimated by multiplying the industry-level IPPS emission factors for each pollutant (for employment) by total establishment employment. As explained in Hettige et al. (1994), this will yield lower bound estimates of true emissions.

[20]Having said that, regressions using just this sample of minimal polluters yield very low pseudo-R^2 statistics. Therefore, the model and the community characteristics are not explaining very much at all about the reported expenditures of these minimal air polluters. These regressions yield similar results to those reported in Table 3, except that income, Democrats, and poverty become insignificant, and doctors and air polluters become statistically significant, with the expected signs. Most of the puzzles remain however.

[21]For example, some of the studies cited in Section 1 focus on just one industry (e.g., hazardous waste processing, steel, pulp and paper). Furthermore, all studies based on TRI data inherently ignore the pollution releases of plants emitting below some threshold.

[22]I estimate that the median plant in this group had over 700 tons of annual emissions of the criteria air pollutants.

Table 4: APA Operating Costs Per Dollar of Output: *Highest Air Polluting Industries*†

	(1)	(2)	(3)
Population density	+0.069**	+0.082**	–0.008
	(0.028)	(0.027)	(0.037)
Homeownership	+1.107**	+0.926**	+0.792**
	(0.279)	(0.280)	(0.312)
Unemployment	–0.015	–0.078	–0.150
	(0.073)	(0.073)	(0.109)
Manufacturing	–0.322**	–0.537**	–0.221**
	(0.076)	(0.077)	(0.090)
Voter turnout	–0.345*	–0.346*	–0.366
	(0.178)	(0.178)	(0.254)
Income	+2.506**	+2.085**	+0.803*
	(0.318)	(0.316)	(0.422)
Sensitive population	+0.003	+0.179	+0.088
	(0.234)	(0.233)	(0.270)
Doctors	–0.031	–0.109**	–0.072
	(0.049)	(0.049)	(0.051)
Democrats	+0.387**	+0.341**	+0.490**
	(0.106)	(0.105)	(0.123)
Poverty	+0.164	+0.147	+0.021
	(0.119)	(0.118)	(0.147)
Education	–0.605**	–0.650**	–0.175
	(0.117)	(0.116)	(0.129)
Home value	–0.126	+0.091	–0.042
	(0.128)	(0.128)	(0.200)
Nonwhite	+0.079**	+0.079**	–0.030
	(0.028)	(0.028)	(0.036)
Foreign-born	–0.084**	–0.071**	+0.002
	(0.035)	(0.035)	(0.053)
Air polluters	–0.203**	–0.042	–0.013
	(0.059)	(0.059)	(0.065)
Metropolitan area	+0.173**	+0.170**	+0.131*
	(0.065)	(0.065)	(0.067)
Industry: Emissions per employee	+1.040**	+0.963**	+0.865**
	(0.022)	(0.023)	(0.025)
Establishment characteristics	–	*yes*	*yes*
State effects	–	–	*yes*
NAAQS-related regulation	–	–	*yes*
Pseudo R^2	0.0292	0.0354	0.0426

†All samples consist of 22,516 observations. All models include year effects. Coefficients are elasticities. Robust Huber-White standard errors are in parentheses. Statistical significance at the 10% and 5% level are indicated by single and double asterisks, respectively.

while the percent foreign-born is actually found to decrease such spending. Neither of these effects is statistically significant in the final specification however, suggesting little evidence of environmental discrimination among these highest polluters.

Population density is found to have a positive effect on APA activity, as might be expected. However, once state and NAAQS-related regulation are controlled for, this effect goes away. High density areas therefore enjoy additional abatement activity at their local plants, but only because they tend to be subject to greater *formal* regulation, *not* because these populaces can muster up more political opposition or because plants are insuring against greater potential damages. The number of other air polluters in the county—meant to capture the potential distraction of regulators—has a negative effect on APA expenditures, but not after one controls for plant-specific characteristics. The implication is that establishments in areas with many other polluting plants per capita, as it turns out, tend to be smaller, younger, less productive, and/or more likely to be "single unit" establishments. Sensitive populations (the young and elderly), the number of doctors, poverty rates, median home value, and unemployment have—for the most part—no effect in any of the regressions in Table 4. The coefficient on unemployment, however, does have the correct (negative) sign and does attain statistical significance at around the 17% level.

Meanwhile, five of these community characteristics have strong, statistically significant effects, of the expected sign, through all three of the regressions in Table 4.[23] Indeed, these five factors are the only ones that are significant in the final specification with full controls (column 3). These robust results include:

(1) Higher rates of homeownership are found to increase APA expenditure, consistent with the notion that constituencies will lobby to protect the value of their property.[24] Other studies have found similar effects. For example, Hamilton (1995) finds that zip codes with a higher percentage of renters are more likely to experience expansions in local hazardous waste facilities. Such residents, he argues, have less attachment to a particular area and therefore may place less value on maintaining its environment. Furthermore, they have fewer "sunk costs" in the area, which means firms may need to compensate them less in the event of damages. Similarly, Arora and Cason (1999) find that toxic emissions are higher in zip codes with higher concentration of renters, at least in the South.

(2) We also see that a heavy presence of manufacturing employment in a county reduces the amount of local APA expenditure, all else being equal. To use the language of Kahn and Matsusaka (1997), the "price" of environmental protection is higher in such areas. Indeed, Gray and Deily (1996) find that steel mills that accounted for a greater proportion of the local labor force faced fewer regulatory actions and inspections.[25] Regulators, they argue, attempt to minimize the political costs of their actions, especially those that threaten the livelihoods of many employees and other local

[23]Note that four of these five are also statistically significant and correctly-signed throughout Table 3 as well.

[24]Interestingly, however, APA activity does not appear to be related to the value of homes.

[25]Helland (1998b) finds similar results with pulp and paper mills.

citizens. A similar effect—for similar reasons—is expected for areas with high local unemployment rates; however, the negative effect in column 3 of Table 4 does not quite attain statistical significance.

(3) Local income levels have a strong positive impact on the pollution abatement activity of these plants. Different and equally plausible interpretations are possible. Kriesel et al. (1996)—using poverty rates instead of income—find that toxic emissions are higher in more impoverished areas and characterize this as "discrimination against poor people." Hamilton (1993, 1995) and Arora and Cason (1999)—had they themselves found these effects—would have offered a Coasian explanation: polluters avoid polluting in areas where potential compensation for damages would be the highest. Meanwhile, Helland (1998a), in his study of inspection targeting, and Kahn and Matsusaka (1997), in their study of voting behavior, see the environment like any other normal good: demand increases with income. And indeed, household surveys show that wealthier individuals favor increased expenditure on the environment (Elliott et al. 1997).

(4) Also in keeping with Kahn and Matsusaka, the political ideology of the populace is found to have an effect on environmental outcomes. In particular, the share of major party votes cast for the Democratic presidential candidate—which is meant to measure a locale's potential support for environmental causes, traditionally thought to be the Democrats' domain (see Elliott et al. 1997)—appears to increase the APA expenditures of these polluting plants. Kahn and Matsusaka (1997) find that this variable explains at least some of the county-level variation in voting outcomes on environmental initiatives in California, and Kahn (1997b) shows that ambient TSP concentrations and TRI emissions are often lower in areas with concentrations of Democrats. Similarly, Earnhart (2004) finds that municipal wastewater treatment facilities in Kansas experience fewer environmental inspections and release more biological oxygen demand (BOD) in areas with higher percentages of Republican voters.

(5) Finally, being in a metropolitan statistical area (MSA) increases APA outlays for these plants, all else being equal. Because I've controlled for many of the factors that are often correlated with cities—including population density, race, homeownership, income, and NAAQS non-attainment—one interpretation of this effect is that plants in these areas are scrutinized more intensely because regulators are more likely to be located nearby (relative to rural, non-metropolitan areas).

Table 5 shows the impact on APA intensity of a one standard deviation increase in these variables from their respective means, holding all other variables constant, using the marginal effects from the regression in column 3 of Table 4. Air pollution abatement is found to be most sensitive to changes in per capita income, with a standard deviation increase translating into a 13.5% increase in APA intensity. A standard deviation increase in votes for Democrats and homeownership rates implies an 11.6% and a 10.6% increase, respectively. Meanwhile, a standard deviation increase in a county's labor force engaged in manufacturing, from a mean of about 26.2%, would

Table 5: Marginal Effects of Covariates on APA Operating Costs Per Dollar of Output†

Homeownership	+10.6%
Manufacturing	–8.0%
Income	+13.5%
Democrats	+11.6%
Metropolitan area	+13.1%

†Percentage change in APA intensity is computed for a one standard deviation change in the respective variable, from the means of the variables, holding all other covariates constant. The effect of metropolitan area is computed for the discrete change (from being outside to inside an MSA). The marginal effects are based on the regression in column 3 of Table 4.

result in an 8.0% decrease in APA intensity.[26] A discrete change from being located in a county outside an MSA to being in an MSA implies a 13.1% increase in air pollution abatement operating costs, all else being equal.

4. Conclusions

The effect of community characteristics on local environmental outcomes—particularly toxic releases and pollution exposure—has received significant examination in recent years. If it is the case that certain populations face greater exposure to toxic and noxious air, this may be due to an extraordinary number of air polluters locating in such areas and/or facilities in those areas do not abate as much of their pollution. This paper focuses on the latter by examining whether spending by manufacturing plants on air pollution abatement—conditional on facility characteristics and various forms of *formal* regulation—depends in any way on the characteristics of the surrounding community, including socioeconomic composition, potential political opposition, economic conditions, and other possible factors.

For the most pollution-intensive industries, larger per capita income is found to *increase* plant-level APA activity, as is a higher degree of homeownership, a greater concentration of Democratic voters, and being located in an MSA.[27] Meanwhile, a greater presence of manufacturing employees is found to *decrease* the pollution abatement expenditure of these plants, suggesting a constituency that is more resistant to additional regulation.[28] Most of these results confirm precedents in the prior

[26]An alternative to the double log functional form is a semi-log specification in which natural logarithms are taken of all variables *except* ones that are percentages (e.g., see Wolverton 2002). In that case, the effect of these three percent-measured explanatory variables—i.e., Democrats, homeownership, and manufacturing—is 12.3%, 14.0%, and -19.2%, respectively.

[27]These same effects are found with the full sample as well.

[28]This paper does not attempt to explain whether regulators regulate more [less] stringently in areas with these characteristics or whether the plants themselves simply behave differently in such areas.

literature. If local populations can indeed affect pollution abatement activity, they can impact the spatial distribution of pollution, thereby creating pollution haven effects.

As interesting as what is found here, is what is *not* found. First, this paper uncovers no evidence that a populace's potential for collective action—as measured by voter turnout—raises the level of local air pollution abatement expenditure. Only Hamilton (1993, 1995) has ever really found this factor to have a positive effect on the local environment. Furthermore, none of the Coasian variables seem to have an effect on APA activity, with the exception of per capita income, which has alternative explanations. Finally, support for environmental injustice is, *at best*, mixed. Using the full sample, APA spending is indeed lower in areas with higher nonwhite populations, once the full set of control variables are in place. The percent of the population that is foreign-born, however, had the opposite effect. In terms of the highest air polluters, outlays are *larger* in nonwhite areas and *smaller* in counties with many foreign-borns—that is until one controls for state and NAAQS-related regulation, which makes even these effects disappear. As this and other examples make clear, controlling for establishment characteristics and formal regulation is vitally important in a study such as this.

References

Arora, Seema and Timothy N. Cason. "Do Community Characteristics Influence Environmental Outcomes? Evidence from the Toxic Release Inventory," *Southern Economic Journal*, 65(4), 691-716, April 1999.

Becker, Randy A. "Air Pollution Abatement Costs under the Clean Air Act: Evidence from the PACE Survey," *Journal of Environmental Economics and Management*, forthcoming (2005).

Becker, Randy A. and J. Vernon Henderson. "Effects of Air Quality Regulations on Polluting Industries," *Journal of Political Economy*, 108(2), 379-421, April 2000.

Becker, Randy A. and J. Vernon Henderson. "Costs of Air Quality Regulation," in *Behavioral and Distributional Effects of Environmental Policy*, Carlo Carraro and Gilbert E. Metcalf (eds.), Chicago: National Bureau of Economic Research and The University of Chicago Press, 2001.

Chay, Kenneth Y. and Michael Greenstone. "Does Air Quality Matter? Evidence from the Housing Market," *MIT Department of Economics Working Paper*, 04-19, February 2004.

Earnhart, Dietrich. "The Effects of Community Characteristics on Polluter Compliance Levels," *Land Economics*, 80(3), 408-432, August 2004.

Elliott, Euel, Barry J. Seldon, and James L. Regens. "Political and Economic Determinants of Individuals' Support for Environmental Spending," *Journal of Environmental Management*, 51(1), 15-27, September 1997.

Gray, Wayne B. and Mary E. Deily. "Compliance and Enforcement: Air Pollution Regulation in the U.S. Steel Industry," *Journal of Environmental Economics and Management*, 31(1), 96-111, July 1996.

Gray, Wayne B. and Ronald J. Shadbegian. "'Optimal' Pollution Abatement – Whose Benefits Matter, and How Much?" *Journal of Environmental Economics and Management*, 47(3), 510-534, May 2004.

Greenstone, Michael. "The Impacts of Environmental Regulations on Industrial Activity: Evidence from the 1970 and 1977 Clean Air Act Amendments and the Census of Manufactures," *Journal of Political Economy*, 110(6), 1175-1219, December 2002.

———. "Estimating Regulation-Induced Substitution: The Effect of the Clean Air Act on Water and Ground Pollution," *American Economic Review*, 93(2), 442-448, May 2003.

Hamilton, James T. "Politics and Social Costs: Estimating the Impact of Collective Action on Hazardous Waste Facilities," *RAND Journal of Economics*, 24(1), 101-125, Spring 1993.

———. "Testing for Environmental Racism: Prejudice, Profits, Political?" *Journal of Policy Analysis and Management*, 14(1), 107-132, Winter 1995.

Helland, Eric. "The Enforcement of Pollution Control Laws: Inspections, Violations, and Self-reporting," *Review of Economics and Statistics*, 80(1), 141-153, February 1998a.

———. "Environmental Protection in the Federalist System: The Political Economy of NPDES Inspections," *Economic Inquiry*, 36(2), 305-319, April 1998b.

Henderson, J. Vernon. "Effects of Air Quality Regulation," *American Economic Review*, 86(4), 789-813, September 1996.

Hettige, Hemamala, Paul Martin, Manjula Singh, and David Wheeler. "The Industrial Pollution Projection System," *World Bank Policy Research Working Paper*, 1431, December 1994.

Kahn, Matthew E. "Particulate Pollution Trends in the United States," *Regional Science and Urban Economics*, 27(1), 87-107, February 1997a.

———. "The Silver Lining of Rust Belt Manufacturing Decline: Killing Off Pollution Externalities," *Center for Economic Studies Discussion Paper*, 97-7, June 1997b.

Kahn, Matthew E. and John G. Matsusaka. "Demand for Environmental Goods: Evidence from Voting Patterns on California Initiatives," *Journal of Law and Economics*, 40(1), 137-173, April 1997.

Kriesel, Warren, Terence J. Centner, and Andrew G. Keeler. "Neighborhood Exposure to Toxic Releases: Are There Racial Inequities?" *Growth and Change*, 27(4), 479-499, Fall 1996.

Levinson, Arik. "An Industry-Adjusted Index of State Environmental Compliance Costs," in *Behavioral and Distributional Effects of Environmental Policy*, Carlo Carraro and Gilbert E. Metcalf (eds.), Chicago: National Bureau of Economic Research and The University of Chicago Press, 2001.

List, John A. and W. Warren McHone. "Measuring the Effects of Air Quality Regulations on "Dirty" Firm Births: Evidence from the Neo- and Mature-Regulatory Periods," *Papers in Regional Science*, 79(2), 177-190, April 2000.

Pargal, Sheoli, and David Wheeler. "Informal Regulation of Industrial Pollution in Developing Countries: Evidence from Indonesia," *Journal of Political Economy*, 104(6), 1314-1327, December 1996.

U.S. Bureau of the Census. *Pollution Abatement Costs and Expenditures, 19__*. Washington, DC: U.S. Government Printing Office, various years.

Wolverton, Ann. "Does Race Matter? An Examination of a Polluting Plant's Location Decision," U.S. Environmental Protection Agency, National Center for Environmental Economics, mimeograph, July 2002.

[7]

Environmental Regulation as Export Promotion: Product Standards for Dirty Intermediate Goods*

Carol McAusland

Abstract

We employ a stylized model of trade in dirty intermediate goods to examine the impacts of product standards on trade volumes and pollution levels. Our focus is on the case with economies of scale in intermediate good production. In this setting, changes in one country's demand for dirty inputs can feed back to affect the quality of intermediate goods offered for sale in international markets. We provide conditions under which tightening product standards in one country can raise both the profits of and exports from that country's final good producers. Greening product standards can have perverse environmental impacts: tightening rules governing emissions arising from the use of dirty inputs may raise rather than reduce local pollution.

KEYWORDS: intermediate goods, trade, pollution, product standards

*My thanks to Charlie Kolstad, Don Fullerton and three anonymous referees for detailed comments on an earlier draft of this paper, and to Sarani Saha for research assistance. All errors are the sole responsibility of the author. Email: mcauslan@econ.ucsb.edu

When a country tightens its environmental regulations, will it drive pollution overseas, hurting its own exports of dirty goods in the process? These are enduring questions in the academic and public debates concerning trade and the environment. So far, theory has indicated that the answer is yes: tightening regulations in one country tends to disadvantage that country's exports of pollution intensive goods. It also tends to reduce polluting activity locally while promoting polluting activity abroad.

To date, most research on this topic has looked only at trade in dirty final goods; environmental problems associated with traded intermediate goods have been almost entirely ignored.[1] This paper presents a stylized model of trade in dirty intermediate goods. Scale economies in intermediate good production serve as the hallmark of the model. In this context, we find that the usual results concerning pollution haven effects may be overturned.[2] Indeed, we derive conditions under which tightening regulation in one country promotes that country's exports of dirty goods and raises its total emissions.

Trade in intermediate goods is far from trivial. A back-of-the-envelope calculation suggests that at least two-thirds of U.S. exports and half its imports are intermediate goods.[3] Moreover, the category of intermediate goods is broad and includes products whose use is inexorably linked to pollution and other environmental problems. For example fertilizer use contributes to ground water contamination, while pesticide use is linked to species loss. Similarly, the use of refined fuels is implicated in much of the air pollution that afflicts urban centers. And just as the lead content of gasoline is determined at the refinery, the emission intensity of many intermediate goods—that is, the emissions that will be generated during *use* of these intermediate goods—is similarly determined at the factory where the intermediate goods are first produced.

Economies of scale are another important characteristic of intermediate goods production. For example, the fixed costs of researching and developing new product lines can be substantial. Consider the agrochemical industry: on average, research-based agrochemical companies spend 10% of their revenues on R&D, and the cost of developing an agrochemical that incorporates a new "active ingredient" was approximately $130 million in the early 1990s (Hartnell 1996). However not all scale economies come in the form of fixed costs. Branstetter (2001) finds strong evidence of intra-national knowledge spillovers in a variety of industries including chemicals, electronics and transportation. Austin (2000) similarly

[1]The exception is Benarroch and Weder (2003), who examine intermediate goods produced by monopolistically competitive upstream firms. They find international trade may lower global pollution with increasing returns to scale in downstream production.

[2]Speaking in general terms, the literature on Pollution Havens has been concerned with two comparative statics. There is the question of whether tighter environmental regulation reduces exports and impacts plant location decisions at the margin. This is distinct from the hypothesis that trade *liberalization* will drive polluting behavior to low standard countries. Following Copeland and Taylor (2004), we refer to these, respectively, as pollution haven *effects* and the pollution haven *hypothesis*. Given that the present paper focuses on the impacts of tightening environmental regulation, then our focus is on pollution haven effects.

[3]In 2003, the two categories Capital Goods and Industrial Supplies alone made up 64% of U.S. exports and 48% of its imports (U.S. Census Bureau 2004). Antweiler and Trefler (2002) similarly use Government of Canada data to conclude that a full two-thirds of US-Canada bilateral trade is in intermediate goods.

finds evidence of substantial knowledge spillovers from biotechnology patents.

Accordingly, this paper develops a partial equilibrium model of trade in intermediate and final goods that highlights scale economies in intermediate good production. The model allows for two distinct forms of increasing returns to scale: analysis begins with the problem of positive external economies of scale, and then instead examines an upstream industry with large fixed costs of production. The paper concludes by assessing how tightening environmental regulation in one country affects trade volumes and pollution levels.

The environmental regulation we consider is an emission cap. In our model emissions arise from use of dirty intermediate goods. Thus the emission cap can come in the form of either a product standard limiting the pollution intensity of intermediate goods—this would be the case with noise and effluent standards for industrial machinery—or in the form of a cap on emission per unit of input, a cap that induces manufacturers to demand cleaner inputs, for example low-sulfur coal.[4] Finally, although our analysis is framed in terms of marginal changes to the emission cap, more restrictive regulations such as outright bans on particular inputs will have similar effects. For example, the European Union (EU) moratorium on new genetically modified (GM) organisms limits the production processes available to manufacturers developing new seed varieties, at least if they want to sell those new varieties in the EU.

How tightening product standards in one country might benefit its producers of final goods depends in large part on market shares. The larger is the fraction of downstream firms that demand clean inputs, then the lower the cost of using clean inputs rather than dirty ones. And if the price gap is big enough, then unregulated firms will use the clean inputs as well. As a consequence, the regulation-induced tastes of the dominant firms will dictate the quality of inputs used industry-wide. In such a case, tightening standards further will have feedback effects impacting all downstream firms, possibly to the net benefit of regulated firms. The suggestion that regulation might promote regulated firms is not new; this is the central claim of the Porter hypothesis. Indeed, a weak version of the Porter hypothesis[5] is that "under stricter environmental regulations, some regulated firms will benefit competitively, at the expense of other regulated firms." (Jaffe et al. 1995, p.154) This itself is just a variant on the Raising Rivals' Cost argument of Maloney and McCormick (1982) and Salop and Scheffman (1983). However the difference in our model is that overseas firms are not formally bound by domestic regulations. Instead, overseas competitors are adversely affected via changes in input markets. In particular, if overseas

[4]Copeland (2001) also considers product standards, examining government incentives to manipulate environmental product standards concerning consumer goods. He similarly finds that green product standards can shift market share toward domestic producers if overseas competitors have higher marginal costs of meeting the tightened regulations. Unlike the present paper, Copeland (2001) only allows for trade in final goods, and finds that environmental regulation can only reduce local pollution.

[5]Porter and van der Linde (1995) make a stronger claim concerning innovation offsets: "By stimulating innovation, strict environmental regulations can actually enhance competitiveness." (p.98) Mohr (2002) investigates this claim in the context of positive external economies of scale. Using a model with learning externalities and multiple equilibria, Mohr finds environmental regulation can serve as a coordinating mechanism that induces atomistic agents to adopt the high-productivity, low-emission technology.

competitors are less-adaptable at employing green inputs, then regulation in one country can shift market share to—and possibly raise the overall profits of—domestic firms.

The above discussion concerns the profits of regulated firms. What about their level of activity overall? We find that regulation may also raise the input use and total output of regulated firms. And this brings us back to the topic of pollution haven effects. When final goods are also traded internationally, more output may mean more exports. In short, environmental regulation may serve as export promotion.

We are not the first to find that environmental regulation might promote exports. This was a central premise of Porter's original argument, though Porter assumed that regulation-induced-innovation would be the driving factor. Simpson and Bradford (1996) examine Porter's assertion using a model of international Cournot duopoly. They find regulation might indeed promote local industry if the resulting innovation generates a sufficiently large shift in the regulated firm's best response function. Modeling competitive firms (and static technologies), Chua (2003) also finds that high pollution taxes may give a country comparative advantage in dirty goods, the necessary condition being that manufacturing and abatement service industries have sufficiently different factor intensity ratios.

Alternately, Copeland and Taylor (1995), Rauscher (1997) and McAusland (2003) all find that environmental regulation can promote net exports if the pollution in question is a by-product of consuming, rather than producing, dirty goods. This is a useful distinction for our purposes. Following McAusland (2003), we label goods whose production generates pollution as *smokestack* goods, and goods that are dirty during use as *tailpipe* goods. In our model, emissions are generated when using intermediate goods to produce final goods. In view of that, we can simultaneously regard the final goods as smokestack goods, and the intermediate inputs as tailpipe goods. And for each interpretation, we find that tightening product standards can have impacts on trade volumes that run counter to the conventional logic. For example, if stricter local effluent standards for industrial machinery promote local use of (potentially imported) machinery, then tailpipe regulation can be seen to promote imports of tailpipe goods. Moreover, if the output of final goods produced using that machinery also rises (and if final goods are exported), this can also be interpreted as a case where smokestack regulation promotes exports of dirty (smokestack) goods.

What about the location of polluting behavior? To the best of our knowledge all scenarios studied to date conclude that tightening environmental regulation in one country reduces the absolute level of polluting behavior occurring within its borders. This is not necessarily so in our model. Even though stricter standards translate to cleaner techniques for producing final goods, the associated shift in market share may dominate. As a result, tighter local regulation may lead to an expansion of the regulated activity in the first place, possibly raising local pollution in absolute terms.

1. Model

We begin by outlining the industrial structure of the model. An upstream industry produces an intermediate good, which firms in the downstream industry then combine with a complementary input to produce final goods. Final goods are then sold to homogenous consumers.

Upstream goods have variable quality, or 'type' e, where e denotes emissions arising from use of one unit of the intermediate good. The variable e may also be interpreted as the emission intensity of the intermediate good, such as machinery that pollutes when used by downstream firms. Downstream firms take the quality offerings of upstream goods as given. These downstream firms take input and output prices as given as well.

The model has two countries, Home and Foreign; let asterisks denote values for Foreign. We assume free trade in all goods. Downstream firms are located in each of the two countries. At this point, locations of the upstream industry and final good consumers are unspecified; Section 2 returns to this issue.

Begin by examining equilibrium in the downstream market. Downstream producers combine the intermediate input x with a complementary input T to produce a final good q in a Cobb-Douglas fashion:

$$q = [g(e)x]^{1-\alpha}T^{\alpha} \tag{1}$$

where $\alpha \in (0,1)$ is a parameter and g is a positive, increasing function with elasticity $\sigma \equiv g'e/g \geq 0$. The function g can be interpreted as the *pollution productivity* of intermediate goods. Downstream firms are heterogenous. In particular, Foreign firms have a production technology with the same form as Home firms, but different parameter values: $q^* = [g^*(e^*)x^*]^{1-\alpha}T^{*\alpha}$ where e^* is the emission intensity of intermediate goods used by Foreign firms—of course e and e^* may be the same in equilibrium—and the function g^* has elasticity $\sigma^* \equiv g^{*\prime}e^*/g^* \geq 0$.

We assume that each country's endowment of the complementary factor is fixed. The reader may wish to interpret T as Home's land endowment; this would be appropriate if final goods were agricultural outputs for example. Note that T and T^* will also serve as partial measures of the scales of the downstream industries in Home and Foreign.

Fixed-factor owners choose inputs of the intermediate good so as to maximize the rent paid to each unit of the fixed factor. Our focus is on the rents paid to a representative land owner, or *firm*. To make things simple, let this representative firm have a factor endowment equal to her country's aggregate endowment T; note though that we will still refer to Home firms as plural when the discussion warrants it. Let P denote the price of final goods and P_{x_e} the price of intermediate inputs of type e. The Home representative firm chooses x to maximize rents Π_T paid to the specific factor T, where

$$\Pi_T = P[g(e)x]^{1-\alpha}T^{\alpha} - P_{x_e}x.$$

Rearranging the first order condition for an interior optimum gives the input demand by Home firms as a function of input and output prices:

$$x(P,P_{x_e}) = \left[\frac{P}{P_{x_e}}[1-\alpha]\right]^{1/\alpha} g(e)^{\frac{1-\alpha}{\alpha}} T;$$

substituting $x(P,P_{x_e})$ into (1) in turn gives

$$q(P,P_{x_e}) = [(1-\alpha)P]^{\frac{1-\alpha}{\alpha}} \left[\frac{g(e)}{P_{x_e}}\right]^{\frac{1-\alpha}{\alpha}} T.$$

Symmetric results obtain for Foreign firms.

Final goods are sold to price-taking consumers who are indifferent about how much pollution is generated during the production of goods they ultimately consume. For simplicity, assume final demand is iso-elastic: $P^{-\varepsilon} = q^{w}$ where $q^{w} = q + q^{*}$ is total world output of downstream goods from the Home and Foreign firms, and ε is a parameter that also serves as the price elasticity of final good demand.

Summing over $q(P,P_{x_e})$ and $q^{*}(P,P_{x_{e^*}})$ and substituting into the inverse demand function gives

$$P = \left[[1-\alpha]^{(1-\alpha)}\Delta^{\alpha}\right]^{-\frac{1}{\varepsilon\alpha+1-\alpha}}$$

$$\text{where } \Delta \equiv \left[\left[\frac{g(e)}{P_{x_e}}\right]^{\frac{1-\alpha}{\alpha}} T + \left[\frac{g^{*}(e^{*})}{P_{x_{e^*}}}\right]^{\frac{1-\alpha}{\alpha}} T^{*}\right]$$

is a price and pollution-productivity weighted sum of the fixed factors available for producing final goods.

We can now solve for Home input use as a function of input price and quality alone. Substituting P into $x(P,P_{x_e})$ gives

$$x = \frac{1}{P_{x_e}}\left[\frac{(1-\alpha)^{\varepsilon}}{\Delta}\right]^{\frac{1}{\varepsilon\alpha+1-\alpha}} \left[\frac{g(e)}{P_{x_e}}\right]^{\frac{1-\alpha}{\alpha}} T \tag{2}$$

and similarly for Foreign firms. Substituting in for x in the production function gives Home output of final goods

$$q = \left[\frac{[1-\alpha]^{\varepsilon}}{\Delta}\right]^{\frac{1-\alpha}{\varepsilon\alpha+1-\alpha}} \left[\frac{g(e)}{P_{x_e}}\right]^{\frac{1-\alpha}{\alpha}} T \tag{3}$$

with a similar expression for Foreign. In the analysis to follow it will be useful to have notation denoting market shares. Define $\lambda \equiv q/[q+q^{*}]$ as Home's share of the final goods market. Equation (3) and its parallel for Foreign gives

$$\lambda = \frac{\left[\frac{g(e)}{P_{xe}}\right]^{\frac{1-\alpha}{\alpha}} T}{\Delta}.$$

Finally, substituting in for x and q into Π_T gives rents paid to the specific factor T available to Home firms:

$$\Pi_T = \frac{\alpha}{1-\alpha}\left[\left[\frac{g(e)}{P_{x_e}}\right]^{\frac{1-\alpha}{\alpha}} T\left[\frac{[1-\alpha]^{\varepsilon}}{\Delta}\right]^{\frac{1}{\varepsilon\alpha+1-\alpha}}\right]. \tag{4}$$

Notably, values of x, q and Π_T all depend on the price and emission intensity of intermediate goods, both directly and indirectly via Δ. These will be determined both by equilibrium in the upstream market and the environmental regulations in each country. We now turn our attention to the latter.

We assume regulation takes the form of product standards on intermediate goods. Let $\bar{e}$ denote the cap on the emission intensity of intermediate goods used within the Home jurisdiction; $\bar{e}^*$ is similarly defined for Foreign. Accordingly, upstream firms can sell intermediate goods in Home only so long as those goods carry emission intensity $e \leq \bar{e}$, and a reduction in $\bar{e}$ represents a tightening of Home regulation. For brevity we consider only the case where $\bar{e} \leq \bar{e}^*$, i.e. where Home is the relatively high standard country. Given that pollution is a by-product of using dirty intermediate goods in final good production, a cap on e can be viewed alternately as a tailpipe regulation for intermediate goods (e.g. cleaner fuels, no GM ingredients), or as a process restriction on smokestack emissions from final goods production (e.g. downstream firms must use scrubbers). Either way, the regulation is within the government's jurisdiction, as the pollution in question is generated within its boundaries. This would be true even if intermediate goods were entirely produced abroad.

1.1 Upstream Market

We analyze two versions of the upstream industry, each characterized by increasing returns to scale in production. In the first version there are positive external economies of scale in the production of intermediate goods of a given quality; as is fitting with external scale economies, assume that upstream firms behave competitively in this example. In the second version adding additional product (quality) lines instead incurs large fixed costs; as large fixed costs favor concentrated production, in this example the upstream good is provided by a monopolist.

In each of the versions to follow, define x_e^w as world production of intermediate goods with quality e. We begin the analysis of each case by treating e and e^* as exogenous, simply to see what prices the upstream industry will charge; this is followed by conditions under which only upstream goods with emission intensity $\bar{e}$ are produced in equilibrium, and how tightening $\bar{e}$ impacts Home's downstream firms.

1.1.1 Positive External Economies of Scale

Let $c(x_e^w)$ denote the unit cost of producing intermediate goods of quality e. Positive external economies of scale are operationalized as follows: $\partial c/\partial x_e^w < 0$ although individual firms treat $c(x_e^w)$ as fixed. In order to simplify matters, further assume that demand is unit elastic ($\varepsilon = 1$) and that no direct cost premium is associated with producing cleaner goods: $\partial c/\partial e = 0$. In this example, upstream producers simply charge $P_{x_e} = c(x_e^w)$ and are indifferent about which quality of goods they supply. Instead, whether any type-$\overline{e^*}$ goods are produced depends entirely on whether downstream firms demand them. The following proposition lays out a sufficient condition under which downstream firms demand only $\bar{e}$-quality intermediate goods.

Proposition 1: *Suppose the upstream industry exhibits positive external economies of scale* ($\partial c/\partial x_e^w < 0$), *unit costs in the upstream industry are independent of quality* ($\partial c/\partial e = 0$), *and final good demand is unit elastic* ($\varepsilon = 1$). *Then all Home and Foreign firms choose to use inputs with quality* $\bar{e}$ *if*

$$g^*(\bar{e})/g^*(\overline{e^*}) > c(x_{\bar{e}}^{\tilde{w}})/c(x_{\overline{e^*}}^{\tilde{w}}), \tag{5}$$

where $x_{\bar{e}}^{\tilde{w}}$ *and* $x_{\overline{e^*}}^{\tilde{w}}$ *are solutions to the system*

$$x_{\bar{e}}^{\tilde{w}} = [1-\alpha]\tilde{\lambda}/c(x_{\bar{e}}^{\tilde{w}}), \quad x_{\overline{e^*}}^{\tilde{w}} = [1-\alpha][1-\tilde{\lambda}]/c(x_{\overline{e^*}}^{\tilde{w}}).$$

$$\tilde{\lambda} = \frac{[g(\bar{e})/c(x_{\bar{e}}^{\tilde{w}})]^{[1-\alpha]/\alpha}T}{[g(\bar{e})/c(x_{\bar{e}}^{\tilde{w}})]^{[1-\alpha]/\alpha}T + [g^*(\overline{e^*})/c(x_{\overline{e^*}}^{\tilde{w}})]^{[1-\alpha]/\alpha}T^*}.$$

Proof: Let $\pi_{j^*}(\tilde{e})$ denote the return paid to Foreign firm j^* with T_{j^*} units of the fixed factor when j^* uses inputs of quality $\tilde{e}$. From (4), when $\varepsilon = 1$ then $\pi_{j^*}(\tilde{e}) = \alpha\lambda_{j^*}(\tilde{e})$ where $\lambda_{j^*}(\tilde{e}) = [g^*(\tilde{e})/P_{x_{\tilde{e}}}]^{[1-\alpha]/\alpha}T_{j^*}/\Delta$ is j^*'s share of the final goods market when j^* uses inputs of quality $\tilde{e}$. Consequently, when faced with the choice between inputs of quality $\bar{e}$ and quality $\overline{e^*}$, firm j^* demands the former whenever $\lambda_{j^*}(\bar{e}) > \lambda_{j^*}(\overline{e^*})$. As firm j^* is atomistic and small, it takes prices and aggregate behavior Δ as exogenous; as input prices equal unit costs then the inequality $\lambda_{j^*}(\bar{e}) > \lambda_{j^*}(\overline{e^*})$ simplifies to $g^*(\bar{e})/c(x_{\bar{e}}^w) > g^*(\overline{e^*})/c(x_{\overline{e^*}}^w)$, or, rearranging, to

$$g^*(\bar{e})/g^*(\overline{e^*}) > c(x_{\bar{e}}^w)/c(x_{\overline{e^*}}^w). \tag{6}$$

Thus, whenever (6) is satisfied, any Foreign firm j^* will demand type $\bar{e}$-quality inputs. The left hand side of (6) is predetermined by $\bar{e}$ and $\overline{e^*}$. Evaluating (2) when $\varepsilon = 1$ shows $x_{\bar{e}}^{\tilde{w}}$ and $x_{\overline{e^*}}^{\tilde{w}}$ are the input demands that will arise if all Home firms use $\bar{e}$-quality inputs, while all Foreign firms use $\overline{e^*}$-quality inputs. Moreover, because Home firms are prevented by law from using inputs with emission intensity greater than $\bar{e}$, and because $\partial c/\partial x_e^w < 0$, then the highest value that the right hand ratio in (6) can take occurs when $x_{\bar{e}}^w = x_{\bar{e}}^{\tilde{w}}$ and $x_{\overline{e^*}}^w = x_{\overline{e^*}}^{\tilde{w}}$. Hence (5) is a sufficient condition for every Foreign firm to demand $\bar{e}$-quality inputs. *Q.E.D.*

Proposition 1 provides a sufficient condition for all downstream firms to use intermediate goods of quality $\bar{e}$. Home firms do so because Home regulation demands it. Foreign firms do so because it is more cost effective. Because Home firms are required by law to use $\bar{e}$-quality inputs, their demand lowers the unit cost, and hence the price, of upstream goods of this type. Foreign firms then face a choice: obtain high productivity by using relatively dirty $\vec{e}^*$-quality inputs, or use relatively inexpensive $\bar{e}$-quality inputs. Sufficient condition (5) simply states that, if *all* Foreign firms were to use $\vec{e}^*$-quality inputs but even this didn't make these goods relatively more cost efficient than $\bar{e}$-quality inputs, then no downstream firm will buy the dirtier inputs. In order for this condition to hold, Home must sufficiently dominate the downstream market, i.e. $\tilde{\lambda}$ must be large. Moreover, $\tilde{\lambda}$ will be large if Home has a sufficiently large endowment of the effective complementary input $[g(\bar{e})/c(x^{\tilde{w}}_{\bar{e}})]^{|1-\alpha|/\alpha}T$, either via raw resource abundance (measured by T) or technology superiority (measured by g). When this endowment is sufficiently large, demand from Home's downstream firms for relatively clean upstream goods (e.g. clean fuels, non-GM seed) will dictate the quality produced by upstream firms, and Home effectively exports its product standards abroad.

We now examine how tightening Home product standards impacts Home's downstream firms if all downstream producers use the same quality inputs. In what follows let ˆ denote percentage changes, e.g. $\hat{q} = dq/q$; moreover, let every occurrence of $\hat{e}$ refer to changes in the Home emission cap, i.e. $\hat{e} \equiv d\bar{e}/\bar{e}$.

Proposition 2: *Suppose the upstream industry exhibits positive external economies of scale* $(\partial c/\partial x^w_e < 0)$, *unit costs in the upstream industry are independent of quality* $(\partial c/\partial e = 0)$, *final good demand is unit elastic* $(\varepsilon = 1)$, *and all downstream firms purchase inputs of the same quality* $\bar{e}$ *(i.e. equation (5) holds). Then tightening Home's product standard (lowering* $\bar{e}$*)*

1. *raises clean input use by Home firms* $(\hat{x}/\hat{e} < 0)$ *and raises rents earned by Home firms* $(\hat{\Pi}_T/\hat{e} < 0)$ *if and only if* $\sigma < \sigma^*$, *and*

2. *raises total output from Home firms* $(\hat{q}/\hat{e} < 0)$ *if and only if* $\sigma < \sigma^*(1-\lambda)(1-\alpha)/[1-\lambda(1-\alpha)]$.

Proof: When all firms use inputs of the same quality $\bar{e}$ then $x^w_{\bar{e}} = x + x^*$. Evaluating (2) and its counterpart for Foreign firms when $\varepsilon = 1$ gives $x^w_{\bar{e}} = [1-\alpha]/c(x^w_{\bar{e}})$ which is independent of $\bar{e}$. Provided this expression has a solution, as assumed, then aggregate input use is independent of quality and so $\hat{P}_{x_{\bar{e}}}/\hat{e} = 0$. Evaluating (2), (3) and (4) when $\varepsilon = 1$ gives $x = [1-\alpha]\lambda/c$, $q = [1-\alpha]^{1-\alpha}\Delta^{\alpha}\lambda$, and $\Pi_T = \alpha\lambda$, while log differentiating reveals $\hat{x}/\hat{e} = \hat{\Pi}_T/\hat{e} = \hat{\lambda}/\hat{e}$ and $\hat{q}/\hat{e} = \alpha\hat{\Delta}/\hat{e} + \hat{\lambda}/\hat{e}$. Log differentiating λ gives $\hat{\lambda}/\hat{e} = [1-\alpha][1-\lambda][\sigma-\sigma^*]/\alpha$ and $\hat{\Delta}/\hat{e} = [\lambda\sigma + (1-\lambda)\sigma^*][1-\alpha]/\alpha$. Provided $\lambda > 0$ then the form of $\hat{\lambda}/\hat{e}$ confirms Part 1; substituting in for $\hat{\lambda}/\hat{e}$ and $\hat{\Delta}/\hat{e}$ and collecting terms to get $\hat{q}/\hat{e} = [\sigma[1-\lambda(1-\alpha)] - \sigma^*(1-\lambda)(1-\alpha)][1-\alpha]/\alpha$ confirms Part 2. *Q.E.D.*

The above proposition gives conditions under which tightening Home's product standard can raise both Home use of the cleaner upstream good and rents paid to the Home specific factor. This requires that Home's productivity be less sensitive to changes in emission intensity than Foreign's, i.e. $\sigma < \sigma^*$. If σ is sufficiently smaller than σ^*, then greening the input also raises Home output of final goods.

Proposition 2 reveals that tighter product standards in Home can promote Home's downstream firms. This outcome requires several conditions be in place. In terms of size, Home must dominate the downstream market so that Home demand for clean inputs dictates the quality of intermediate goods used by all downstream firms; as noted following Proposition 1, this requires that Home have either a sufficiently large endowment of the complementary factor T, and/or a sufficiently high level of technological superiority in using intermediate goods (g large).[6] Then, if Home tightens its product standards, this will induce upstream suppliers to green their products so that they can keep selling goods to Home's downstream firms. Ordinarily this greening of intermediate goods should hurt downstream firms because it will impact their productivity. But if Foreign downstream production technology is sufficiently less adaptable than is Home's, then Foreign firm's will suffer a larger reduction in output. This will shift market share in favor of Home firms; indeed it is straightforward to show $\text{sgn}[\hat{\lambda}/\hat{e}] = \text{sgn}[\sigma - \sigma^*]$. If Foreign technology is *sufficiently* less adaptable—recall Parts 1 & 2 of Proposition 2—then the shift in market share more than fully offsets losses to Home firms from lower pollution productivity: Home's downstream industry will be promoted, rather than hindered, by a stricter Home product standard. Although seemingly perverse, given the context this result should not be surprising. This is simply a version of Raising Rivals' Costs, in which stricter environmental regulation can benefit a subset of firms if their production technologies are more amenable to using green inputs. What is unique about the case we present is that, *de jure*, the tightened product standard applies only to Home's downstream firms. But because of interactions in the upstream market, the Home product standard has *de facto* influence throughout the downstream industry. As mentioned earlier, the large presence of Home firms in the downstream market allows Home effectively to export its product standards to Foreign. As a result, tightening those Home standards can have a Raising Rivals' Cost effect on Foreign firms even though, legally, Foreign firms are exempt.

Should we expect to see this in practice? Recall the example of the EU's moratorium on new GM organisms.[7] This moratorium has effectively prevented any GM seed, other than Bt maize, from being planted in the EU. And whether intentional or not, this moratorium may well influence the pattern of future seed development, as upstream producers may

---

[6]The conditions that g be large but σ be relatively small are not inherently inconsistent: the variable g measures the *level* of Home pollution productivity while σ measures the *sensitivity* of that productivity to changes in emission intensity.

[7]The EU moratorium may also violate World Trade Organization (WTO) rules. For example, in 2003 the United States, Canada, Egypt and Argentina filed complaints with the WTO, arguing that the EU moratorium is an unfair barrier to trade. They argue there is no scientific evidence that consuming GM organisms harms human health or the environment, and so the ban in unwarranted. This case is still pending.

be less willing to develop costly GM product lines if they are unsellable in the large EU market. Moreover, such a shift in product development will likely have heterogeneous impacts on downstream producers. GM crops are believed to be better suited for use in large scale agricultural enterprises, such as are common in the United States, and so a global shift away from developing genetically modified seed would likely disadvantage U.S. agriculture disproportionately.

1.1.2 Fixed Costs

We now consider the case where increasing returns in the upstream market instead take the form of fixed costs associated with producing distinct product lines. Let $c(e)$ denote the marginal cost of producing upstream goods of quality e, where $\mu \equiv -(\partial c/\partial e)(e/c) \geq 0$; we assume no other economies of scale in this case (i.e. $\partial c/\partial x_e^w = 0$). The presence of fixed costs suggests that the equilibrium number of upstream firms will be small. We consider the limiting case with only a single upstream firm acting as a monopolist, selling to both Home and Foreign.

Taking e and e^* as given for the moment, the monopolist chooses prices of the intermediate inputs it sells to maximize its variable profits $\Pi_V = [P_{x_e} - c(e)]x + [P_{x_{e^*}} - c(e^*)]x^*$, where x and x^* satisfy (2) and a symmetric expression for Foreign input use. Differentiating Π_V with respect to P_{x_e} and $P_{x_{e^*}}$, setting each derivative equal to zero and solving as a system of simultaneous equations yields

$$P_{x_e} = \frac{c(e)}{1-\alpha}\frac{\varepsilon}{\varepsilon-1}, \quad P_{x_{e^*}} = \frac{c(e^*)}{1-\alpha}\frac{\varepsilon}{\varepsilon-1}. \tag{7}$$

This is a variant on the familiar pricing rule by which a monopolist sets a selling price that is a constant markup over marginal cost. As is common in models with market power, intermediate good prices are positive only if $\varepsilon > 1$ (i.e. if demand for downstream goods is elastic) and we impose this restriction throughout the remainder of this example. Substituting (2) and (7) into Π_V and collecting terms gives

$$\Pi_V(e,e^*) = a_1 a_2 \Lambda(e,e^*)^{\frac{\alpha[\varepsilon-1]}{[\varepsilon\alpha+1-\alpha]}},$$

$$\text{where } \Lambda(e,e^*) \equiv \left[\frac{g(e)}{c(e)}\right]^{\frac{1-\alpha}{\alpha}} T + \left[\frac{g^*(e^*)}{c(e^*)}\right]^{\frac{1-\alpha}{\alpha}} T^*,$$

$$a_1 \equiv \frac{[\varepsilon\alpha+1-\alpha][1-\alpha]}{\varepsilon}, \text{ and } a_2 \equiv \left[\frac{[\varepsilon-1][1-\alpha]^2}{\varepsilon}\right]^{\frac{[1-\alpha][\varepsilon-1]}{\varepsilon\alpha+1-\alpha}}.$$

This establishes the groundwork for the following proposition.

Proposition 3: *Assume fixed costs F of producing distinct upstream product lines, marginal production costs are independent of quantity* ($\partial c/\partial x_e^w = 0$)*, and final good demand is elastic* ($\varepsilon > 1$)*. Then the upstream monopolist supplies only* $\bar{e}$*-quality intermediate goods if and only if both*

(8) $$\frac{c(\bar{e})}{c(\bar{e}^*)} < \left[\frac{Tg(\bar{e})^{\frac{1-\alpha}{\alpha}} + T^*g^*(\bar{e})^{\frac{1-\alpha}{\alpha}}}{T^*g^*(\bar{e}^*)^{\frac{1-\alpha}{\alpha}}}\right]^{\frac{\alpha}{1-\alpha}}$$

(9) $$\text{and} \quad \frac{F}{a_1a_2} > \Lambda(\bar{e},\bar{e}^*)^{\frac{\alpha[\varepsilon-1]}{\varepsilon\alpha+1-\alpha}} - \Lambda(\bar{e},\bar{e})^{\frac{\alpha[\varepsilon-1]}{\varepsilon\alpha+1-\alpha}}.$$

Proof: We first establish *which* quality the monopolist would offer if it supplied intermediate goods of only one quality. If the monopolist offers only $\bar{e}$-quality intermediate goods then both Home and Foreign can participate in the downstream market and the monopolist earns variable profits

$$\Pi_V(\bar{e},\bar{e}) = a_1a_2c(\bar{e})^{-\frac{[\varepsilon-1][1-\alpha]}{\varepsilon\alpha+1-\alpha}}\left[Tg(\bar{e})^{\frac{1-\alpha}{\alpha}} + T^*g^*(\bar{e})^{\frac{1-\alpha}{\alpha}}\right]^{\frac{\alpha[\varepsilon-1]}{\varepsilon\alpha+1-\alpha}}.$$

If it instead offers goods with emission intensity $\bar{e}^*$, then the Home firm is excluded from the market, and so the monopolist's profits are $\tilde{\Pi}_V(\bar{e}^*)$, where

$$\tilde{\Pi}_V(\bar{e}^*) = a_1a_2c(\bar{e}^*)^{-\frac{[\varepsilon-1][1-\alpha]}{\varepsilon\alpha+1-\alpha}}\left[T^*g^*(\bar{e}^*)^{\frac{1-\alpha}{\alpha}}\right]^{\frac{\alpha[\varepsilon-1]}{\varepsilon\alpha+1-\alpha}}.$$

Thus the upstream monopolist offers an input satisfying Home's stricter regulation if $\Pi_V(\bar{e},\bar{e}) > \tilde{\Pi}_V(\bar{e}^*)$, or, equivalently, condition (8) holds.

Next suppose condition (8) is met. The monopolist will produce only $\bar{e}$-quality goods if $\Pi_V(\bar{e},\bar{e}) > \Pi_V(\bar{e},\bar{e}^*) - F$, which is equivalent to condition (9). *Q.E.D.*

Proposition 3 lays out conditions under which the monopolist produces only relatively clean ($\bar{e}$) inputs. As specified in (9), this outcome requires first that the fixed costs of offering multiple product lines be sufficiently large. The second condition is that, if producing only one variety of intermediate goods, producing $\bar{e}$-quality intermediate goods is more profitable to the monopolist. Inspecting (8) reveals the monopolist chooses to produce clean inputs if $[Tg(\bar{e})^{[1-\alpha]/\alpha} + T^*g^*(\bar{e})^{[1-\alpha]/\alpha}]/T^*g^*(\bar{e}^*)^{[1-\alpha]/\alpha}$ is sufficiently large. Interpret this condition as follows. If the monopolist offered only $\bar{e}^*$-quality goods, Home firms would be excluded from the market and only Foreign firms would purchase the monopolist's products; Foreign firms, however, would have relatively high productivity. If the monopolist instead offered only clean inputs, both Home and Foreign firms could participate in the downstream market. Whether this raises global demand for the monopolist's product enough to compensate for the monopolist's higher unit costs depends, among other things, on the size of Home's downstream industry. Specifically, the larger is Home's fixed factor endowment T, or Home's productivity when using clean inputs (measured by $g(\bar{e})$) then the larger is the right hand ratio in (8) and the more likely the monopolist produces only $\bar{e}$ quality goods.

Let us now examine how tightening Home's environmental regulation will impact Home's downstream firms if all downstream firms use the same quality inputs.

Proposition 4: *Suppose there are fixed costs of producing distinct upstream product lines, marginal production costs are independent of quantity* ($\partial c/\partial x_e^w = 0$)*, final good demand is elastic* ($\varepsilon > 1$)*, and all downstream firms use inputs of quality* $\bar{e}$ *(i.e. conditions (8) and (9) hold). Then lowering* $\bar{e}$ *promotes Home's downstream firms if* $\sigma^* - \sigma$ *is sufficiently large. In particular, lowering* $\bar{e}$

1. *raises input use by Home firms* ($\hat{x}/\hat{e} < 0$) *if and only if*

(10) $$\sigma^* - \sigma > \frac{\alpha[(1-\alpha)(\varepsilon-1)\sigma + \varepsilon\mu]}{(1-\alpha)(1-\lambda)},$$

2. *raises output from Home firms* ($\hat{q}/\hat{e} < 0$) *if and only if* $\sigma^* - \sigma > \alpha\varepsilon[\sigma+\mu]/[(1-\alpha)(1-\lambda)]$, *and*

3. *raises rents earned by Home firms* ($\hat{\Pi}_T/\hat{e} < 0$) *if and only if* $\sigma^* - \sigma > \alpha[\varepsilon-1][\sigma+\mu]/[1-\lambda]$.

Proof: Part 1. Substitute (7) into (2), set $e = e^* = \bar{e}$ and log differentiate with respect to $\bar{e}$. This gives

(11) $$\frac{\hat{x}}{\hat{e}} = \frac{\frac{1-\alpha}{\alpha}[(\varepsilon\alpha + 1 - \alpha - \lambda)\sigma - [1-\lambda]\sigma^*] + \varepsilon\mu}{\varepsilon\alpha + 1 - \alpha}$$

Isolating $\sigma^* - \sigma$ reveals $\hat{x}/\hat{e} < 0$ if and only if condition (10) holds.

Proofs of Parts 2 and 3 have identical form as the proof of Part 1. *Q.E.D.*

Propositions 2 and 4 are derived for quite different versions of the upstream industry: Proposition 2 applies to a competitive upstream industry with external economies of scale, while Proposition 4 applies to an upstream monopolist with internal fixed costs. But the results appear quite robust. Comparing the two propositions reveals that regulations' impact depends pivotally on the gap between σ and σ^* in each case: if Foreign pollution productivity is sufficiently sensitive to the emission intensity of intermediate goods, then a stricter product standard in Home may benefit Home firms. In this latter example we have also allowed for upstream marginal costs that vary with emission intensity: $\mu \geq 0$. This generates some, but not full, restraint on Home's ability to aid its firms via strict regulation. Now when $\bar{e}$ is lowered, Home firms are also disadvantaged by higher input costs: $dP_{x_e}/d\bar{e} = c'[dP_{x_e}/dc] \geq 0$. But still, if the gap between Foreign and Home productivity responses is sufficiently large—recall Parts 1-3 of Proposition 4—we will again see the shift in market share to Home firms dominate, and Home regulation promotes Home downstream firms.

2. Pollution Havens

Section 1 above examines the impact of stricter product standards in one country on that country's downstream producers. We now address how this affects trade volumes and, ultimately, pollution levels.

As noted earlier, given that emissions are generated during the production of final goods, a reduction in $\bar{e}$ translates to stricter Home smokestack regulation. Let β denote the fraction of global consumers of final goods who reside in Home. Then Home exports of smokestack goods equal $E = q - \beta q^w$. Differentiating gives $dE/d\bar{e} = dq/d\bar{e} - \beta dq^w/d\bar{e}$, and so $dE/d\bar{e}$ depends on how both Home and global supply of final goods vary with $\bar{e}$. Consider two polar examples: $\beta = 0$ and $\beta = 1$. When $\beta = 0$ all final goods consumers are offshore and so $\text{sgn}[dE/d\bar{e}] = \text{sgn}[\hat{q}/\hat{e}]$, where the conditions for $\hat{q}/\hat{e} < 0$ are as laid out in Propositions 1-4. Alternately, if $\beta = 1$, then all final goods consumers reside in Home and so $E = -q^*$, which is decreasing in $\bar{e}$ if $\sigma^* > \sigma$.[8] In sum, under the conditions by which tightening Home's product standard promotes Home output of final goods and hinders Foreign output—recall Propositions 1-4—then when final goods are traded, stricter Home (smokestack) regulation promotes Home exports of dirty (smokestack) goods if the gap between σ and σ^* is sufficiently large. In short, regulation can be good for exports.

Next examine net imports of tailpipe goods, denoted M. As $\bar{e}$ caps emissions associated with use of the dirty intermediate good, we can also interpret $\bar{e}$ as the stringency of Home tailpipe regulation. Again, we offer two polar examples. If all upstream production is located offshore, then $M = x$, and so tighter Home tailpipe regulation promotes Home tailpipe imports under the conditions for $\hat{x}/\hat{e} < 0$ as laid out in Propositions 1-4. Alternately, suppose all upstream production occurs in Home; then $M = -x^*$ and so $dM/d\bar{e} = -dx^*/d\bar{e}$, which is negative if $\sigma^* > \sigma$. Again we see that, under the conditions laid out in Propositions 1-4, when Foreign firms are sufficiently less adaptable to using green inputs (i.e. σ^* sufficiently greater than σ), then stricter Home tailpipe regulation can promote net imports of tailpipe goods.

Finally, consider the impact of a tighter Home product standard on Home's pollution level. Define $Z = ex$ as Home's emissions. If tightening regulation promotes Home input use, then even though stricter product standards translate to cleaner techniques of production, the increase in the level of polluting activity may dominate. We can be more concrete. Differentiate $Z = ex$ and convert to percentage changes to get $\hat{Z}/\hat{e} = 1 + \hat{x}/\hat{e}$. Substituting in for $\hat{x}/\hat{e}$ gives

[8] With either form of upstream scale economies, if all downstream firms use the same quality intermediate goods and $\lambda > 0$, then it is straightforward to show that $\sigma^* > \sigma$ is a sufficient condition for $dq^*/d\bar{e} > 0$ and $dx^*/d\bar{e} > 0$.

$$\frac{\hat{Z}}{\hat{e}} = 1 + \frac{[1-\alpha][1-\lambda]}{\alpha}[\sigma - \sigma^*] \tag{12}$$

in the case of positive external economies of scale and, using (11),

$$\frac{\hat{Z}}{\hat{e}} = 1 + \frac{1-\alpha}{\alpha[\varepsilon\alpha + 1 - \alpha]}\left[\frac{\alpha}{1-\alpha}\varepsilon\mu + [\alpha[\varepsilon - 1] + 1 - \lambda]\sigma - [1-\lambda]\sigma^*\right] \tag{13}$$

in the case of upstream monopoly and fixed costs. In general these expressions are of ambiguous sign. If $\sigma = \sigma^*$, for example, then in each case $\hat{x}/\hat{e}$ is positive, and so $\hat{Z}/\hat{e}$ is positive as well. If instead $\sigma = \mu = 0$, then whether Home pollution rises or falls with stricter Home regulation depends on relative parameter sizes: in the case of upstream external economies of scale, $\hat{Z}/\hat{e} < 0$ if $\sigma^* > \alpha/[(1-\alpha)(1-\lambda)]$; in the case of upstream monopoly and fixed costs, $\hat{Z}/\hat{e} < 0$ if $\sigma^* > \alpha[\varepsilon\alpha + 1 - \alpha]/[(1-\alpha)(1-\lambda)]$.

Examining these relations reveals the magnitude of σ^* is pivotal once again. The reason is the same as before: the more sensitive is Foreign productivity to the emission intensity of intermediate goods, the greater the shift in market share that accompanies a reduction in $\bar{e}$. Thus, even though Home's downstream producers use cleaner techniques of production, a sufficiently large σ^* means that the scale of Home activity rises by even more, and Home's total emissions rise.

3. Caveats and Extensions

We conclude our analysis by pointing out what this paper does not do. As mentioned, this paper employs a stylized model of trade and intermediate goods so as to highlight some potentially perverse outcomes associated with tightening environmental regulation in downstream countries. In essence, the goal of this paper is to provide a counter-example regarding the impacts of environmental regulation on trade volumes and pollution levels. And as is often true when trying to make a simple case, many issues have been left out along the way.

For instance, this paper considers only one type of environmental regulation: product standards. However the flavor of results would be essentially unchanged if, for example, governments instead employed emission taxes. With emission taxes, the main difference would be that highly-taxed downstream firms would still be eligible to participate in the downstream market if intermediate goods were relatively dirty. However, as high emission taxes induce firms to demand cleaner inputs, the upstream industry may still choose to supply the downstream market with only relatively clean intermediate goods if the highly-taxed firms make up a sufficiently large portion of the downstream market. Just as in our model, this will be true if the fixed costs of additional product lines are large and/or large cost savings are associated with producing multiple units of the same type. If raising taxes leads to even cleaner upstream goods, then environmental regulation may still have perverse impacts on market share and trade volumes if downstream firms have heterogenous capacity to make use of green inputs.

This analysis has also treated the level of environmental regulation in each country as exogenous. Although this is appropriate for highlighting possible impacts of regulation on trade volumes, government motives when setting product standards are also worthy of study. Sections 1 and 2 derive conditions under which a government can simultaneously raise rents received by its fixed factors and protect its local environment; the downside is higher consumer prices and possibly higher costs for upstream producers. This suggests that government incentives to manipulate product standards for strategic purposes may vary across countries depending on industrial structure, technology, and rate of consumer participation. Moreover, we employ a partial equilibrium analysis: we do not allow for factor reallocations across sectors as might reasonably accompany regulatory changes. If only because the magnitude of trade in intermediate goods is so large, all these issues warrant further examination.

4. Conclusion

This paper highlights possible impacts of emission caps, and product standards in particular, on economic and environmental variables. Our focus is a country whose purchasing power in input markets is large, large enough that its environmental regulation influences the quality choices of upstream suppliers. This context gives rise to possibilities that may seem perverse at first glance: local environmental regulation can raise local rents and *promote* exports. But these outcomes make perfect sense if downstream firms have heterogenous capacity to use environmentally friendly inputs. Thus, if one country's regulations induce suppliers to green their products, this may confer advantage on some firms at the expense of others. This is simply an example of Raising Rivals' Costs in an international context, where governments that are unable legally to manipulate product standards outside their jurisdictions instead use their firms' market share to dictate input quality.

The environmental impacts of these strategic standards may also be perverse. If production technologies are identical worldwide, tightening local product standards unambiguously lowers local pollution. But if technologies are heterogenous, there are conditions under which shifting market share dominates the greening of actual inputs, such that tightening local product standards *raises* local pollution. In short, tighter regulation may promote polluting behavior, making the regulated region more, not less, of a pollution haven.

Although our model is admittedly stylized, we argue that it has implications both for empirical research and policy. Trade in intermediate goods accounts for a substantial portion of global trade. Our results suggest scope for the presumed relationship between regulatory stringency and export-competitiveness to fail when intermediate goods are involved. We provide an example in which smokestack regulation promotes smokestack exports, while tailpipe regulation promotes tailpipe imports; this serves as a counter-example that violates the conventional wisdom that environmental regulation hinders exports. At minimum our results suggest researchers should think twice before lumping final and intermediate goods together when using international data. Where the purchasing power of downstream users influences the quality of intermediate goods produced by the international market, the differences between trade in intermediate versus final goods may indeed be non-trivial.

As regards policy, we argue that governments should remain vigilant about ensuring a scientific basis for product standards imposed both locally and abroad. Exporting governments are already wary of overseas product standards that protect local firms at the expense of competing imports. Our results suggest that governments should also be suspicious of overseas product standards that influence the quality and variety decisions of upstream suppliers, both with regards to current product varieties and with respect to R&D into new product lines. Just as health officials in developing countries lament that medical research is misdirected towards developing new drugs for affluent consumers in rich countries, similarly the product standards applied in large countries can influence the shape and direction of research and development of new intermediate products. Where those new products reflect the tastes and needs of users in a subset of countries, users in smaller, fringe markets may be made worse off.

References

Antweiler, Werner and Daniel Trefler, Increasing Returns and All That: A View from Trade, *American Economic Review* **92** (2002), 93-119.

Austin, David H., Patents, Spillovers and Competition in Biotechnology, Resources for the Future Discussion Paper 00-53 (2000).

Benarroch, Michael and Rolf Weder, Intra-Industry Trade in Intermediate Products, Pollution and Internationally Increasing Returns, *mimeo*, University of Winnipeg (2003).

Branstetter, Lee G., Are Knowledge Spillovers International or Intranational in Scope? Microeconometric Evidence from the U.S. and Japan, *Journal of International Economics* **53** (2001), 53-79.

Chua, Swee, Does Tighter Environmental Policy Lead to a Comparative Advantage in Less Polluting Goods? *Oxford Economic Papers* **55** (2003), 25-35.

Copeland, Brian R., Trade and the Environment: Product Standards in a National Treatment Regime, *mimeo*, University of British Columbia (2001).

—— and M. Scott Taylor, Trade and the Environment: A Partial Synthesis, *American Journal of Agricultural Economics* **77** (1995), 765-771.

—— and ——, Trade, Growth and the Environment, *Journal of Economic Literature* **42** (2004), 7-71.

Hartnell, Gaynor, The Innovation of Agrochemicals: Regulation and Patent Protection, *Research Policy* **25** (1996), 379-395.

Jaffe, Adam B., Steven R. Peterson, Paul R. Portney and Robert N. Stavins, Environmental Regulation and the Competitiveness of U.S. Manufacturing: What Does the Evidence Tell Us? *Journal of Economic Literature* **33** (1995), 132-163.

Maloney, Michael T. and Robert E. McCormick, A Positive Theory of Environmental Quality Regulation, *Journal of Law and Economics* **25** (1982), 99-123.

McAusland, Carol, Trade, Politics, and the Environment: Tailpipe vs. Smokestack, University of California at Santa Barbara, Department of Economics Working Paper 8-03 (2003).

Mohr, Robert D., Technical Change, External Economies, and the Porter Hypothesis, *Journal of Environmental Economics and Management* **43** (2002), 158-168.

Porter, Michael E. and Claas van der Linde, Toward a New Conception of the Environment-Competitiveness Relationship, *Journal of Economic Perspectives* **9** (1995), 97-118.

Rauscher, Michael, *International Trade, Factor Movements, and the Environment,* Oxford: Clarendon Press, 1997.

Salop, Steven C. and David T. Scheffman, Raising Rivals' Costs, *American Economic Review* **73** (1983), 267-271.

Simpson, D.R. and R.L. Bradford III, Taxing Variable Cost: Environmental Regulation as Industrial Policy, *Journal of Environmental Economics and Management* **30** (1996), 282-300.

United States Census Bureau, Press Release: U.S. International Trade in Goods and Services: July 2004 (September 10, 2004).

[8]

Unilateral Emission Reductions and Cross-Country Technology Spillovers*

Rolf Golombek and Michael Hoel

Abstract

With limited participation in an international climate agreement, standard economic analysis suggests that a unilateral action taken by a group of countries in order to reduce its emissions is likely to be undermined by increases in emissions from other countries (carbon leakage). While analyses of carbon leakage typically have regarded the technology in each country as given, abatement technologies are in reality endogenous, and thus technology development may be affected by environmental policies. We demonstrate that with endogenous technologies and technology diffusion between countries, it is no longer obvious that reduced emissions in some countries will increase emissions in other countries. We identify cases in which reduced emissions in some countries might reduce emissions also in other countries.

KEYWORDS: transboundary pollution, unilateral environmental action, R&D investments, technology spillovers

*Research support of the Research Council of Norway under the programme SAMSTEMT is gratefully acknowledged. Rolf Golombek, Frisch Centre, Gaustadalleen 21, N-0349 Oslo, Norway. Michael Hoel, Department of Economics, University of Oslo, P.O. Box 1095 Blindern, N-0317 Oslo, Norway.

With transboundary environmental pollution it is necessary to have some kind of coordination of policies among the involved countries in order to achieve a first-best efficient outcome. If the number of involved countries is large enough, however, it is well known that the free-rider incentives of each country may undermine the possibility of reaching a collectively good outcome, see e.g. Barrett (2003). It therefore seems unlikely that an efficient cooperative outcome will be achieved in the near future for important global environmental problems such as the climate problem.

At present (August 2004), it is not clear whether the Kyoto agreement will enter into force. If Russia ratifies the agreement, it is likely to enter into force. The Kyoto agreement will, however, cover less than a third of global greenhouse gas emissions. Even if emissions are cooperatively allocated among the countries ratifying the Kyoto agreement, emissions from the Kyoto group as a whole and from other large emitters such as the US will be determined in a non-coordinated manner.

In a situation with less than full cooperation among the countries involved, the effect of a unilateral action taken by one country (or a group of countries) in order to reduce its emissions is a crucial issue. Standard economic analysis suggests that such a unilateral emissions reduction is likely to be undermined by increases in emissions from other countries. There are several reasons for such a response from other countries. One reason given in Hoel (1991) is that the environmental damage cost functions are strictly convex in the sum of emissions from all countries. For given emissions from other countries, it is individually rational for each country to equate its marginal abatement cost with its marginal environmental cost. If emissions are reduced in a particular country (e.g. due to a change in the environmental concern in that country), marginal environmental costs will go down in all other countries. Each country will therefore adjust its emissions upwards, so that marginal abatement costs again are equal to their marginal environmental damage costs.

In the context of the climate problem, increased emissions from other countries as a response to an emissions reduction in one country is often called carbon leakage. In addition to the reason given above, there are several other reasons for carbon leakage: The most obvious way carbon emissions in a country may be affected by abatement policies in another country is through the prices of fossil fuels. If a country reduces its demand for fossil fuels by introducing a carbon tax, international fuel prices will decline. Lower fuel prices will increase fuel demand in other countries, and thus increase their CO_2 emissions. A second mechanism of carbon leakage is via the prices of energy intensive tradable goods: if the use of fossil fuels is reduced through a carbon tax in the sectors producing energy intensive tradable goods, production of these goods will be reduced in the country which introduces the carbon tax, thereby resulting in reduced CO_2 emissions from this country. However, the reduced supply of energy intensive tradable goods in this country will increase the international price of these goods, thus increasing the supply in other countries. Emissions from other countries therefore increase.

Issues related to carbon leakage have been studied in e.g. Bohm (1993), Golombek et al. (1995), Hoel (1994, 1996, 2001), Rauscher (1997; Section 5.8), Gurtzgen and Rauscher (2000) and Mathiesen and Mæstad (2004). This literature discusses how policies introduced in one country, or in a group of countries, with the aim of reducing carbon emissions may increase carbon emissions in other countries, thus weakening the effect of the policies. Numerical simulations quantifying the importance of carbon leakage include Pezzey (1992), who concludes that the carbon leakage might be quite strong, while e.g. Golombek and Bråten (1994), Felder and Rutherford (1993), and Perroni and Rutherford (1993) give numerical simulations indicating that the effects are modest.

In all of the literature referred to above, the technology in each country is regarded as given. However, abatement technologies are in reality endogenous, and technology development may be affected by environmental policies and other policies. Several recent articles have studied interactions between endogenous technological change and environmental policy, see e.g. Jaffe *et al.* (2002) and Löschel (2002) for overviews. With endogenous technologies, it is no longer obvious that reduced emissions in one country will increase emissions in other countries. Assume e.g. that a country that becomes more concerned about the environment (modeled as a positive shift in its environmental damage function) increases its R&D expenditures in order to reduce its abatement costs. If there are cross-country technology spillovers, this may also reduce abatement costs in other countries. Even if the environmental concern in other countries is unchanged, this cost reduction may induce other countries to reduce their emissions. Effects of this type via technology interdependencies have previously been discussed by e.g. Heal (1994) and Barrett (2003). However, these authors have not analyzed how the technology interdependencies may be rooted in cross-country spillovers of knowledge created by investments in R&D.

The present paper gives a formal analysis of the arguments above, and in particular identifies cases in which reduced emissions in some countries might reduce emissions also in other countries. Although we have the climate problem in mind, the level of abstraction of the model we use makes it relevant also for other transboundary environmental problems. The specific features of carbon leakage discussed above related to international trade of fuel and fuel intensive goods are not present in our simple model. However, our model does have the property that for given technologies, reduced emissions in one country may increase emissions in other countries. This property of our model is captured by introducing convex environmental damage functions. This convexity may be seen as a simplified way to represent the property of carbon leakage that would be present in a richer model specifically designed for the climate problem.

In Section 1 we present a simple static two-country model where countries are identical except for their environmental concern. In each country the technology level depends both on R&D expenditures domestically and in the other country. We model this technology spillover by a "standard" linear equation, which we generalize in Section 4. Since our model is static, there is no time lag between R&D expenditures and the effects on technology levels. Uncertainty related to the effects of R&D expenditures on technology levels is also ignored.

Section 2 presents the initial equilibrium. In Section 3.1 we consider the case where countries initially are identical, and where R&D expenditures are positive in both countries. Starting from this symmetric equilibrium, preferences change in one of the countries.[1] Such a preference change could be interpreted as a genuine preference change by all of the citizens in the country, or by so many citizens that the political majority changes. Alternatively, the preference change could represent new scientific knowledge demonstrating that the consequences of pollution for this country are worse than previously believed. For the climate problem an example could be that sea level will raise more than one previously thought, with correspondingly larger damages for a low-lying country. We show that for the case considered in Section 3.1, abatement and R&D expenditure increase in the country where the preferences change, whereas abatement and R&D expenditures decline in the other country.

In Sections 3.2 and 3.3 we assume that countries are different. In the "dirty" country, environmental concern is so low that R&D expenditures are zero, while these expenditures are positive in the "clean" country. Hence, the dirty country might represent a poor country, which, due to its low income, has low valuation of environmental damages. We show that the effects of increased environmental concern in one country depend on the country in which this preference change takes place. A shift in the environmental damage function of the dirty country increases abatement in the dirty country, whereas abatement and R&D expenditures decline or remain unchanged in the clean country. On the other hand, a shift in the environmental damage function of the clean country increases abatement and R&D expenditures in the clean country, whereas the effect on abatement in the dirty country is ambiguous.

In Section 4 we consider a more general equation determining technology spillovers than in the previous sections. We show that the results in Sections 3.2 and 3.3 are not sensitive to the properties of technology spillovers, whereas some of the results in Section 3.1 are sensitive. In particular, abatement and R&D expenditures may increase in the country where preferences do not change.

Since investments in R&D are decisions of a more long-run nature than abatement decisions, in Section 5 we examine a two-stage game with R&D being determined prior to abatement instead of all variables being determined simultaneously as in Sections 3-4. We show that in the two-stage game each country has an incentive to increase R&D relative to the one-stage game discussed in Section 3. On the other hand, in the two-stage game there are cases in which increased concern for the environment in one country has an ambiguous effect on abatement whereas the corresponding cases have an unambiguous effect in the one-stage game (Sections 3-4).

In Section 6 we analyze the consequences of intellectual property rights that make it possible for the country that develops a technology to prevent the other country from using it. If the other country is allowed to use the technology, it has to compensate the country that developed the technology. We show that with identical countries

---

[1] As an alternative to a preference change, abatement in one country could be exogenous (and shifted). We prefer the present approach because in our model changes occur in fundamentals (preferences) rather than in variables that are choice variables (abatement). Note that the alternative approach would not change any of our propositions.

increased concern for the environment in one country generally has ambiguous effects on abatement and R&D expenditures. On the other hand, if one country is dirty, abatement increases in both countries and R&D expenditures increase in the clean country.

Finally, Section 7 sums up our findings, points at welfare effects and concludes.

1. The model

We consider a model with two countries, "home" and "foreign". We use small letters for home country variables, and capital letters for foreign country variables. The abatement cost functions of the two countries are identical and continuous, and are given by $\Omega(a, y)$ and $\Omega(A, Y)$, where a and A are abatement levels and y and Y are technology levels. We assume that $\Omega(0, y) = 0$ for all y, and that (for $a > 0$) $\Omega_a = \partial\Omega/\partial a > 0$, $\Omega_{aa} > 0$, $\Omega_y < 0$, $\Omega_{yy} > 0$, $\Omega_{ay} < 0$ and $\Omega_{aa}\Omega_{yy} - (\Omega_{ay})^2 > 0$ (i.e. the Ω-function is strictly convex). We also assume that $\Omega_a(0, y) = 0$ for any y. This assumption implies that both countries will have positive abatement levels as long as they have *some* concern for the environment.

The technology level in the home country depends on R&D in the home country, x, and in the foreign country, X. However, technology diffusion is not perfect: Only part ($\gamma < 1$) of the R&D expenditures undertaken in the foreign country is beneficial for the home country (and vice versa).[2] The technology levels of the home and foreign countries are thus given by

$$y = x + \gamma X \tag{1}$$

$$Y = X + \gamma x \tag{2}$$

In (1) and (2) we have assumed an additive structure of technology spillovers, that is, the technology level of a country depends on the sum of R&D expenditures undertaken in all countries, corrected by the technology diffusion parameter γ. This way of modeling spillovers can be found in a wide range of theoretical and empirical contributions analyzing spillovers within countries, and goes back at least to Spence (1984). It is also used in most of the literature on climate policy in the context of interactions between countries and endogenous technology development, see e.g. van der Ploeg and de Zeeuw (1994), Xepapadeas (1995), Ben Youssef (2003), Buonanno et al. (2003), Buchner and Carraro (2004) and Rosendahl (2004).

Although spillovers often are modeled as in (1) and (2), it is not obvious that this is the best way of modeling technology spillovers between firms and countries. Cohen

[2] In Sections 1-3 γ is assumed to be exogenous. The possibility that the country developing a new technology may actively prevent other countries from using this technology through intellectual property rights is discussed in Section 6.

and Levinthal (1989) have argued that the ability of a firm to learn from other firms may depend on its own R&D effort. Graevenitz (2002) discusses the policy implications of whether one models spillovers additively as in (1) or in a similar way as Cohen and Levinthal suggest. We shall return to alternative ways of modeling technology spillovers in Section 4.

Business as usual emissions are identical in the two countries, and normalized to 1. Net emissions are thus $1-a$ in the home country and $1-A$ in the foreign country.

Each country faces three types of costs; abatement costs, R&D expenditures and environmental damage costs. The sum of the first two costs is $\Omega(a,y)+x$ for the home country and $\Omega(A,Y)+X$ for the foreign country. In both countries environmental damage costs depend on the sum of emissions, which we denote $\sigma\,(=2-a-A)$. In the home country environmental damage costs are $e(\sigma)+f\sigma$, where *f* is a shift parameter that is equal to zero in the initial equilibrium. In the foreign country environmental damage costs are $E(\sigma)$. In Sections 3-4 we consider the case of linear damage functions as well as the case in which *e* and *E* are strictly convex, while in Sections 5 and 6 we restrict ourselves to the case of linear damage functions.

2. Equilibrium

In the non-cooperative equilibrium, each country chooses its abatement level and R&D expenditures in order to minimize its own total costs, taking the abatement level and R&D expenditures in the other country as given. Hence, the home county minimizes $\Omega(a,y)+x+e(2-a-A)+f\cdot(2-a-A)$, taking *A* and *X* as given, where *y* is given by (1). The foreign country solves a similar minimization problem.

Below we distinguish between two cases. In the first case countries are identical, and there exists an interior equilibrium (*x* and *X* are positive) given by:

(3) $$\Omega_a(a,y)=e'(2-a-A)+f$$

(4) $$-\Omega_y(a,y)=1$$

(5) $$\Omega_A(A,Y)=E'(2-a-A)$$

(6) $$-\Omega_Y(A,Y)=1$$

where $f=0$ initially. The six equations (1)-(6) determine the equilibrium values of our six endogenous variables *a, A, x, X, y, Y*.

In the second case environmental damage functions differ between the countries. With different countries we shall refer to one country as the dirty country, that is, the country with the lowest valuation of environmental damage, and the other country as the clean country. Moreover, we assume that in the dirty country the equilibrium value of e' (or E') is so low that instead of the interior equilibrium condition (4) we get a corner solution given by $x=0$ (or $X=0$) and $-\Omega_y(a,\gamma X)\leq 1$[3] (or $-\Omega_Y(A,\gamma x)\leq 1$).

The case of identical countries is examined in Section 3.1. We shall call the country in which the preference shift takes place the home country, and the other country is called the foreign country. We then study effects of increased valuation of the environment when countries differ. First, we study effects of increased valuation in the dirty country (Section 3.2) and then in the clean country (Section 3.3). Again, in both sections the country in which the shift takes place we shall call the home country. Hence, in Section 3.2 we assume that the home country is dirty, whereas the home country is clean in Section 3.3.

3. Increased environmental concern

3.1 Identical countries

In this section we assume that countries are identical initially, and we also assume positive R&D expenditures in both countries initially. We investigate the effects of increased concern for the environment in one country, denoted the home country.

For the case in which the environmental damage cost functions are linear, we immediately see from (5) and (6) that the foreign county wants the same abatement level and same technology level as prior to the preference change. However, since the home country wants a higher technology level (and more abatement) after the shift, R&D expenditures in the foreign country must go down in order to keep the technology level in the foreign country unchanged.[4] In the case of increasing marginal environmental costs, the foreign country wants to abate less when the home country abates more. This reduces the incentives for R&D expenditures in the foreign country even more, that is, R&D expenditures decline in the foreign country.

The reasoning above is confirmed by a formal derivation given in the Appendix. We thus have the following proposition:

Proposition 1: *With linear technology spillovers and identical countries, increased concern for the environment in the home country raises abatement and R&D expenditures in that country, whereas R&D expenditures in the foreign country decline. If the damage functions are linear, abatement in the foreign country is unchanged. Otherwise (i.e. with a strictly convex damage function in the clean country) abatement in the foreign country decreases.*

---

[3] For any given positive pair (A,X) it is not possible to find a non-negative pair (a,x) satisfying the interior equilibrium conditions (1)-(4) when e' is suffciently low.

[4] In a dynamic model with irreversible R&D investments, increased environmental concern in the home country would leave the foreign country with a better technology than it would have chosen had it known of this preference change at the time of the R&D investment.

3.2 Home country is dirty

When the home country is dirty, the initial equilibrium is given by (3), (5) and (6) (with $y = \gamma X$ and $Y = X$). Assume now that the environmental damage function of the dirty country (that is, the home country) shifts. Differentiation of (3), (5) and (6) with respect to f ($f = 0$ initially) yields

$$(7) \qquad \frac{da}{df} = \frac{-(\Omega_{AA} + E'')\frac{\Omega_{YY}}{\Omega_{YA}} + \Omega_{AY}}{N} > 0$$

$$(8) \qquad \frac{dA}{df} = \frac{\frac{\Omega_{YY}}{\Omega_{YA}} E''}{N} \leq 0$$

$$(9) \qquad \frac{dX}{df} = \frac{-E''}{N} \leq 0$$

where N is positive.[5] A shift in the environmental damage function of the dirty country will increase abatement in that country. If $E'' > 0$, increased abatement in the dirty country will reduce marginal costs of environment in the clean country, and thus abatement will be reduced in the clean country. Lower abatement in the clean country makes investments in R&D less profitable in the clean country, that is, R&D expenditures are reduced.

If the environmental damage function in the clean country is linear[6], increased abatement in the dirty country does not change marginal environmental costs of the clean country. Hence, in the clean country abatement and R&D expenditures do not change. See (8) and (9) for $E'' = 0$. We have thus shown:

Proposition 2: *With linear technology spillovers and no R&D expenditures in the dirty country, increased concern for the environment in the dirty country increases abatement in the dirty country. If the damage function in the clean country is linear, neither abatement nor R&D expenditures in the clean country change. Otherwise, abatement and R&D expenditures are reduced in the clean country.*

---

[5] $N = \Omega_{aa}[(E'' + \Omega_{AA})\frac{-\Omega_{YY}}{\Omega_{YA}} + \Omega_{AY}] - E''\Omega_{ay}\gamma + e''(\frac{\Omega_{AA}\Omega_{YY}}{-\Omega_{YA}} + \Omega_{AY}) > 0$

[6] The damage function in the dirty country could be linear or strictly convex.

3.3 Home country is clean

In the previous section we studied effects of increased environmental concern in the dirty country if there is no R&D expenditure in that country. We now turn to the case where the home country is clean, and study effects of increased environmental concern in the clean country. Like in the previous section there is no R&D expenditure in the dirty country. The initial equilibrium is given by (3), (4) and (5) (with $y = x$ and $Y = \gamma x$). With $f = 0$ initially, differentiating the system w.r.t. f gives

$$\frac{da}{df} = \frac{\dfrac{\Omega_{yy}}{-\Omega_{ya}}(\Omega_{AA} + E'')}{N} > 0 \tag{10}$$

$$\frac{dx}{df} = \frac{\Omega_{AA} + E''}{N} > 0 \tag{11}$$

$$\frac{dA}{df} = \frac{-(\gamma\,\Omega_{AY} + E''\dfrac{\Omega_{yy}}{-\Omega_{ya}})}{N} \tag{12}$$

where $N > 0$, see footnote 4. A shift in the environmental damage function of the clean country raises abatement in that country. Because increased abatement raises the profitability of R&D expenditures, these expenditures increase in the clean country. Increased R&D expenditures in the clean country raises the technology level of the dirty country through spillovers, which tends to increase abatement in the dirty country ($\frac{-\gamma\,\Omega_{AY}}{N} > 0$ in (12)) because abatement costs have decreased in the dirty country. However, increased abatement in the clean country lowers marginal environmental damage in the dirty country, which tends to decrease abatement in that country ($\frac{E''\dfrac{\Omega_{yy}}{\Omega_{ya}}}{N} < 0$ in (12)). Hence, the effect on abatement in the dirty country is ambiguous.

With linear damage function in the dirty country, marginal environmental costs of the dirty county do not change as abatement increases in the clean country. Hence, there is only a positive effect on abatement in the dirty country through technology spillovers, that is, abatement and thus technology level are increased in the dirty country. To sum up:

Proposition 3: *With linear technology spillovers and no R&D expenditures in the dirty country, increased concern for the environment in the clean country increases* *abatement and R&D expenditures in the clean country. If the damage function in the dirty country is linear, abatement in the dirty country increases. Otherwise, the effect on abatement in the dirty country is ambiguous.*

4. Technology complementarity

In the previous sections we studied unilateral emissions reductions when technology spillovers between countries were linear. As mentioned in the Introduction, however, a more generalized approach may be preferable. In the present section we shall assume that the technology level of the home country y is given by the function $\Phi(x, X)$. The partial derivatives of this function with respect to x and X are denoted by Φ_1 and Φ_2, respectively. We assume that $\Phi_i > 0, i = 1,2,$ and $\Phi_1 > \Phi_2$ for $x = X$ as technology diffusion is not perfect. For the foreign country the technology level is given by $Y = \Phi(X, x)$. We assume that Φ is homogenous of degree 1. Moreover, for each country the marginal benefits of R&D expenditures undertaken in the other country depend positively on the magnitude of its own R&D expenditures, that is, $\Phi_{12} > 0$. We will refer to the latter assumption as technology complementarity.

Note that as Φ is homogenous of degree 1, Φ_1 is homogenous of degree 0, that is, $\Phi_{11}x + \Phi_{12}X = 0$. From our assumption of technology complementarity it now follows that $\Phi_{ii} < 0, i = 1,2$, that is, both own R&D expenditures and spillovers from the other country have "decreasing returns". Finally, in the case of identical countries $(x = X)$, we have $\Phi_{11} + \Phi_{12} = 0$. Below we examine the different cases studied in Section 3 under the assumption of technology complementarity.

4.1 Identical countries

With technology complementarity, the initial equilibrium is given by (with $f = 0$ initially)

(13) $$\Omega_a(a, \Phi(x, X)) = e'(2 - a - A) + f$$

(14) $$-\Omega_y(a, \Phi(x, X))\Phi_1(x, X) = 1$$

(15) $$\Omega_A(A, \Phi(X, x)) = E'(2 - a - A)$$

(16) $$-\Omega_Y(A, \Phi(X, x))\Phi_1(X, x) = 1$$

Consider first the case of linear environmental cost functions. From Section 3.1 we know that for the special case of linear spillovers, the foreign country's desired technology level (and abatement level) is unaffected by the preference shift in the home country. However, as R&D expenditures increase in the home country as a response to the preference shift, we see from (16) that this is no longer true in the present case. Due to technology complementarity, the foreign country will want a higher technology level as x increases. (It is, however, not obvious whether this implies larger R&D expenditures in the foreign country.) Moreover, a higher technology level raises the abatement level in the foreign country (ceteris paribus). If marginal environmental costs are increasing (not constant as above), higher abatement in the home country tends to crowd out abatement in the foreign country. Combined with the previous result we thus expect the effect on abatement in the foreign country to be ambiguous.

In order to check the above intuition, we have differentiated (13) - (16) with respect to f. The details are given in the Appendix, and the results are as follows:

Proposition 4: *With technology complementarity and identical countries, increased concern for the environment in the home country increases abatement and R&D expenditures in that country. The effect on abatement and R&D expenditures in the foreign country are ambiguous. However, if the damage functions are linear, abatement in the foreign country is increased.*

From Propositions 1 and 4 (for identical countries), we see that a shift in the environmental damage function in the home country has similar effects in the home country in the two cases (linear technology spillovers vs. technology complementarity). In the foreign country the effects differ: Under strictly convex damage functions, abatement and R&D expenditures are reduced under linear spillovers whereas under technology complementarity the impact on abatement and R&D expenditures are ambiguous. Similarly, with linear damage functions, abatement in the foreign country is unchanged under linear spillovers, whereas abatement in the foreign country increases under technology complementarity. These differences reflect that under technology complementarity it is more profitable to invest in R&D, which tends to increase abatement as costs of abatement are reduced when the technology level is increased.

4.2 Different countries

In Section 3.2 (home country is dirty and undertakes no R&D) we had $y = \gamma X$ and $Y = X$. The only change from Section 3.2 to the present section is that these two equations are replaced by $y = \Phi(0, X)$ and $Y = \Phi(X, 0)$, respectively. The derivatives of y and Y with respect to X are thus changed from γ and 1 in Section 3.2 to Φ_2 and Φ_1 (which are both positive) in the present section. Going through the derivations in Section 3.2, it is straightforward to see that the three inequalities derived are unaffected (but the sizes of these derivatives will of course depend on the spillover function). Similarly, the only change from Section 3.3 (home country is clean) to the present section is that $y = x$ and $Y = \gamma x$ are replaced by $y = \Phi(x, 0)$ and $Y = \Phi(0, x)$, respectively. Going through the derivations in Section 3.3, it is straightforward to see that the three inequalities derived are unaffected. We thus have the following proposition:

Proposition 5: *The results in Proposition 2 and 3 are valid also under technology complementarity.*

5. A two-stage game

In the previous sections we assumed that all variables, that is, abatement and R&D expenditures in both countries, were determined simultaneously. One could, however, argue that since investments in R&D are decisions of a more long-run nature than abatement decisions, it would be more reasonable to model the interactions between the countries as a two-stage game where R&D expenditures are determined in the first stage and abatement levels in the second stage. We shall briefly consider this alternative in the present Section, under the assumption that technology spillovers are linear. Because we obtained unambiguous results in Section 3 under the assumption of linear damage functions (whereas the case of convex damage functions led to some ambiguous results), we focus on the case of linear damage functions in both countries in order to easily identify the effects of changing the order of moves.

In the second stage, x and X (and hence y and Y) are given. In this stage the home county thus minimizes $\Omega(a,y)+x+e(2-a-A)+f\cdot(2-a-A)$ w.r.t. a, taking A, x and y as given. The foreign country solves a similar minimization problem. This gives us equations (3) and (5). With linear damage functions ($e''=E''=0$) equation (3) defines abatement in the home country as a function of the technology level in this country and the shift parameter f, whereas equation (5) defines abatement in the foreign country as a function of the technology level in the foreign country:

(17) $$a=a(y,f)$$

(18) $$A=A(Y)$$

where $a_y=\partial a/\partial y>0$, $a_f=\partial a/\partial f>0$ and $A'>0$.

In the first stage the home country takes into account that its R&D expenditures will influence abatement in the foreign country via (18). Similarly, the foreign country takes into account that its R&D expenditures will influence abatement in the home county via (17).[7] Using (2), it is therefore clear that minimization of $\Omega(a,y)+x+e(2-a-A)+f\cdot(2-a-A)$ for the home country gives (instead of (4))

[7] By the envelope theorem, none of the countries need explicitly take account of the fact that its own abatement will be influenced by its own R&D expenditures.

(19) $$-\Omega_y(a,y)+\gamma A'(Y)\left[e'(2-a-A)+f\right]=1$$

The corresponding minimization problem for the foreign country, using (1), gives (instead of (6)):

(20) $$-\Omega_Y(A,Y)+\gamma a_y(y,f)E'(2-a-A)=1$$

The second terms of the l.h.s. of (19) and (20) are positive, implying that R&D levels are higher in the present case than in the case where all variables are determined simultaneously. The interpretation is that when R&D expenditures are determined prior to abatement, the home country takes into account that increased R&D expenditures will, through spillovers (γ), increase the technology level of the other country, and thus lead to more abatement in the other country (A'), which is valuated by the home country according to its marginal environmental costs ($e'+f$). The foreign country has a similar incentive to increase its R&D expenditures. With higher R&D expenditures in both countries (relative to the case in which all variables are determined simultaneously), technology levels will be higher in both countries, and thus abatement will also be higher in both countries. We thus have the following proposition:

Proposition 6: *With linear technology spillovers and linear damage functions, technology and abatement levels are higher in both countries when R&D expenditures are determined prior to abatement than when R&D expenditures are determined simultaneously with abatement. If a country has positive R&D expenditures, these expenditures are higher when they are determined prior to abatement than when R&D expenditures are determined simultaneously with abatement.*

In order to see how our endogenous variables are altered by a change in the shift parameter f we differentiate (19) and (20) with respect to f (where $f=0$ initially), after inserting (1), (2), (17) and (18). With identical countries the changes in the endogenous variables will in general depend on all the second order derivatives of $a(y,f)$ and $A(Y)$. Because $a(y,f)$ and $A(Y)$, which are derived from (3) and (5), depend on the first order derivative of the abatement cost function and the damage function, the changes in the endogenous variables will depend on third order derivatives of the abatement cost function (the damage function is assumed to be linear in the present section, that is, all third order derivates of the damage function are zero). Without making further assumptions on the abatement cost functions, these third order derivatives may be either positive or negative (or zero). Hence, in the case of identical countries the effects of increased environmental concern are ambiguous.

Consider next the case in which one of the countries is dirty. Let the first-order-condition of the home and foreign country w.r.t. R&D expenditures (i.e. (19) and (20)) be written as $l(x,X,f)=0$ and $L(X,x,f)=0$, respectively.[8] When the home country is dirty, $x=0$ and $l(0,X,f)>0$. From $L(X,0,f)=0$ it follows that the effect on R&D (in the foreign country) due to increased concern for the environment in the home country is

$$\frac{dX}{df}=\frac{-L_f}{L_X}=\frac{\gamma E' a_{yf}}{L_X} \tag{21}$$

where $L_X>0$ from the second-order condition. Using (13) and remembering that $e''=0$ in the present section, it can be shown that $a_{yf}>0$ if a rightward shift, measured horizontally, in the marginal abatement cost curve is larger the higher up on this curve we are initially. Hence, if $a_{yf}>0$ then R&D expenditures in the foreign country increase if the damage function in the dirty country increases. More R&D expenditures will increase the technology level in both countries, which tends to increase abatement in each country, see (17) and (18). The positive effect on abatement in the home country is enhanced due to the direct effect of the shift ($a_f>0$). We thus have the following proposition:

Proposition 7: *Consider the two-stage game where R&D expenditures are determined prior to abatement, and there are linear technology spillovers and linear damage functions. If the home country is dirty (i.e. has no R&D expenditures) and* $a_{yf}>0$*, increased concern for the environment in the dirty country increases R&D expenditures in the clean country, and abatement in both countries increases.*

Comparing Proposition 2 (in the case of linear damage functions) with Proposition 7, we see that in both cases increased concern for the environment in the dirty country leads to more abatement in this country. While abatement and R&D expenditures in the foreign country do not change when all variables are determined simultaneously (Proposition 2), these two variables change when R&D expenditures are determined prior to abatement (provided $a_{yf}\neq 0$).

When the home country is clean, the effect on R&D (in the home country) due to increased concern for the environment follows from differentiating $l(x,0,f)=0$ w.r.t. x:

$$\frac{dx}{df}=\frac{-l_f}{l_x}=\frac{\gamma A' - \Omega_{ya} a_f}{l_x}>0 \tag{22}$$

---

[8] More precisely, $l(x,X,f)=1+\Omega_y(a,y)-\gamma A'(Y)\left[e'(2-a-A)+f\right]$ and $L(X,x,f)=1+\Omega_Y(A,Y)-\gamma a_y(y,f)E'(2-a-A)$.

where $l_x > 0$ from the second-order condition. Increased R&D expenditures will increase technology levels, and hence tend to increase abatement in both countries. Moreover, the direct effect of increased environmental concern is more abatement in the home country. We thus have the following proposition:

Proposition 8: *In a two-stage game where R&D expenditures are determined prior to abatement, linear technology spillovers, linear damage functions and no R&D in the dirty country, the results in Proposition 3 for the case of linear damage functions are valid.*

6. Intellectual property rights

So far, we have assumed that once a technology is developed by one of the countries, it is freely available also to the other country, although less beneficial than to the country which developed the technology (since $\gamma < 1$ in the linear case and a similar assumption holds in the case of technology complementarity). In this section we analyze the consequences of intellectual property rights that make it possible for the country that develops a technology to prevent the other country from using it.

Formally, we model this is follows. As in Sections 3-4, R&D expenditures and abatement levels are chosen simultaneously in the two countries. After these variables are chosen, each county can decide whether or not to let the other country use the technology it has developed. We restrict ourselves to the case of linear spillovers. Consider the home county (a similar argument applies to the foreign country). If it decides that the foreign country may use the technology it has developed, the technology level in the foreign country will be $Y = X + \gamma x$ as before. If, however, the home country decides that the foreign country may *not* use the technology it has developed, the technology level in the foreign country will instead simply be $Y = X$.

Since the technology available to the foreign country is a by-product of the technology the home country has developed for its own use, its cost is zero once the technology is developed. For the home country, welfare is therefore unaffected by whether or not the foreign country is allowed to use the technology the home country has developed.[9] For the foreign country, receiving the technology reduces its costs from $\Omega(A, X)$ to $\Omega(A, X + \gamma x)$. The total gain from letting the foreign county use the technology developed by the home country is thus $\Omega(A, X) - \Omega(A, X + \gamma x)$. Clearly, an efficient bargain between the two countries will imply that the foreign country is

[9] In a richer model, this would not necessarily be true as a country's competitiveness could depend on what technologies are available in other countries. Moreover, in a two-stage model of the type considered in the previous section, abatement in the foreign country would depend on whether or not the foreign country had access to the technology developed by the home country. In this case the home country would benefit from transferring the technology it had developed to the foreign country, even if it was not compensated for this technology transfer.

allowed to use the technology, and that the home country receives something in return for allowing this use. This compensation may take different forms, but in the simple model used here it is natural to specify this as money transfer. We shall assume that the money transfer to the country developing the technology is a fixed share β of the total gains, where $0<\beta<1$. Denoting the money transfer to the home country by r we thus have

(23) $$r=r(A,X,x)=\beta[\Omega(A,X)-\Omega(A,X+\gamma x)]$$

For the foreign country, the compensation it receives for the technology it develops and lets the home country use is denoted by R:

(24) $$R=R(a,x,X)=\beta[\Omega(a,x)-\Omega(a,x+\gamma X)]$$

We shall assume that the negotiated transfers r and R are correctly foreseen at the time when abatement levels and R&D expenditures are determined. With these assumptions, the change in the optimization problem of the home country is that it now minimizes $\Omega(a,y)+x+e(2-a-A)+f\cdot(2-a-A)-r(A,X,x)+R(a,x,X)$ while it previously was assumed to minimize $\Omega(a,y)+x+e(2-a-A)+f\cdot(2-a-A)$. A similar change applies to the foreign country. For the case of an interior optimum, using (23) and (24) it is straightforward to verify that the equilibrium conditions (3)-(6) are changed to

(25) $$\Omega_a(a,x+\gamma X)+\beta[\Omega_a(a,x)-\Omega_a(a,x+\gamma X)]=e'+f$$

(26) $$-\Omega_y(a,x+\gamma X)+\beta\left[\Omega_y(a,x+\gamma X)-\Omega_y(a,x)-\gamma\Omega_Y(A,X+\gamma x)\right]=1$$

(27) $$\Omega_A(A,X+\gamma x)+\beta[\Omega_A(A,X)-\Omega_A(A,X+\gamma x)]=E'$$

(28) $$-\Omega_Y(A,X+\gamma x)+\beta\left[\Omega_Y(A,X+\gamma x)-\Omega_Y(A,X)-\gamma\Omega_y(a,x+\gamma X)\right]=1$$

Notice that for $\beta=0$, then (25)-(28) are equivalent to (3)-(6).

To see the difference between the case of $\beta=0$ and $\beta>0$, consider the home country (where an equivalent result holds for the foreign country). The term in square brackets in (25) is positive for $X>0$. The interpretation of this term is that the higher is abatement in the home country, the more it will have to pay for the technology developed in the foreign country (since the home country's cost reduction from this technology is larger the higher is its abatement). The isolated effect of this positive term is to make abatement lower in the home country.

The square term in (26) is also positive. This is because if $X > 0$, increased R&D in the home country will reduce the payment to the foreign country (as the cost reduction in the home country due to technology transfer from abroad is lower the more technology it has developed itself). In addition, higher R&D in the home country will increase the payment it will receive from the foreign country. The isolated effect of this positive term is to make R&D expenditures higher in the home country.

The direct effects of the new terms in (25)-(28) compared with (3)-(6) are thus to reduce abatement and to increase R&D. Further, since $\Omega_{ay} < 0$, lower abatement tends to reduce the incentives for R&D, and higher R&D tends to increase abatement. Thus it is not obvious how the equilibrium values of abatement and R&D levels change when β is increased from $\beta = 0$. It can be verified by differentiating (25)-(28) w.r.t. β that changes in abatement and R&D expenditures are ambiguous unless we make additional assumptions on the abatement cost function. There is one exception. If one country is dirty, say, the foreign country, the term in square brackets in (25) is zero. There is thus no factor tending to make abatement lower for the clean country in this case. The term that gives higher R&D expenditures for the clean country remains positive (ceteris paribus). In this case the result is therefore that abatement and R&D expenditures are higher in the clean country when $\beta > 0$ (intellectual property rights) than when $\beta = 0$.

6.1 Identical countries

We next consider the effect of increased concern for the environment in the home country. Consider first the case of equal countries. Recall that for $\beta = 0$ the effect of an increase in f is that a and x increase, and that X goes down so much that Y remains unchanged (see Proposition 1 for linear damage functions). Moreover, A remains unchanged as a response to the increase in f. When $\beta > 0$, there are additional effects via the terms in square brackets in (25)-(28). In particular, when abatement in the home country increases, the incentive for R&D in the foreign country is increased via the term in square brackets in (28), thus increasing X (ceteris paribus). On the other hand, an increase in x will - via the term in square brackets in (27) - tend to reduce A. Since lower abatement tends to reduce the incentives for R&D, and higher R&D tends to increase abatement, it is not obvious what the effect of the additional terms are on the equilibrium values of abatement and R&D levels in the foreign country.

The square brackets in (25)-(28) depend on differences between first-order derivatives. Therefore, the differentiated system will contain several terms for which the signs depend on differences between second-order derivatives. These terms can only be signed if we impose assumptions on the third order derivatives. Hence, without further assumptions on the abatement cost function the effects of increased environmental concern in one country are ambiguous. However, with quadratic abatement cost function, where all third order derivatives are zero, it can be shown that abatement increases in both countries.[10]

6.2 Different countries

Consider first the case where the home country is dirty, i.e. $x = 0$. The effect of an increase in f is as before that abatement in this county increases. When $\beta = 0$, this had no consequence for abatement and R&D in the clean country, cf. Proposition 2 for linear damage functions. When $\beta > 0$, there is an effect on the clean (i.e. foreign) country via the term in square brackets in (28): The increase in a increases this term, giving higher R&D expenditures in the clean country. This increase in R&D expenditures in the clean country raises abatement in this country, see the Appendix for a proof.

Turning to the case in which the foreign country is dirty (i.e. $X = 0$), (28) is now not part of the model. Because the term in square brackets in (27) is independent of abatement in the home country, the effects on the foreign country of changes in the home country are similar to those in the case of no intellectual property rights. Hence, Proposition 3 is still valid (for linear damage functions). The results in Section 6.2 are summarized in the following Proposition:

Proposition 9: *With linear technology spillovers, intellectual property rights and one dirty country, increased concern for the environment in one country will increase abatement in both countries, whereas R&D expenditures in the clean country are increased.*

7. Concluding remarks

A number of studies suggest that the effect of a unilateral action taken by a group of countries in order to reduce its emissions is likely to be undermined by increases in emissions from other countries (carbon leakage). While analyses of carbon leakage typically have regarded the technology in each country as given, abatement technologies are in reality endogenous, and thus technology development may be affected by environmental policies. An important lesson from our paper is that the *isolated* effect of technology diffusion from one country to another is decreased emissions in the other country due to lower costs of abatement in the other country. However, the *net* effect on abatement in the other country may still be negative (carbon leakage). First, improved technology in the other country due to spillovers from the first country might have an impact on R&D expenditures in the other country. In addition, as mentioned in the Introduction, increased abatement in the first country (following from increased environmental concern in this country) might directly reduce abatement in the

[10] These results are available upon request.

other country. We have thus demonstrated that with endogenous technologies and technology diffusion between countries, it is no longer obvious that reduced emissions in one country will increase emissions in other countries. In fact, in our two-country model there are cases where the opposite is true, so that increased environmental concern in one country might lead to lower emissions in both countries.

More specifically, we have showed that if countries differ, and the unilateral action is undertaken by countries with the highest valuation of environmental damage, abatement in the other countries will increase if there are no R&D expenditures in the countries with the lowest valuation of environmental damage and the environmental damage function of these countries is linear. Yet, under other assumptions a unilateral action will be undermined by decreased abatement in other countries. See Table 1 for an overview of our main results.

Increased environmental concern in the home country has welfare effects in the foreign country. For any given policy in the foreign country, this country is better off the higher is abatement and the higher are R&D expenditures in the home country. Since the foreign country maximizes its own welfare given the choices made by the home country, welfare in the foreign country moves in the same direction as abatement and R&D expenditures in the home country, provided these two home country variables move in the same direction. For the one-stage game without intellectual property rights (Sections 3-4) we see that when the home country becomes more concerned about the environment, abatement is increased in this country. R&D expenditures in the home country also increase, or are unchanged for the case in which the home country is dirty (no R&D expenditures). Hence, when the concern for the environment increases in the home country, welfare increases in the foreign country under these cases.

Table 1: Abatement effects in the foreign country following from increased environmental concern in the home country.

	Identical countries	*Home is dirty*	*Home is clean*
Linear technology spillovers			
Linear damage function	0	0	+
Strictly convex damage function	-	-	?
Technology complementarity			
Linear damage function	+	0	+
Strictly convex damage function	?	-	?
Two-stage game			
Linear damage function	?	+[a]	+
Intellectual property rights			
Linear damage function	?[b]	+	+

[a] Provided $a_{sf} > 0$. [b] Positive if all third-order derivatives are zero.

Since we are considering a change in preferences, it is not obvious how to measure welfare effects in the home country. However, a detailed discussion of welfare effects in the home country is given in an earlier version of the present paper, see Golombek and Hoel (2004). There we argue that when the concern for the environment increases in the home country, the welfare effects in the home country are ambiguous. The effects depend on whether the two countries initially are identical, whether environmental damage cost functions are linear, and whether technology spillovers are linear.

Throughout the present paper we have assumed that there is no international environmental agreement. However, our paper might shed some light on the effect on global emissions if only some countries participate in an international climate agreement. In the Introduction we noted that the dirty country might represent a poor country (or a group of poor countries). As a rule of thumb, developing countries have not signed the Kyoto agreement. Hence, if the Kyoto agreement leads to more abatement and R&D expenditures among the Protocol members ("clean country"), and there are technology spillovers to developing countries, abatement in developing countries (that is, poor countries) may increase, see Proposition 3 and 5. Hence, the present concern among some OECD countries that non-signatory developing countries will undermine the Kyoto agreement might be overstated.

Appendix

We first consider the general case of technology complementarity (Proposition 4), and then examine the special case of linear technology spillovers (Proposition 1).

Proof of Proposition 4:

Differentiating (13) - (16) w.r.t. f, and substituting for da and dA, yields the system

$$\begin{aligned} Gdx + HdX &= df \\ Hdx + GdX &= 0 \end{aligned} \tag{29}$$

where $G = e''(\Phi_1 + \Phi_2)\frac{\Omega_{yy}}{-\Omega_{ya}} + \Phi_1(\frac{\Omega_{aa}\Omega_{yy}}{-\Omega_{ya}} + \Omega_{ay}) - \frac{\Omega_{aa}\Omega_y}{\Omega_{ya}}\frac{\Phi_{11}}{\Phi_1} > 0$ and

$H = e''(\Phi_1 + \Phi_2)\frac{\Omega_{yy}}{-\Omega_{ya}} + \Phi_2(\frac{\Omega_{aa}\Omega_{yy}}{-\Omega_{ya}} + \Omega_{ay}) - \frac{\Omega_{aa}\Omega_y}{\Omega_{ya}}\frac{\Phi_{12}}{\Phi_1}$. Note that the two first terms in G and H are positive. The third term in G is positive, whereas the third term in H is negative. Hence, the sign of H is ambiguous, but $G > H$ and due to Φ being homogenous of degree 1 we also have $|G| > |H|$.[11]

Solving (29) yields

$$\frac{dx}{df} = \frac{G}{K} > 0 \tag{30}$$

$$\frac{dX}{df} = \frac{-H}{K} \tag{31}$$

where $K = G^2 - H^2 > 0$. Moreover, using (4) and (6) we find

$$\frac{da}{df} = \frac{[2\Omega_y(\frac{\Omega_{AA}\Omega_{YY}}{-\Omega_{YA}} + \Omega_{AY}) + \Omega_y\Omega_{yy}\frac{\Omega_{AA}}{-\Omega_{YA}}]\frac{\Phi_{11}}{-\Omega_{YA}}(1 + \frac{\Phi_2}{\Phi_1})}{K} +$$

$$\frac{\frac{\Omega_{yy}}{-\Omega_{YA}}(\frac{\Omega_{AA}\Omega_{YY}}{-\Omega_{YA}} + \Omega_{AY})(\Phi_1^2 - \Phi_2^2)}{K} + \tag{32}$$

$$\frac{e''(\Phi_1 + \Phi_2)\frac{\Omega_{yy}}{\Omega_{ya}^2}[\Omega_{yy}(\Phi_1 - \Phi_2) + 2\Omega_y\frac{\Phi_{11}}{\Phi_1}]}{K} > 0$$

$$\frac{dA}{df} = \frac{\Omega_Y\Phi_{11}(1 + \frac{\Phi_2}{\Phi_1}) + e''(\Phi_1 + \Phi_2)\frac{\Omega_{yy}}{\Omega_{ya}^2}[\Omega_{YY}(-\Phi_1 + \Phi_2) + 2\Omega_Y\frac{\Phi_{12}}{\Phi_1}]}{K} \tag{33}$$

---

[11] If $H > 0$ then $|G| > |H|$ since $G > H$. Assume $H < 0$. If $|H| \geq |G|$ then $-e''(\Phi_1 + \Phi_2)\frac{\Omega_{yy}}{-\Omega_{ya}} - \Phi_2(\frac{\Omega_{aa}\Omega_{yy}}{-\Omega_{ya}} + \Omega_{ay}) + \frac{\Omega_{aa}\Omega_y}{\Omega_{ya}}\frac{\Phi_{12}}{\Phi_1} \geq e''(\Phi_1 + \Phi_2)\frac{\Omega_{yy}}{-\Omega_{ya}} + \Phi_1(\frac{\Omega_{aa}\Omega_{yy}}{-\Omega_{ya}} + \Omega_{ay}) - \frac{\Omega_{aa}\Omega_y}{\Omega_{ya}}\frac{\Phi_{11}}{\Phi_1}$, which implies that $\frac{\Omega_{aa}\Omega_y}{\Omega_{ya}}\frac{(\Phi_{11} + \Phi_{12})}{\Phi_1} \geq 2e''(\Phi_1 + \Phi_2)\frac{\Omega_{yy}}{-\Omega_{ya}} + (\Phi_1 + \Phi_2)(\frac{\Omega_{aa}\Omega_{yy}}{-\Omega_{ya}} + \Omega_{ay})$. In the latter inequality the right hand side is positive, whereas the left hand side is zero since Φ is homogenous of degree 1 and the two countries are identical. Hence, the latter inequality is not fulfilled, that is, $|G| > |H|$.

In *(33)* the first term is positive, whereas the second term is negative, that is, the effect on abatement in the foreign country is ambiguous. Note that with linear damage functions ($e''=0$) the nominator in *(33)* is positive, and hence abatement in the foreign country increases.

Proof of Proposition 1:

In the special case of liner spillovers we have $\Phi_1=1$, $\Phi_2=\gamma$ and $\Phi_{ij}=0$ for all i and j. The derivatives are then simplified to

(34) $$\frac{dx}{df}=\frac{\left(\frac{\Omega_{aa}\Omega_{yy}}{-\Omega_{ya}}+\Omega_{ay}\right)-e''\frac{\Omega_{yy}}{\Omega_{ya}}(1+\gamma)}{L}>0$$

(35) $$\frac{dX}{df}=\frac{-\left(\frac{\Omega_{aa}\Omega_{yy}}{-\Omega_{ya}}+\Omega_{ay}\right)\gamma+e''\frac{\Omega_{yy}}{\Omega_{ya}}(1+\gamma)}{L}<0$$

(36) $$\frac{da}{df}=\frac{\frac{\Omega_{yy}}{-\Omega_{ya}}\left[\left(\frac{\Omega_{aa}\Omega_{yy}}{-\Omega_{ya}}+\Omega_{ay}\right)(1-\gamma^2)+e''\frac{\Omega_{yy}}{\Omega_{ya}}(\gamma^2-1)\right]}{L}>0$$

(37) $$\frac{dA}{df}=\frac{-e''\left(\frac{\Omega_{yy}}{\Omega_{ya}}\right)^2(1-\gamma^2)}{L}\le 0$$

where $L=\left[\left(\frac{\Omega_{aa}\Omega_{yy}}{-\Omega_{ya}}+\Omega_{ay}\right)-e''\frac{\Omega_{yy}}{\Omega_{ya}}(1+\gamma)\right]^2-\left[\left(\frac{\Omega_{aa}\Omega_{yy}}{-\Omega_{ya}}+\Omega_{ay}\right)\gamma-e''\frac{\Omega_{yy}}{\Omega_{ya}}(1+\gamma)\right]^2>0$.

Proof of Proposition 9:

We want to examine how increased concern for the environment in the home country affects abatement and R&D expenditures in both countries under intellectual property rights (and linear damage functions) when one country is dirty.

When the home country is dirty, differentiation of (25), (27) and (28) gives

$$\frac{da}{df} = \frac{\Omega_{aa}\Omega_{yy} - \Omega_{ay}^2 + \beta\gamma^2\Omega_{aa}\Omega_{yy}}{\Omega_{AA}P} \tag{38}$$

$$\frac{dX}{df} = \frac{-\beta\gamma\Omega_{ya}}{P} \tag{39}$$

$$\frac{dA}{df} = \frac{-\Omega_{AY}}{\Omega_{AA}}\frac{dX}{df} \tag{40}$$

where

$$\begin{aligned} P = {} & [\Omega_{aa}(a,\gamma X) + \beta(\Omega_{aa}(a,0) - \Omega_{aa}(a,\gamma X))][\frac{-\Omega_{YA}^2}{\Omega_{AA}} + \Omega_{YY}] + \gamma^2\beta^2\Omega_{ay}^2 + \\ & \gamma^2\beta\{[\Omega_{aa}(a,\gamma X) + \beta(\Omega_{aa}(a,0) - \Omega_{aa}(a,\gamma X))]\Omega_{yy} - \Omega_{ay}^2\} \end{aligned} \tag{41}$$

The first term in (41) is positive due to a second order condition and the second term is also positive. In the last (third) term the square brackets are positive due to a second order condition, but this is not sufficient to ensure that the whole third term is positive. We find it reasonable to impose (regularity) assumptions that ensure that $\frac{da}{df} > 0$. Because the numerator in (38) is positive, we will assume that $P > 0$. This is the case if, for example, all third-order derivatives are zero, or more generally, if the square brackets in the last term in (41) are non-negative. Under $P > 0$ abatement and R&D expenditures in the foreign country increase.

Finally, assume that the home country is clean. Differentiation of (25)-(27) gives

$$\begin{aligned} \frac{da}{df} = {} & \frac{[\Omega_{AA} + \beta(\Omega_{AA}(A,0) - \Omega_{AA}(A,\gamma x))]\Omega_{yy} + \gamma^2\beta^2\Omega_{AY}^2 +}{[\Omega_{AA} + \beta(\Omega_{AA}(A,0) - \Omega_{AA}(A,\gamma x))]Q} \\ & \frac{\gamma^2\beta\{\Omega_{YY}[\Omega_{AA} + \beta(\Omega_{AA}(A,0) - \Omega_{AA}(A,\gamma x))] - \Omega_{AY}^2\}}{[\Omega_{AA} + \beta(\Omega_{AA}(A,0) - \Omega_{AA}(A,\gamma x))]Q} \end{aligned} \tag{42}$$

(43) $$\frac{dx}{df} = \frac{-\Omega_{ya}}{Q}$$

(44) $$\frac{dA}{df} = \frac{-\gamma(1-\beta)\Omega_{AY}}{\Omega_{AA} + \beta(\Omega_{AA}(A,0) - \Omega_{AA}(A,\gamma x))} \frac{dx}{df}$$

where

$$Q = [\Omega_{AA} + \beta(\Omega_{AA}(A,0) - \Omega_{AA}(A,\gamma x))](\Omega_{aa}\Omega_{yy} - \Omega_{ay}^2) + \gamma^2\beta^2\Omega_{aa}\Omega_{AY}^2$$
$$+\gamma^2\beta\Omega_{aa}\{\Omega_{YY}[\Omega_{AA} + \beta(\Omega_{AA}(A,0) - \Omega_{AA}(A,\gamma x))] - \Omega_{AY}^2\}$$

None of these derivatives can be signed without imposing more assumptions. If regularity assumptions are imposed in order to ensure that $\frac{da}{df} > 0$, then $\frac{dx}{df} > 0$ and $\frac{dA}{df} > 0$.

References

Barrett, S. (2003), *Environment & Statecraft. The Strategy of Environmental Treaty-Making*. Oxford University Press.

Ben Youssef, S. (2003), "Transboundary pollution, R&D spillovers and international trade," Nota di lavaro 39.2003, Fondazione Eni Enrico Mattei.

Bohm, P. (1993), "Incomplete international cooperation to reduce CO_2 emissions: alternative policies," *Journal of Environmental Economics and Management* ***24***, 258-71.

Buchner, B. and C. Carraro (2004), "Economic and environmental effectiveness of a technology-based climate protocol," Nota di lavaro 61.2004, Fondazione Eni Enrico Mattei.

Buonanno, B., Carraro, C. and M. Galeotti (2003), "Endogenous induced technical change and the costs of Kyoto," *Resource and Energy Economics* ***25***, 11-34.

Cohen, W. M. and D. A. Levinthal (1989), "Innovation and learning: the two faces of R&D," *The Economic Journal* ***99***, 569-596.

Felder, S. and T. F. Rutherford (1993), "Unilateral CO_2 reductions and carbon leakage: the consequences of international trade in oil and basic materials," *Journal of Environmental Economics and Management* **25**, 162-176.

Golombek, R. and J. Bråten (1994), "Incomplete international agreements: Optimal carbon taxes, market failures and welfare effects," *The Energy Journal* **15**, 141-165.

Golombek, R. and M. Hoel (2004), "Unilateral emission reductions when there are cross-country technology spillovers," Working paper from Department of Economics, University of Oslo.

Golombek, R., Hagem, C. and M. Hoel (1995), "Efficient Incomplete International Climate Agreements," *Resource and Energy Economics* **17**, 25-46.

Graevenitz, G. von. (2002), "Spillovers reconsidered: how optimal R&D subsidies depend on the spillover process," Mimeo, Department of Economics, University College London.

Gurtzgen, N. and M. Rauscher (2000), "Environmental policy, intra-industry trade and transfrontier pollution," *Environmental and Resource Economics* **17**, 59-71.

Heal, G. (1994), Formation of international environmental agreements, in C. Carraro (ed.): *Trade, Innovation, Environment.* Kluwer Academic Publishers, ch. 3.2, pp. 301-322.

Hoel, M. (1991), "Global Environmental Problems: The effects of unilateral actions taken by one country," *Journal of Environmental Economics and Management* **20**, 55-70.

Hoel, M. (1994), "Efficient Climate Policy in the Presence of Free Riders," *Journal of Environmental Economics and Management 27 (3),* 259-274.

Hoel, M. (1996), "Should a carbon tax be differentiated across sectors?" *Journal of Public Economics* **59**, 17-32.

Hoel, M. (2001), "International trade and the environment: how to handle carbon leakage," in H. Folmer, H.L. Gabel, S. Gerking and A. Rose (eds.): *Frontiers of Environmental Economics*, Edward Elgar, Cheltenham (UK), ch. 7, pp. 176-191.

Jaffe, B., G. Newell, R. Stavins (2002), "Environmental policy and technological change," *Environmental and Resource Economics* **22** (special issue), 41-69.

Löschel, A. (2002), "Technological change in economic models of environmental policy: A survey," *Ecological Economics* ***43**, 105-126.*

Mathiesen, L. and O. Mæstad (2004), "Climate policy and the steel industry: achieving global emission reductions by an incomplete climate agreement," *The Energy Journal*, forthcoming.

Perroni, C. and T. F. Rutherford (1993), "International trade in carbon emission rights and basic materials: general equilibrium calculations for 2020," *Scandinavian Journal of Economics* **95**, 257-278.

Pezzey, J. (1992), "Analysis of unilateral CO_2 control in the European Community," *The Energy Journal* **13**, 159-172.

Ploeg, F. van der and A. de Zeeuw (1994), "Investment in Clean Technology and Transboundary Pollution control," in Carraro, C., *Trade, Innovation, Environment*. 229-240. Netherlands, Dordrecht: Kluwer Academic Publishers.

Rauscher, M. (1997), *International Trade, Factor Movements and the Environment*, Clarendon Press, Oxford.

Rosendahl, K. E. (2004), "Cost-effective environmental policy: implications of induced technological change," *Journal of Environmental Economics and Management*, forthcoming.

Spence, M. (1984), "Cost reduction, competition, and industry performance," *Econometrica* **52**, 101-122.

Xepapadeas, A. (1995), "Induced technical change and international agreements under greenhouse warming," *Resource and Energy Economics* **17**, 1-23.

[9]

Trade Liberalization and Pollution Havens*

Josh Ederington, Arik Levinson, and Jenny Minier

Abstract

U.S. Presidential Executive Order 13141 commits the United States to a careful assessment and consideration of the environmental impacts of trade agreements. The most direct mechanism through which trade liberalization would affect environmental quality in the U.S. is through the composition of industries. Freer trade means greater specialization, increasing the concentration of polluting industries in some countries and decreasing it in others. We begin by documenting the substantial shift in U.S. manufacturing toward cleaner industries from 1972 to 1994. We then use annual industry-level data on imports to the U.S. to examine whether this compositional shift can be traced to the significant trade liberalization that occurred over the same time period, and we conclude that no such connection exists. A shift toward cleaner industries has also occurred among U.S. imports, and we find no evidence that pollution-intensive industries have been disproportionately affected by the tariff changes.

KEYWORDS: Pollution Haven, PACE, Environment

*Josh Ederington and Jenny Minier are at University of Kentucky Economics Department, Lexington KY 40506. Arik Levinson is at Georgetown University Economics Department, Washington DC 20057. The authors would like to thank Anna Maria Mayda and Rodney Ludema for helpful discussions.

In recent decades, while United States manufacturing output increased, total pollution from manufacturers declined. Much of the decline has been due to a large shift in the composition of the U.S. manufacturing sector away from polluting industries and toward cleaner industries. During the same period, U.S. tariffs on imported manufactured goods declined. These concurrent trends have led many to conclude that trade liberalization – "globalization" – has *caused* the shift in U.S. manufacturing to clean industries at the expense of environmental quality in developing countries. In this paper we seek empirical support for that claim, using data on tariffs, U.S. imports, pollution abatement costs, and industries' relative pollution intensities. While much research has been devoted to studying the effect of environmental regulations on trade, we know of no other direct empirical tests of whether trade liberalization has exacerbated the tendency for pollution-intensive industries to relocate to developing countries.

This concern, that trade liberalization leads to a transfer of polluting industries to developing countries, has been part of the impetus for widespread protests against the World Trade Organization. It has been the motivation for environmental addenda to trade agreements such as the North American Free Trade Agreement (NAFTA), and it is the subject of President Clinton's Executive Order 13141, "Environmental Review of Trade Agreements." EO13141 requires that the United States "factor environmental considerations into . . . its trade negotiating objectives," and that the U.S. Trade Representative and the Chair of the Council on Environmental Quality oversee analyses of the "environmental impacts of trade agreements."

Despite the intuitive appeal of the idea, trade liberalization will not necessarily lead to a shift toward cleaner goods production in U.S. manufacturing. Indeed, using country-level data, recent work by Antweiler, Copeland and Taylor (2001) and Copeland and Taylor (2003) has shown that increases in the volume of trade can result in increased pollution emissions in the U.S. From this they infer that trade liberalization could actually shift the composition of U.S. manufacturing toward *more* polluting industries. Our work differs from theirs in that we examine the compositional effect of tariff reductions directly using industry-level data, while Antweiler *et al.* and Copeland and Taylor infer the compositional effect indirectly from country-level correlations between pollution emissions and trade. As we show, our results support their inference.

Trade liberalization can affect the environment through several mechanisms, such as interjurisdictional competition to lower standards, transfer of pollution abatement technology, cross-border spillovers, or changes to the overall scale of economies. But it seems to us that the most direct effect of trade liberalization on the environment would be through the composition of industries. Trade liberalization leads to specialization, and countries that specialize in less pollution-intensive goods will have cleaner environments.

In this paper we assess the degree to which the reduction of U.S. tariffs on imported goods accounts for the changing composition of U.S. manufacturing toward cleaner industries. We do this in three ways. First, in section 1 of the paper, we examine the pollution content of U.S. manufactured goods, imports, and exports, in the spirit of

Kahn (2003), to see whether imports have replaced the pollution-intensive goods no longer produced in the U.S. Second, in section 2 we use a regression approach similar to Grossman and Krueger (1993) to see whether the tariff reductions of the past several decades have exacerbated the "pollution haven" effect. If this is the case, then broad-based tariff reductions could have caused the compositional shift observed in the data. Finally, in section 3 we investigate whether the compositional shift in U.S. industries could be due to asymmetric trade liberalization (i.e., larger U.S. tariff reductions in the more pollution-intensive industries). Specifically, we use our regression coefficients to predict the changes in the composition of U.S. industries that are due to past tariff reductions (as well as the changes that would be the result of setting current tariff levels to zero).

We find that that the tariff reductions over the past several decades cannot account for the change in the composition of U.S. manufacturing toward cleaner industries. First, imports from other countries do not appear to have replaced the domestic production of pollution-intensive goods in the U.S. Second, we find no evidence that pollution-intensive industries are disproportionately sensitive to tariff reductions. (In fact, just the opposite appears likely.) Finally, while it is true that tariff reductions have been greater for polluting industries, that difference explains only a small part of the overall shift in U.S. manufacturing.

1. The Pollution Content of U.S. Manufacturing

The various effects of trade on environmental quality can be divided into three components: how trade affects the overall *scale* of the economy; how trade affects the *techniques* of production, and how trade affects the *composition* of industries.[1] The various parties concerned about trade liberalization's effect on the environment appear to concentrate on the composition effect: specifically, the concern that trade liberalization will result in the creation of pollution havens (countries that specialize in polluting industries). Hence we investigate mainly the composition effect and not the technique effect. We begin by examining the degree to which the composition of U.S. manufacturing has shifted toward clean industries, and how much of that shift can be explained by changes in the composition of imports and exports.

In order to isolate this composition effect, we need a metric with which to label various industries as being relatively "clean" or "dirty." To that end, we rely on the World Bank's "Industrial Pollution Projection System" (IPPS).[2] The IPPS reports the amount of each of 14 pollutants, in pounds per million dollars of value added, that were generated from each of 459 four-digit SIC codes in 1987. These data are summarized in Table 1. They represent a snapshot of the production techniques as of 1987.

Figure 1a demonstrates the degree to which the U.S. industrial composition has shifted toward less-polluting industries. The bold line plots an index of total real

[1] See, for example, Copeland and Taylor (2003).

[2] See Hettige *et al.* (1994).

manufacturing output. The recessions of the mid-1970s, early 1980s, and early 1990s are apparent, and overall manufacturing grew by 57 percent between 1972 and 1996.

Table 1: Pounds of Pollution per Dollar Value Added 1987

		IPPS Coefficients [lbs/$]		
Pollutant	**Industries (1)**	**Mean (2)**	**Std. dev. (3)**	**Max (4)**
Air pollution				
Particulates	448	1.64	8.79	147
CO	448	2.67	14.5	202
SO2	448	2.47	10.78	140
NOx	448	1.63	6.23	67
VOC	448	1.38	4.77	85
PM10	448	0.89	7.64	108
Water pollution				
BOD	321	0.0011	0.0066	0.068
TSS	321	0.0069	0.053	0.70
Toxics				
Air	434	1.31	5.31	77
Land	434	1.64	7.28	83
Water	434	0.204	2.208	43
Metals				
Air	317	0.020	0.091	1.24
Land	317	0.592	4.50	74
Water	317	0.0035	0.310	0.53

Source: World Bank Industrial Pollution Projection System (Hettige *et al.*, 1994).

The dotted lines in Figure 1a plot the predicted total emissions of sulfur (SO_2), suspended solids in water, and hazardous waste, three of the 14 pollutants tracked by the IPPS. The predicted pollution levels are calculated by multiplying each 4-digit manufacturing industry's real annual value added by the industry's IPPS coefficient, and then summing across industries to get total predicted pollution for each year. By holding the IPPS coefficients constant as of 1987, we are in effect freezing the technique of production, and only looking at changes in pollution due to changes in the scale and composition of manufacturing. If industrial growth were balanced, and every four-digit industry grew by the same 57 percent as did overall manufacturing, then predicted pollution would coincide perfectly with manufacturing growth, and grow 57 percent. The dotted pollution lines would overlie the bold manufacturing line. But because relatively pollution-intensive industries shrank, while clean industries grew, the total amount of predicted pollution grew by much less than 57 percent over this time period. Predicted sulfur grew 21 percent, suspended solids fell 15 percent, and hazardous waste grew 35 percent, all much less than the growth of overall manufacturing.[3]

Figure 1a. Trends in Manufacturing, Pollution, and Tariffs

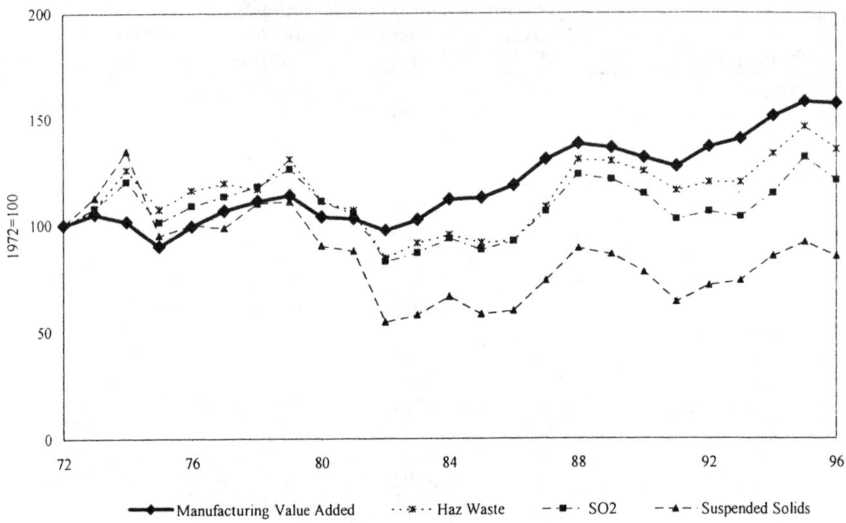

Table 2 summarizes the data behind Figure 1. The first column presents the growth in real U.S. manufacturing value added from 1972 to 1994 (51 percent), compared with the predicted growth in each of the 14 pollutants, where the predictions are based on the scale and composition of manufacturing, holding technique constant.[4]

Figure 1a and Table 2 demonstrate the degree to which the U.S. manufacturing sector shifted toward cleaner industries. During this same time period, U.S. tariffs fell: from 1974 to 1994, the average tariff on manufactured imports to the U.S. fell by over 50 percent, from over 8 percent to less than 4 percent. Given the steep drop in tariffs, combined with the dramatic shift toward cleaner industries, we understand how casual observers could conclude that the two trends are related. Indeed, this inference makes sense intuitively, since trade liberalization could lead to increased specialization, which in turn could result in the U.S. producing cleaner goods.

[3] Thirteen of the 14 pollutants tracked by the IPPS grew by less than manufacturing. The only exception was biological oxygen demand (BOD), which had predicted growth of 69 percent. Of this growth, 49 percent came from two industries: "chemical preparations n.e.c." (SIC 2899) and "paperboard mills" (SIC 2631). These industries grew by 68 and 54 percent, respectively, and had 57 and 32 times the mean IPPS coefficient for BOD.

[4] Table 2 uses data only up until 1994, which is the last year of the pollution abatement cost data we use later in the paper.

Table 2: Predicted Pollution Changes 1972-1994

Pollutant	US Manufacturing (1)	Imports to the US (2)	Imports from non-OECD (3)	US Exports (4)
Particulates	13.3	103.6	109.4	169.4
CO	1.9	96.7	113.2	158.6
SO2	14.8	110.3	119.3	183.5
NO2	22.8	114.7	117.2	205.2
VOC	36.4	171.0	177.3	223.3
pm10	-6.3	51.3	66.0	124.0
BOD	71.5	95.9	95.3	120.7
TSS	-14.7	66.0	100.7	152.6
Air toxics	47.8	196.6	197.2	264.3
Water toxics	34.7	148.4	151.5	273.7
Solid waste toxics	33.2	159.8	162.7	242.1
Metals to air	-2.3	89.5	109.9	217.5
Metals to land	-1.0	81.5	91.6	207.8
Metals to water	19.8	111.2	135.9	234.9
Percent change in manufactured products	51.1	317.9	344.4	268.8

If it is true that the composition shift of U.S. manufacturing toward clean industries has been caused by tariff reductions, one might expect that the composition of imports will have shifted toward dirty industries. Figure 1b investigates the pollution composition of imports. The bold line in Figure 1b plots the real value of U.S. imports, which grew by 318 percent from 1972 to 1994. The dotted lines plot the predicted pollution content of those imports, calculated in the same way as in Figure 1a. We multiply the real value of each 4-digit industry's imports by its IPPS coefficient and then aggregate across industries. If the composition of imports remained the same, then the dotted lines would overlie the bold imports line (i.e., grow by 318 percent). The fact that the dotted lines lie below the imports line indicates that the composition of imports to the U.S. has also shifted toward clean industries.

The second column of Table 2 summarizes the data behind Figure 1b. Real imports grew 318 percent, while the predicted pollution composition of those imports grew by markedly less. Of the 14 pollutants tracked by the IPPS, none grew by even two-thirds as much as real imports. Figure 1b and Table 2 thus demonstrate that cleaner U.S. manufacturing composition is not offset by more polluting imports. Rather, the composition of imports has also become cleaner.

One potential counterargument to our observation that the composition of imports has become cleaner is that the concerns about tariff reductions exacerbating pollution havens is based largely on imports from developing countries, not from developed countries. To see whether the composition of imports from relatively poorer countries has become more pollution-intensive, in Figure 1c we conduct the same exercise for imports from non-OECD countries only. The bold line plots the real value of imports from non-OECD countries. The dotted lines again plot pollution predicted from the IPPS coefficients, multiplied by their respective 4-digit industries' imports, and aggregated across industries. Again, pollution predicted by the scale and composition effects, holding technique constant, grew by far less than overall imports, even from poorer countries. Table 2 summarizes these data in column (3). Real imports from non-OECD countries grew by 344 percent from 1972 to 1994, but the composition of those imports shifted toward cleaner industries. As a result, the pollution predicted from those imports grew by far less than 344 percent.

Figure 1b. Trends in Imports and Pollution

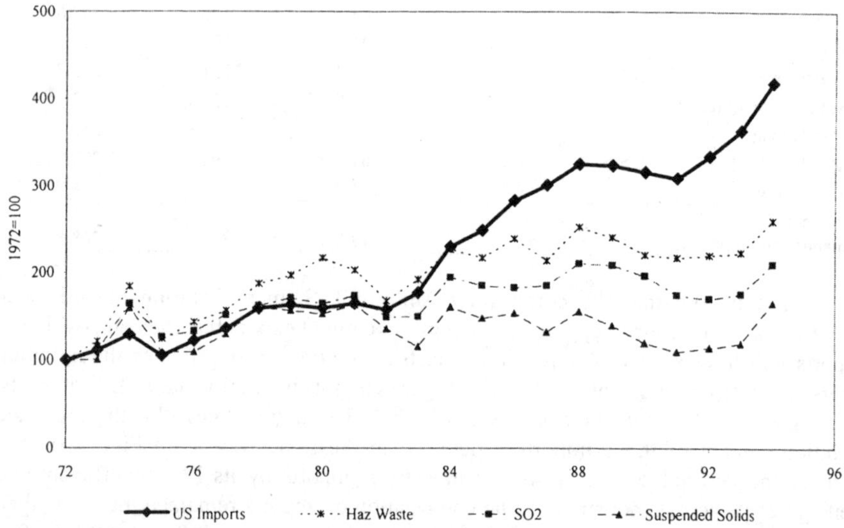

So far we have demonstrated that the industrial composition of U.S. imports has become less pollution-intensive, and that the missing polluting industries have not been replaced by imports. A remaining piece of the puzzle, then, is to ask what happened to the pollution composition of U.S. exports. If the composition of U.S. *consumption* has remained steady, and the composition of U.S. manufactures and imports have become cleaner, then the U.S. must be exporting the excess clean manufactured goods.

Figure 1d plots the growth in real U.S. exports, and the predicted pollution content of those exports, in the same manner as Figures 1a-1c. Here the predicted pollution lines also lie below the real exports line, but the shift is much less stark. Column (4) of Table 2 presents the data behind Figure 1d. Real manufacturing exports grew by 269 percent, and most of the predicted pollutants grew by slightly less. The composition of U.S. exports has shifted toward cleaner goods, but this shift is less dramatic than the shift that occurred in U.S. manufacturing or imports.

Figure 1c. Imports from Non-OECD and Pollution

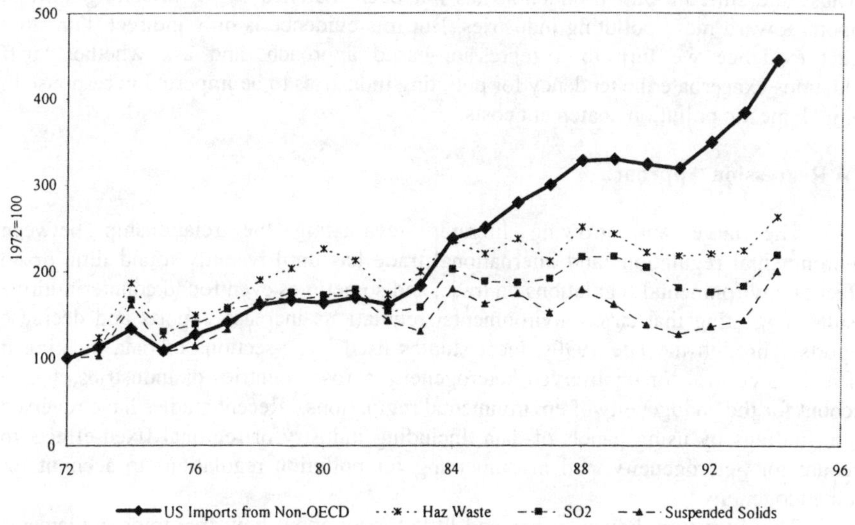

Figure 1d. U.S. Exports and Pollution

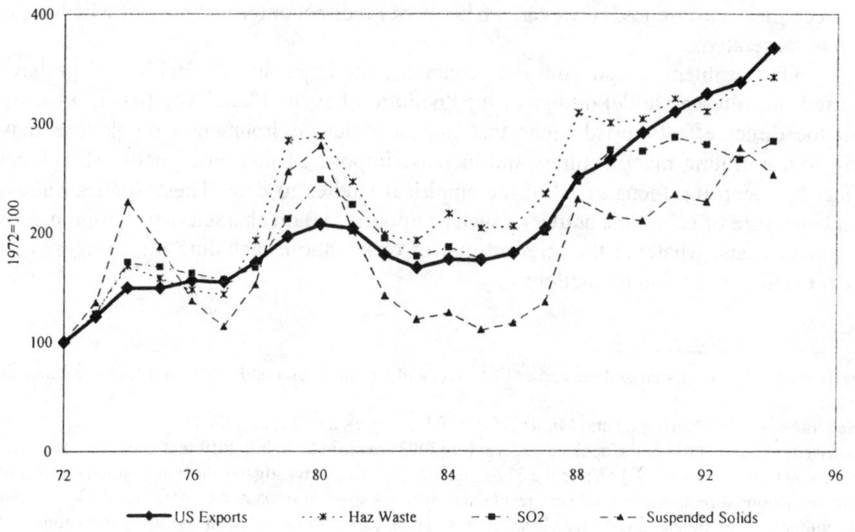

Together, Figure 1 and Table 2 provide suggestive evidence that the shift in U.S. manufacturing toward cleaner industries has not been caused by trade liberalization, because the shift in U.S. production has not been matched by an offsetting shift in imports toward more polluting industries. But this evidence is only indirect. For more direct evidence we turn to a regression-based approach, and ask whether tariff reductions exacerbate the tendency for polluting industries to be imported in response to rising domestic pollution abatement costs.

2. A Regression Approach

The large and growing literature evaluating the relationship between environmental regulations and international trade has until recently found little or no effect of environmental regulations on trade, and sometimes even found counterintuitive results suggesting that strict environmental regulations increase exports and decrease imports. Through the late 1990s, most studies used cross-sections of data, making it difficult to control for unobserved heterogeneity across countries or industries, and to account for the endogeneity of environmental regulations.[5] Recent studies have reversed these findings by using panels of data, including industry or regional fixed-effects to account for heterogeneity, and instrumenting for pollution regulations to account for their endogeneity.[6]

This literature, however, has had little to say about how free trade agreements are likely to affect the environment, the focus here. As mentioned previously, U.S. tariffs on imports have fallen, and the composition of U.S. manufacturing has shifted toward cleaner goods in recent decades, leading many observers to infer a causality between these two trends. However, we know of no direct empirical tests of whether the trends are related.[7]

One problem is that both the economics literature and the public debate have blurred two alternative definitions of the "pollution haven effect." The first is what we call the direct effect: jurisdictions that impose strict environmental regulations may drive out polluting manufacturers and increase imports of polluting goods. This direct effect has been the focus of all of the empirical studies to date. These studies regress some measure of economic activity, (such as imports, M) on characteristics of industries or jurisdictions, whatever the relevant unit of observation, including the stringency of the prevailing pollution regulations:

---

[5] Jaffe *et al.* (1996) summarized the earlier literature, and Brunnermeier and Levinson (2004) provides an update.

[6] See, for example, Ederington and Minier (2003) and Levinson and Taylor (2004).

[7] Antweiler *et al.* (2001) and Copeland and Taylor (2003) explore a slightly different question. They ask how *trade* (rather than *tariffs*) affects the environment. Our work investigates the compositional effect of tariff reductions directly (using industry level data), while Antweiler *et al.* and Copeland and Taylor infer the compositional impact indirectly (from country-level correlations between pollution-emissions and trade). Nevertheless, we both reach the same conclusion: trade liberalization has not resulted in a compositional shift in the U.S. toward cleaner industries.

(1) $$M_{it} = \beta_1 P_{it} + \beta_2 T_{it} + \mathbf{X}'_{it}\gamma + d'_i\boldsymbol{\alpha} + v'_t\boldsymbol{\delta} + e_{it}$$

where P_{it} is a measure of pollution regulations, T_{it} is a measure of trade restrictions, $\mathbf{X}_{it}$ is a set of industry characteristics, and d_i and v_t are cross-section and time dummies. In this context, the direct effect of pollution regulations (*P*) on imports (*M*) is captured by the coefficient β_1 in equation (1). Estimation of a negative coefficient on β_1 is taken as evidence for the existence of a pollution haven effect (i.e., more stringent environmental standards result in more imports).

Whether such pollution haven effects exist has taken on great importance in current policy debates, as the existence of such effects is a potential problem in negotiating trade agreements. Specifically, anti-globalization protests (and Executive Order 13141), are concerned that international trade agreements, by lowering tariffs broadly, may enable polluting industries to avoid strict environmental laws by relocating overseas and thus assist in the creation of pollution havens. However, estimation of β_1 tells us nothing about how trade liberalization will influence the composition of industries and hence pollution patterns. Indeed, if equation (1) is correctly specified, then trade agreements and tariff liberalization will have no effect on the composition of industries unless the trade agreements are asymmetrically implemented and result in disproportionately greater tariff reductions in the more pollution-intensive industries.[8]

Many environmental activists, however, are concerned with what we call the indirect effect: the effect of trade liberalization on the direct (pollution-haven) effect. Implicit in this concern is the assumption that pollution-intensive industries are more sensitive to tariff reductions than other industries, and thus a broad-based U.S. tariff reduction will result in a compositional shift with relatively large increases in U.S. imports of pollution-intensive goods. Technically, if the direct effect is $\partial M/\partial P$, then the indirect effect is $\partial^2 M/(\partial P \partial T)$. Empirically, this indirect effect can be measured by adding an interactive term to equation (1) to get

(2) $$M_{it} = \beta_1 P_{it} + \beta_2 T_{it} + \beta_3 \overline{P}_i T_{it} + \mathbf{X}'_{it}\gamma + d'_i\boldsymbol{\alpha} + v'_t\boldsymbol{\delta} + e_{it}$$

where β_3 captures the indirect effect of pollution regulations on the sensitivity of trade to tariffs.

Note that we have interacted T_{it} with $\overline{P}_i$, the *average* pollution abatement costs for industry *i* across all time periods. This means that we are asking whether tariff changes have a larger effect on imports for industries whose average pollution abatement costs are larger. The independent effect of average pollution abatement costs on imports is captured by the industry dummies. We could, of course, have asked a

---

[8] The coefficient on tariffs, β_2, is constant across industries and is assumed to be independent of the pollution intensity of the industry. Thus, a symmetric tariff reduction on all manufacturing industries will have an equal effect on all industries, and thus no effect on the composition of manufacturing.

slightly different question, by interacting T_{it} with P_{it}, which would estimate the degree to which tariffs have a larger effect on imports for industries whose pollution abatement costs rose more. Or, we could have interacted $\overline{T_i}$ with P_{it}, which would estimate the degree to which increases in pollution abatement costs have a larger effect on industries with higher tariffs. We believe, however, that equation (2) best captures the concerns of Executive Order 13141 and the WTO protests.[9]

We estimate versions of equation (2) in which the dependent variable (M_{it}) is annual imports to the U.S., by four-digit manufacturing SIC code. We have two main motivations. First, we believe that this interactive effect (β_3) is an important test of the effect of trade agreements on the environment. One of the chief mechanisms by which trade will change environmental quality in various countries is by changing the distribution of industries. Countries that host increasingly polluting industries as a consequence of liberalized trade will become more polluted, while countries whose industrial compositions become less polluting will become cleaner. If a trade agreement lowers tariffs on polluting industries, and raises tariffs on clean industries, the environmental consequences will perhaps be obvious. More realistically, if a trade agreement lowers tariffs across the board, the environmental consequences will depend on whether the polluting industries or the clean industries are more responsive to the tariff reductions: in other words β_3 in equation (2).

The policy debate seems to presume that this indirect effect is negative for the U.S. – that lowering tariffs will increase imports more in those industries that have stricter pollution regulations. This inference assumes that the U.S. is at a comparative disadvantage in industries facing stringent environmental regulations and thus, a reduction in tariff barriers will result in a relatively greater import surge in these more regulated industries. We, however, see no particular reason for this to be true as comparative disadvantages can arise from many factors, not just environmental costs.[10] We test for the interaction anyway, because we believe it to be the underlying assumption of so much public debate.

The second important motivation for our empirical strategy is that if equation (2) is the correct specification, then equation (1) contains an important missing covariate that is clearly correlated with pollution abatement costs. The indirect pollution haven effect is thus not only important in its own right, as the focus of policy concerns, but it may also provide an additional explanation for why early attempts to measure the direct pollution haven effect have been biased downwards.

---

[9] In section 2.3 below we discuss a sensitivity test, running a version of equation (2) interacting T_{it} with P_{it}, rather than with $\overline{P_i}$.

[10] A similar point is made by Copeland and Taylor (2003), in which they point out that increased trade could shift dirty industries to high-environmental standard (high-income) countries if such countries have a comparative advantage in dirty goods production (due, for example, to relative capital abundance, or labor costs necessary to compensate workers for poor conditions).

2.1 Data

Because only the United States has collected pollution abatement cost data for a significant period of time, researchers studying pollution havens and trade barriers have focused on U.S. imports and exports. This paper, like many before it, starts by following Grossman and Krueger (1993) and regressing manufacturing imports by industry on industry characteristics, as in equation (1). Data on imports to the U.S. come from the *NBER Trade Database*, documented in Feenstra *et al.* (2002). Our dependent variable (M_{it}) is imports divided by the value of shipments, taken from the *NBER Manufacturing Productivity Database* (Bartelsman *et al.*, 2000).[11]

The industry characteristics used include environmental costs, tariffs, capital intensity, and human capital intensity. Environmental costs come from the Pollution Abatement Costs and Expenditures (PACE) survey (U.S. Department of Commerce). We use the data from 1978 to 1994, the last year the PACE data were collected, and focus on pollution abatement operating costs.[12] The PACE data were not collected in 1987, and were not disaggregated by four-digit SIC code in 1979, so we exclude those two years. Our measure of an industry's environmental costs (P_{it}) is pollution abatement operating costs (U.S. Dept. of Commerce) divided by total materials costs (Bartelsman *et al.*, 2000).

Tariffs by four-digit SIC code are taken from Feenstra *et al.* (2002), where the tariff rate for each industry is duties paid divided by the customs value of imports. We converted the 1989-94 data to the 1972 SIC codes using the Bartelsman *et al.* concordance.

Finally, each industry is characterized by its capital intensity and human capital intensity. Like Grossman and Krueger we measure capital intensity as one minus payroll's share of value added:

$$\text{capital intensity} = 1 - \frac{\text{payroll}}{\text{value added}} \tag{3}$$

Human capital intensity is then the total payroll less what would have been paid to workers had each earned the average wage of an 18-year-old worker with less than a high school education, all divided by the value added in the industry:

$$\text{human capital intensity} = \frac{\text{payroll - (low-skill wage)} \times \text{(worker hours)}}{\text{value added}} \tag{4}$$

[11] It should be noted that we strictly follow Grossman and Krueger (1993) in using gross imports (not net imports) as our dependent variable. This is due to the fact that we are estimating the effect of unilateral trade liberalization by the U.S. on its imports (i.e., we do not have foreign tariff data to measure the corresponding effect of foreign tariff liberalization on U.S. exports).

[12] After 1987 the data switch from the 1972 SIC codes to the 1987 SIC codes. We used the concordance in the *NBER Manufacturing Productivity Database* to reallocate pollution costs to 1972 industry definitions.

Income data come from the *Current Population Survey*, May supplemental surveys.

Table 3 presents summary statistics of these data. There are 394 industries, and 15 years of data (1978-1994, less 1979 and 1987 when the PACE data are unavailable). Many 4-digit SIC codes are not reported in the 4-digit PACE data, partly to prevent disclosure of confidential business information. We have omitted these missing observations, resulting in a sample size of 4409.[13] The average U.S. manufacturing industry imported 17 percent of the value shipped by domestic producers, divided about evenly between OECD countries and the rest of the world. The average industry spent 1.1 percent of total materials costs on pollution abatement operating costs, ranging from zero to 18 percent. Tariffs averaged 4.0 percent, ranging from zero to over 40 percent. The question we ask is whether those industries whose tariffs dropped most saw larger increases in imports if their environmental costs were high.

Table 3: The Data

	Means (std. dev.) (1)	**Max (2)**
Dependent variable:	0.169	14.4
gross imports / value shipped	(0.414)	
From OECD countries	0.085	12.1
	(0.261)	
From non-OECD countries	0.084	7.1
	(0.231)	
Environmental cost	0.011	0.176
	(0.015)	
Tariff	0.040	0.41
	(0.040)	
Tariff × Average	0.000346	0.00675
environmental cost	(0.000608)	
Human capital	0.275	0.99
	(0.092)	
Physical capital	0.612	0.95
	(0.120)	
Observations	4,409	
Number of industries	394	

[13] We have run all of the specifications below with the missing environmental cost data treated as zeros, with no change in the substantive conclusions. Coefficients' magnitudes change, but not the pattern of signs nor statistical significances.

2.2 Empirics

Table 4 begins with a basic fixed-effect regression of import penetration, by industry, on industry characteristics from 1978 to 1994. It is a panel-data version of Grossman and Krueger (1993), as represented by equation (1). Industries whose environmental costs increased also saw their imports increase, and industries whose tariffs increased saw imports decrease.[14] Equally sensibly, higher capital intensity and human capital intensity are associated with lower imports. At the bottom of Table 4 we have calculated the elasticity of imports with respect to tariffs (-0.44) as a useful benchmark to compare with later specifications and samples. It suggests that a 10 percent increase in tariffs is associated with a 4 percent decline in imports.

Table 4: Baseline Regressions

Dependent variable: gross imports / value shipped	**Baseline** (1)	**With interaction** (2)
Environmental cost	1.45*	1.86*
	(0.28)	(0.29)
Tariff	-1.84*	-2.17*
	(0.14)	(0.16)
Tariff × Avg. env. cost		50.6*
		(12.4)
Human capital	-0.74*	-0.71*
	(0.22)	(0.22)
Physical capital	-0.44*	-0.42*
	(0.18)	(0.18)
Observations	4,409	4,409
Number of industries	394	394
R-squared	0.91	0.91
Elasticity of imports with respect to tariffs	-0.436	-0.382
Chi-sq. from Hausman test of fixed versus random effects	14.3	124.7

Notes to Table: The regressions are estimated with year and industry fixed effects, and cover the period 1978-94 (1979 and 1987 are omitted due to missing data). The dependent variable is gross imports divided by value shipped.

* statistical significance at the 5% level.

In column (2) of Table 4 we include an interaction term, as in equation (2). Our interpretation of public rhetoric, and EO13141, is that the coefficient (β_3) on this interactive term is expected to be negative. However, in practice the coefficient (50.6) is

[14] The fact that the coefficient on the environmental cost variable is positive and statistically significant is consistent with recent literature that uses time series data and fixed effects, and departs from previous efforts (such as Grossman and Krueger) that relied on a single cross section.

positive and statistically significant. Polluting industries appear to be *less* sensitive to tariff reductions rather than more sensitive, and trade liberalization does not exacerbate the pollution haven effect.[15]

If we take the point estimates from column (2) seriously, and calculate the elasticity of import penetration with respect to environmental costs at the average level of tariffs, that elasticity is -0.38. Though this number is not markedly different from the tariff elasticity when the interaction term is omitted (-0.44), the important point is that the interaction term is positive, so that as we consider industries with lower pollution abatement costs, this estimated elasticity shrinks in absolute value.

We conclude from Table 4 that in general, tariff reductions from 1978 to 1994 did not significantly increase imports more from polluting industries than from clean industries. In fact, the opposite is more likely to be true: if anything, trade liberalization has shifted U.S. industrial composition toward dirtier industries, by increasing imports of polluting goods by less than clean goods.

In previous work, we explored the possibility that environmental regulations have different effects on trade with developing countries than they do on trade with developed countries (Ederington *et al.*, 2005). It seems plausible that the same forces are at work here. Trade agreements may alter the pollution intensity of trade between the U.S. and developing countries, but not with other developed countries.[16]

To test whether trade agreements have increased imports in pollution-intensive industries from developing countries, in Table 5 we run versions of equations (1) and (2) separately for OECD and non-OECD countries. The coefficients are smaller than for the whole sample because the scale of the dependent variable is smaller: imports from OECD divided by U.S. value shipped. Hence comparisons can be made only by examining the relevant elasticities. In column (1) of Table 5, without the interaction term, the elasticity of imports with respect to tariffs (-0.15) is smaller for OECD countries than for all imports. Most importantly, as with the full sample, the interactive term in column (2) of Table 5 is still positive and statistically significant.

The real test, however, comes with imports from non-OECD countries, where the concern about pollution havens is strongest. Once again the interactive term between tariffs and pollution costs remains positive and statistically significant. The point estimate (31.7) implies that tariff reductions have had a *smaller* effect on imports from non-OECD countries in the more pollution-intensive industries. The interacted coefficient in column (4) has exactly the opposite sign to what we perceive to be the conventional wisdom. Rather than the U.S. manufacturing sector becoming cleaner at the expense of the manufacturing sector in developing countries, column (4) suggests

---

[15] This result, that polluting industries are less sensitive to tariff reductions, is consistent with previous work (Ederington, Levinson and Minier, 2005), in which we show that polluting industries are less geographically footloose than clean industries. This would make them less sensitive to changes in tariffs as well as to changes in pollution regulations.

[16] Intuitively, if the U.S. is at a comparative disadvantage in dirty goods production (due to high regulatory standards) against developing countries but not other developed countries, then trade liberalization might only result in large increases in polluting imports from the developing world.

that the U.S. manufacturing sector has become more polluting as a consequence of tariff reductions.[17]

Table 5: Trading Partners' Environmental Standards

Dependent variable:	OECD		Non-OECD	
gross imports / value shipped	(1)	(2)	(3)	(4)
Environmental cost	1.03*	1.19*	0.33†	0.58*
	(0.14)	(0.15)	(0.20)	(0.21)
Tariff	-0.31*	-0.44*	-1.52*	-1.72*
	(0.07)	(0.08)	(0.10)	(0.12)
Tariff × Avg. env. cost		18.9*		31.7*
		(6.3)		(8.8)
Human capital	-0.24*	-0.23*	-0.51*	-0.49*
	(0.11)	(0.11)	(0.16)	(0.16)
Physical capital	-0.23*	-0.22*	-0.24*	-0.23*
	(0.09)	(0.09)	(0.13)	(0.13)
Observations	4,409	4,409	4,409	4,409
Number of industries	394	394	394	394
R-squared	0.94	0.94	0.85	0.85
Elasticity of imports with respect to tariffs	-0.146	-0.109	-0.723	-0.653

Notes to Table: The dependent variable in each regression is gross imports divided by value shipped to specified trading partners (OECD countries in columns 1 and 2, non-OECD in columns 3 and 4). All regressions include year and industry fixed effects.

* Statistical significance at the 5% level. † Statistical significance at the 10% level.

2.3 Democracy, Endogeneity, and Other Concerns

A number of readers have speculated that trade liberalization might have a larger effect on imports from dictatorships than from democracies. The intuition may not be explained by any formal trade theory, but rather by the notion that an oppressive government has a comparative advantage in that citizens with fewer civil liberties cannot vote for environmental policies and workplace safety laws that raise production costs. We have estimated a version of Table 4 separately for dictatorships and democracies, as defined by the Freedom House (2004) index of political rights, which is commonly used in empirical studies involving democracy. Freedom House assigns countries an index value from 1-7. The typical classification is that countries indexed 1-2 are "free," and those indexed 6-7 are "not free."

Appendix Table A contains a version of the basic specifications. Column (1) uses only imports from "non-dictatorships," with Freedom House indices from 1 to 5. The results are essentially the same as for the full specifications in Table 4. Column (2)

[17] This result is actually consistent with those of Antweiler et al., (2001) and Copeland and Taylor (2003) who, using a different technique, estimate that trade integration has shifted dirty-goods industries to high-standard (high-income) countries. (See fn. 7.)

uses only imports from "dictatorships," with Freedom House indices of 6 or 7. Here the direct effects of environmental costs are small and insignificant, but the direct effect of tariffs and the indirect effect of tariffs on high-environmental-cost industries are about the same as for the full sample. In other words, separating out the dictatorships does not change the substantive conclusion that trade liberalization does not increase imports of polluting industries any more than clean industries.

Another concern raised by many analysts of the pollution haven literature, ourselves included, is that regressions of trade flows on pollution costs suffer from endogeneity problems. It may be that countries use environmental policy as a strategic substitute for tariffs (Ederington and Minier, 2003). Or, it could be that the proxy for regulatory stringency, expenditures on pollution abatement, is a function of trade, if trade changes the composition of sub-industries within each 4-digit SIC code (Levinson and Taylor, 2004).

While these issues are not the focus of this paper, in column (3) of Appendix Table A we have tried employing the instruments from Levinson and Taylor (2004). That paper constructs an instrument based on the geographic dispersion of industries throughout the U.S. The idea is that industries concentrated in states whose environmental standards increased most will have experienced the highest run-up in environmental costs. So long as the geographic dispersion of industries throughout the U.S. is unrelated to import penetration, the geographic dispersion will be a good instrument for an industry's environmental regulatory costs. (See Levinson and Taylor for details.) The current paper represents a slightly different application: four-digit industries rather than three-digit industries, the dependent variable is gross rather than net imports, the regulatory proxy is pollution abatement cost over the cost of materials rather than value added, and the data run from 1974-1994, rather than just 1977-1986. Nevertheless, column (3) of Appendix Table A estimates a version in which P_{it} is instrumented using the Levinson and Taylor instruments. The environmental cost coefficient (2.86) is larger than in Table 4, but is statistically insignificant in this case. (We have fewer years of data and industries.) The interaction term, which is the focus of this paper, is smaller (13.7), and statistically insignificant, but still positive. So the basic finding from Table 4 survives. Industries whose environmental compliance costs are largest are not more responsive to tariff reductions.

Because of the concerns about endogeneity, we do not want to interpret the results in Tables 4 and 5 causally. Rather, the estimates in Tables 4 and 5 are conditional correlations that tell us that imports have increased more in industries whose tariffs have fallen the most, but that this conditional correlation is not larger for industries that are more pollution-intensive. Trade liberalization does not seem to be associated with increased imports from relatively pollution-intensive industries.

Finally, some commentators have argued that the correct version of equation (2) would interact tariffs T_{it} with current pollution costs (P_{it}), rather than with average pollution costs ($\overline{P_i}$). In column (4) of Appendix Table A we estimate that alternative model. All of the resulting coefficients are statistically indistinguishable from those in equation (2) except for the coefficient on P_{it} alone. This is because most of the variation in P_{it} is across industries rather than within industries over time, the cross-industry variation is soaked up by industry fixed effects, and including the interaction $P_{it}T_{it}$ means there are *two* right-hand-side variables trying to capture the effect of time-series variation in pollution costs.

In short, the basic result holds for a variety of alternative specifications (separating imports from dictatorships, instrumenting for pollution costs, and including a time-varying interactive term). Imports of pollution-intensive goods do not appear to be more sensitive to tariff reductions than imports of clean goods.

3. Predicted Pollution Changes

Tables 4 and 5 examine whether pollution-intensive industries are more responsive to broad-based tariff reductions than clean industries, and find just the opposite. Thus our estimates imply that symmetric tariff reductions in the U.S. would have actually shifted the composition of U.S. manufacturing toward dirtier industries. There is, however, an alternative mechanism through which the trade agreements could make U.S. manufacturing cleaner. It might be that trade agreements, either past or future, are skewed toward pollution-intensive industries. If tariffs fell *more* for polluting industries than for clean industries from 1978 to 1994, then trade liberalization might be said to have caused a shift of polluting industries overseas, even if the interactive coefficient β_3 in equation (2) is zero. Alternatively, if current tariffs are higher for polluting industries than for clean industries, then future trade agreements that lower all tariffs to zero might be expected to shift polluting industries overseas.

The first thing to note is that on average, industries whose tariffs fell further had higher pollution abatement costs than industries whose tariffs rose, or fell by less. Specifically, the raw correlation between the change in tariffs from 1978 to 1994 and average pollution costs is -0.19, which is statistically significantly different from zero. While this correlation is not strong, it lends plausibility to the claim that tariff reductions might have caused the shift in U.S. manufacturing composition, because the reductions were skewed toward polluting industries. That said, those industries that experienced the very largest tariff reductions are not among the most polluting industries.[18]

To explore the magnitudes of the correlations between environmental costs and tariff reductions, in Table 6 we predict the declines in U.S. pollution resulting from U.S. tariff reductions from 1978 to 1994 using the coefficients of Table 4. In column (1) of Table 6 we list the estimated change in the emissions of the 14 pollutants modeled by the World Bank's Industrial Pollution Projection System (IPPS) due to scale effects and compositional shift in U.S. industries. These are the same data used to generate Figure 1, but for the shorter time period for which pollution abatement cost data are available.

[18] The five industries that have experienced the largest tariff reductions are SIC 2252 "Hosiery, not elsewhere classified," SIC 2322 "Mens' & boys' underwear & nightwear," SIC 2075 "Some fat/oil not cottonseed," SIC 2342 "Bras, girdles, and allied garments," and SIC 2341 "Womens' & childrens' outerwear."

To generate the numbers in column (1), we used the IPPS coefficients to predict the amount of pollution generated in each industry in 1978 and then aggregated across all industries. We did the same for 1994 and then reported the percentage change between 1978 and 1994. The numbers differ because industries with varying pollution intensities for different pollutants grew by different amounts during the period.

As in Figure 1, while the real value of manufacturing output increased by 36 percent from 1978 to 1994, the predicted pollution either rose by much less, or declined outright. This is because the manufacturing industries have shifted toward those that produce less pollution. In other words, while manufacturing output increased, predicted airborne particulate emissions declined by 6.4 percent because those four-digit industries that produce the most particulates declined over the period, while those that produced the least increased their output.

Table 6: Predicted Pollution Changes

Pollutant	Percent change in pollution 1978-1994 (1)	Percent change relative to real manufacturing growth (2)	Predicted percent change in pollution 1978-1994 (3)	Percent change from 1994 if tariffs set to zero. (4)
Air pollution				
Particulates	-6.4	-31.1	-3.0	-1.6
CO	-19.3	-40.6	-1.2	-1.0
SO2	-12.7	-35.8	-1.6	-1.5
NOx	-5.8	-30.7	-2.3	-2.2
VOC	+4.5	-23.1	-3.0	-3.0
PM10	-17.5	-39.3	-2.0	-1.3
Water pollution				
BOD	+26.9	-6.6	-1.1	-3.8
TSS	-29.2	-47.9	-1.4	-1.6
Toxics				
Air	+12.8	-17.0	-3.2	-2.3
Land	-3.5	-29.0	-1.4	-2.5
Water	-4.4	-29.7	-1.1	-1.4
Metals				
Air	-21.4	-42.2	-1.7	-1.4
Land	-20.9	-41.8	-1.4	-1.0
Water	-10.5	-34.1	-1.1	-0.9

Source: authors' calculations from World Bank Industrial Pollution Projection System, the NBER Manufacturing Industry Productivity Database (Bartelsman *et al.*, 2000), NBER Tariff Data (Feenstra *et al.*, 2002).

To represent the pollution changes in column (1) of Table 6 comparably with the predictions from the regression coefficients, in column (2) we report the predicted change relative to the change in real manufacturing growth. For example, the decline in particulates of 6.4 percent represents a 31 percent decline relative to the 36 percent growth in manufacturing. ([1-0.064]/[1.36]-1 = -0.31) This amounts to examining the composition effect alone, eliminating both the scale and technique effects.

In column (3) we use what we know about the changes in tariffs over this time period to ask whether the fact that tariff reductions were steeper in more polluting industries could have caused this shift in U.S. manufacturing toward cleaner industries, presumably at the expense of other countries' environmental quality. We used the tariff coefficients from column (2) of Table 4, along with the actual industry-level tariffs and environmental costs, to predict the changes in imports (as a share of value shipped) that resulted from U.S. tariff reductions. We then multiplied this result by the actual value shipped in each industry, to get the predicted change in imports. We then assumed as a benchmark that changes in imports are offset dollar-for-dollar with domestic production, so we can multiply the import change by the IPPS coefficients to get predicted changes in pollution emanating from each industry.[19] Finally, we aggregated these pollution changes across all industries, and divided this Figure total by the total estimated pollution as of 1978, to get the predicted percentage changes reported in column (3).

As column (3) shows, predicted pollution declined only slightly due to the varying declines in industry-specific tariffs. Given the degree to which tariffs declined over this time period, the predicted effect of this on imports, and the predicted pollution associated with each industry, imports are sufficient to replace between 1 and 3 percent of pollution generated by U.S. manufacturing. This change is small relative to the overall shift in U.S. manufacturing toward cleaner industries demonstrated in column (2). Though tariff reductions have been steeper in more pollution-intensive industries, that can explain only a small part of the shift of U.S. manufacturing toward cleaner industries.

There is, however, one last interpretation of the environmental opposition to trade liberalization that we can examine with these data. Observers could be concerned that *current* tariff levels offer more protection to polluting industries than to clean industries, and that *future* trade agreements that lower all U.S. tariffs to zero could cause U.S. polluting industries to relocate overseas more than relatively clean industries. In column (3) of Table 5 we predict the change in U.S. manufacturing pollution that would result from a drop in tariffs from their 1994 levels down to zero. For each four-digit SIC code we calculate the level of the 1994 tariffs (from Feenstra *et al.*, 2002). We then use the tariff and interactive coefficients from Table 4 to predict the change in imports that would result from reducing these 1994 tariffs to zero. We then assume as a benchmark that those imports replace U.S. production, and use the IPPS pollution coefficients to generate the resulting change in each of the 14 pollutants. As column (4) shows, this results in very small changes in pollution relative to actual changes over the 1978 to 1994 period. Once again, we find little basis for the fear that tariff reductions themselves will generate large shifts in the composition of U.S. manufacturers away from polluting industries.

[19] Note that this exaggerates the pollution changes, because the IPPS coefficients are calculated per dollar of *value added*, and we multiply by *value shipped*.

4. Conclusions

The "trade and environment" debate has captured public interest, from the street protests outside WTO meetings to President Clinton's executive order requiring that trade agreements be accompanied by environmental impact statements. The general consensus seems to be that tariff barriers protect U.S. manufacturers that would otherwise move overseas to avoid increasingly strict U.S. environmental standards, and that trade agreements, by lowering these tariffs, will disproportionately affect the most polluting industries.

This story is, on the surface, consistent with recent U.S. experience that has seen the U.S. manufacturing sector shift toward cleaner industries. As we show in this paper, while manufacturing's real value added grew by 51 percent from 1972 to 1994, that growth occurred mostly among the least polluting industries, causing the estimated emissions of most of the 14 pollutants tracked by the World Bank's IPPS to rise by much less than 51 percent, or even to decline. This shift has occurred simultaneously with a general reduction in U.S. tariff barriers due primarily to the success of GATT negotiations.

We show, however, that these two trends appear unrelated. We find no evidence that domestic production of pollution-intensive goods in the U.S. is being replaced by imports from overseas. In addition, we find that symmetric tariff reductions in the U.S. have most likely induced a compositional change toward dirtier industries (not cleaner) among U.S. manufacturing. Finally, while polluting industries have experienced larger tariff reductions than other industries in the U.S., these differences in tariff reductions do not appear to be of sufficient magnitude to explain more than a small fraction of the shift in U.S. manufacturing toward cleaner industries.

Appendix

Appendix Table A: Extensions

Dependent variable: gross imports / value shipped	**Imports from non-dictatorships (1)**	**Imports from dictatorships (2)**	**2SLS version 1977-1987 (3)**	**With $P_{it} \times T_{it}$ interaction (4)**
Environmental cost	1.60*	0.08	2.86	0.19
	(0.23)	(0.15)	(3.67)	(0.34)
Tariff	-0.97*	-1.15*	-0.55	-2.15*
	(0.13)	(0.08)	(0.42	(0.15)
Tariff × Avg. env. cost	24.02*	25.7*	13.7	
	(9.78)	(6.3)	(32.3)	
Tariff ×Annual env. cost				54.1*
				(8.2)
Human capital	-0.56*	-0.15	-0.84*	-0.76*
	(0.18)	(0.11)	(0.34)	(0.22)
Physical capital	-0.36*	-0.10	-0.48[†]	-0.46*
	(0.14)	(0.09)	(0.27)	(0.18)
Observations	4,409	4,393	1731	4,409
Number of industries	394	393	307	394

Notes to Table: The regressions are estimated with year and industry fixed effects, and cover the period 1978-94 (1979 and 1987 are omitted due to missing data). The dependent variable is gross imports divided by value shipped. "Dictatorships" are defined as counties with Freedom House (2004) indices of 6 or 7 on a scale from 1 to 7. Column (4) uses instrumental variables as in Levinson and Taylor (2004).
[*] statistical significance at the 5% level. [†]Statistical significance at the 10% level.

References

Antweiler, Werner, Brian Copeland, and M. Scott Taylor. 2001. "Is Free Trade Good for the Environment?" *American Economic Review* 91(4): 877-908.

Bartelsman, Eric J., Randy A. Becker, and Wayne B. Gray. 2000. "The NBER-CES Manufacturing Industry Database," www.nber.org/nberces/nbprod96.htm. A prior version is documented as NBER Technical Paper #205.

Brunnermeier, Smita and Arik Levinson. 2004. "Examining the Evidence on Environmental Regulations and Industry Location," *Journal of the Environment and Development* 13(1): 6-41.

Copeland, Brian R. and M. Scott Taylor. 2003. *Trade and the Environment: Theory and Evidence*. Princeton University Press.

Ederington, Josh and Jenny Minier. 2003. "Is Environmental Policy a Secondary Trade Barrier? An Empirical Analysis," *Canadian Journal of Economics* 36(1): 137-154.

Ederington, Josh, Arik Levinson, and Jenny Minier. 2005. "Footloose and Pollution-free," *Review of Economics and Statistics*, 87(1) forthcoming.

Feenstra, Robert C., John Romalis, and Peter K. Schott. 2002. "U.S. Imports, Exports, and Tariff Data, 1989-2001," NBER Working Paper #9387.

Frankel, J.A. and D. Romer. 1999. "Does Trade Cause Growth?" *American Economic Review*, 89(3): 379-99.

Freedom House. 2004. *Freedom in the World: The Annual Survey of Political Rights and Civil Liberties.* New York, Rowman & Littlefield Publishers, Inc.

Grossman, Gene M. and Alan B. Krueger. 1993. "Environmental Impacts of a North American Free Trade Agreement," in *The Mexico-U.S. Free Trade Agreement*, Peter M Garber, ed., Cambridge, MA: MIT Press. Reprinted in Robert Percival, and Dorothy C. Alevizatos, eds., *Law and the Environment: A Multidisciplinary Reader.* Philadelphia: Temple University Press, 1997.

Hettige, Hemamala, Robert E.B. Lucas, and David Wheeler. 1992. "The Toxic Intensity of Industrial Production: Global Patterns, Trends, and Trade Policy," *American Economic Review Papers and Proceedings* 82(2): 478-481.

Hettige, Hemamala, Paul Martin, Manjula Singh, and David Wheeler. 1994. "The Industrial Pollution Projection System," World Bank Policy Research Working Paper #1431 (Part 1).

Jaffe, Adam B., Steven R. Peterson, Paul R. Portney, and Robert N. Stavins. 1995. "Environmental Regulations and the Competitiveness of U.S. Manufacturing: What does the Evidence Tell Us?" *Journal of Economic Literature* 33(1):132-63.

Kahn, Matthew E. 2003. "The Geography of U.S. Pollution Intensive Trade: Evidence from 1959 to 1994," *Regional Science and Urban Economics* 33(4): 383-400.

Levinson, Arik, and M. Scott Taylor. 2004. "Unmasking the Pollution Haven Effect," NBER working paper #10629.

U.S. Department of Commerce, various years. *Current Industrial Reports: Pollution Abatement Costs and Expenditures*, 1972-92.

[10]

Testing for Pollution Havens Inside and Outside of Regional Trading Blocs*

Matthew E. Kahn and Yutaka Yoshino

Abstract

This paper examines the composition of bilateral manufacturing trade in terms of its pollution intensity. A growing share of world trade takes place between nations who are members of the same regional trade agreement (RTA). We analyze how the formation of RTAs affects the composition of trade. We focus on whether richer or poorer nations specialize in exporting dirty goods. Using bilateral trade data over the years 1980 to 1997 for 128 nations for 34 manufacturing industries, we document that low-, middle-, and high-income nations differ with respect to their income elasticity in exporting dirty goods. Outside of RTA blocs, we find general support for the pollution haven hypothesis. As a nation's income rises, its exports of dirty goods decrease relative to its exports of clean goods. When we compare, low-income, middle-income and high-income nations, we find that middle-income nations have the largest dirty trade income elasticity. Participation in a RTA slightly weakens the pollution haven effect observed outside of regional trading blocs. We also document cross-RTA heterogeneity in the composition of trade across 13 major RTAs in our sample.

KEYWORDS: trade, pollution havens, regional trading blocs

*Matthew E. Kahn is an Associate Professor of International Economics at the Fletcher School, Medford, MA 02155 (Matt.kahn@tufts.edu). Yutaka Yoshino is a Ph.D. student in the Department of Economics at the University of Virginia, Charlottesville, VA 22904 (yfy2d@virginia.edu) and a consultant at the World Bank, Washington, D.C. We thank Don Fullerton, two anonymous referees and seminar participants at Stanford University for useful comments. This paper was written while Kahn was visiting Stanford University. We thank Christopher Magee for kindly sharing data with us.

Two competing theories might explain bilateral pollution-intensive trade flows. The "pollution haven hypothesis" (PHH) posits that the poorer nations will become more specialized in exporting pollution-intensive products, because they are less likely to enact stringent domestic environmental regulation. On the other hand, the "factor endowment hypothesis" (FEH) predicts that the richer nations, who are endowed with more capital, will become more specialized in pollution-intensive exports because these industries tend to be capital intensive (Copeland and Taylor 2003, 2004).

This paper examines the composition of bilateral manufacturing trade in terms of its pollution intensity. All else equal, do exports of pollution-intensive products or "dirty goods" increase or decrease as an exporter's national per-capita income increases? Relative to trade in less pollution-intensive products or "clean goods", is the income elasticity of exports of dirty goods larger or smaller?

We examine trading patterns for nations who are and are not members of the same regional trade agreement (RTA). Over the last 30 years, RTAs have proliferated around the world. As of October 2003, 189 agreements were filed with the World Trade Organization (WTO). Since the establishment of the WTO in 1994, over 130 RTAs have been signed and filed with the WTO, compared to 124 that were signed in the much longer period between 1946 and 1994.[1] These agreements include free trade agreements, customs unions, preferential agreements, and services agreements.[2] Over this time period, a growing share of world trade is taking place within these blocs (Schiff and Winters 2003).[3] We test whether the formation of such blocs affects the composition of bilateral trade. For example, if bloc members harmonize their environmental regulations, then we would expect that the FEH effect would dominate the PHH effect within the bloc.

Despite the growing importance of within-RTA trade, the trade and environment literature has not theoretically modeled how RTA formation affects the composition of trade flows nor empirically estimated the size of these effects.[4] In this paper, we seek to address both of these issues. In the theory section, we build an intra-industry trade model that highlights the conditions such that the poorer trading partner will emerge as the pollution haven. In addition, we examine how RTA participation affects this outcome.

---

[1] See the WTO official web site (www.wto.org). The WTO members are required to notify the organization concerning any RTAs they participate in.

[2] A free trade agreement liberalizes trade among its members, while each member has its own tariff policy toward the non-member countries. A customs union liberalizes intra-bloc trade and establishes a common tariff policy toward non-members. A preferential agreement is an agreement for a developed country (or a group of developed countries) that provides unilateral tariff concessions to a developing country or group of developing countries to assist the latter's economic development.

[3] Through lowering within-bloc trade barriers, RTA participation may cause nations to trade more (Frankel and Rose 2002). Alternatively, intra-bloc trade may grow by diverting flows from extra-bloc trade.

[4] Panagariya and Suthiwart-Narueput (1997) theoretically model the impact of trading bloc formation. However, they analyze the welfare implications of the interaction between bloc formation and the environment, based on the predetermined direction of pollution-intensive trade among countries inside and outside of the bloc, rather than modeling how bloc formation affects the direction of trade.

To investigate the patterns of clean and dirty trade flows empirically, we use bilateral trade data over the years 1980 to 1997 for 128 nations for 34 manufacturing industries. We document that low-, middle-, and high-income nations differ with respect to their income elasticity in exporting dirty goods. Outside of RTA blocs, richer nations export less of dirty goods relative to cleaner goods as they grow richer. Under the assumption that richer nations have more stringent regulations than poorer nations and have more capital, estimates of the income elasticity of dirty trade exports provide a test of whether the PHH effect dominates the FEH effect. We find that middle-income nations have a larger dirty trade income elasticity than high-income nations. At the same time, low-income nations are found to have a smaller dirty trade elasticity than middle-income nations. Together, these results provide support for both the PHH and FEH hypotheses. Middle-income nations are likely to specialize in dirty production because these nations have more capital than poorer nations and less stringent environmental regulations than richer nations.

RTA bloc participation has a small statistically significant effect on the income elasticity of dirty trade. The sign and size of this effect depends on the exporting nation's income and on the specific RTA's characteristics. For the average nation that participates in a RTA, its dirty exports income elasticity increases relative to its outside-bloc dirty exports income elasticity. For nations whose per-capita GDP is between $2,000 and $10,000, we find that RTA participation reduces its dirty exports income elasticity relative to its outside-bloc dirty exports income elasticity. We recognize that RTAs differ along a number of dimensions. Some RTAs such as the European Union are comprised of rich nations while other RTAs feature significantly greater income inequality within the bloc. To measure the consequences of these differences, we estimate an econometric model that allows us to examine how participation in thirteen different major RTA blocs affects the income elasticity of dirty exports. Across the thirteen RTA blocs in our sample, we find significant heterogeneity with respect to how bloc participation affects the income elasticity of dirty trade. In blocs that encourage within-bloc environmental regulatory convergence, we find empirical evidence supporting the FEH hypothesis. While many of our econometric estimates of RTA effects are statistically significant, their economic magnitude is surprisingly small.

Our study contributes to two parts of the trade and environment literature, one that examines the pollution content of trade flows and the other that investigates the environmental consequences of regional economic integration. Starting with Tobey's (1990) study, researchers have analyzed the implications of domestic environmental policies on trade flows (Ederington and Minier 2003, Grether and de Melo 2003, Kahn 2003, Levinson and Taylor 2004, and van Beers and van den Bergh 2000). Relatively little research has explicitly considered the environmental implications of trading bloc formation. Xu and Song (2000) and Maskus (2003) investigate dirty industry trade patterns while focusing on the economic integration of specific regions. Our study is unique in its emphasis on the composition of trade flows both within and across a broad variety of blocs including trading blocs among developed nations (North-North blocs), among developing nations (South-South blocs), and with a mixture of developed and developing nations (North-South blocs).

In the next section, we present a theoretical model of bilateral trade to show the conditions such that the poorer nation in the trading bloc could become the pollution haven. Section 2 presents our data sources. Section 3 discusses our empirical framework and results. Section 4 provides greater detail on the implications of our results for testing for pollution havens both inside and outside of regional trading blocs. Section 5 concludes.

1. A Model of Bilateral Trade in a Polluting Industry

1.1 Basic Structure

In this section, we provide a theoretical framework for analyzing the existence of pollution havens in trading blocs. We first examine sectoral bilateral trade flows for goods whose pollution intensities differ. We then look at how trading bloc formation between country pairs affects the bilateral trading patterns for different levels of pollution intensity.

We consider a stylized three-country trade model, in which each of the three countries, *i*, *j*, and *k*, has two sectors: agriculture (*A*) and manufacturing (*M*). The manufacturing sector is composed of a continuum of sub-sectors, indexed by $z \in [0,1]$. Each country has three basic types of production factors: labor (*L*), capital (*K*), and the environment (*E*). Capital in our model is used only in manufacturing. It consists of capital specific to individual sub-sectors, where $C(z)$ denotes the amount of capital that is specific to sub-sector z. The three countries are symmetric in *L*, but differ with respect to the aggregate level of capital *K*.[5] For simplicity, we assume that the three countries have the same distribution of aggregate capital among sub-sectors so that the ratio $s(z) \equiv C(z)/K$ is identical across the three countries for all z. The equilibrium size of *E*, that is supplied elastically, is determined endogenously in the model.

Sector *A* produces homogeneous goods in a perfectly competitive setting with a constant-return-to-scale technology, using only labor as an input. One unit of labor input produces one unit of agricultural product. This implies that the domestic price of agricultural product in Country i will be equal to the marginal cost, that is the wage rate w_i.

On the other hand, each manufacturing sub-sector z produces horizontally differentiated products in monopolistically competitive settings with increasing-return-to-scale technologies. The model's basic structure is a type of intra-industry trade model developed by Helpman and Krugman (1985) and others based on the monopolistic competition model of differentiated products in Dixit and Stiglitz (1977).[6]

---

[5] Later in Subsection 1.2, we will assign which country has the larger *K*.

[6] The product differentiation framework is appropriate for analyzing bilateral trade volumes in manufacturing sectors with a significant quantity of intra-industry trade between countries. As discussed

The development of a variety in sub-sector z requires one unit of specific capital as a fixed input. The factor price of specific capital is denoted as $\pi(z)$.[7] One unit of production of an individual variety requires one unit of labor input. The production process generates pollution as a by-product. Following the trade and environment literature, we characterize pollution emissions E as a factor of production. Pollution tax τ is the unit price producers have to pay to the government to emit pollution. While τ is given exogenously in this model, we follow the usual assumption that τ is increasing with the level of income. Furthermore, following Copeland and Taylor (1994), we incorporate a pollution abatement technology in the production function by characterizing the production function based on labor and environmental factors in a Cobb-Douglas form.[8] The exponent on environment in the production function is denoted as $\theta(z)$ and increases with z. This indicates that z is ordered according to the pollution intensity in production. We assume $\theta(z) \in [\underline{\theta}, \overline{\theta}]$ with $0 < \underline{\theta} < \overline{\theta} < 1$. The total cost of producing the output level $Y(z)_i$ in sub-sector z is therefore:

$$(1) \qquad TC(z)_i = \pi(z)_i + (\tau_i)^{\theta(z)} (w_i)^{1-\theta(z)} Y(z)_i$$

The representative consumer in each country has a utility function, that exhibits the two-tier budgeting process. The upper tier has the following form:

$$(2) \qquad U_i = \beta \int_0^1 \alpha(z) \ln X(z)_i dz + (1-\beta) \ln X_i^A - f(E_i)$$

where $X(z)_i$ is the sub-utility for sub-sector z's varieties, that will be defined below; X_i^A is the amount of agricultural consumption; and the last term $f(E_i)$ represents the negative utility from pollution generated by the domestically produced varieties with $f' > 0$. In equation (2), $\alpha(z)$ is sub-sector z's national income budget share and has the property $\int_0^1 \alpha(z) dz = 1$. The sub-utility for z's varieties takes a CES form, as:

$$(3) \qquad X(z)_i = \left\{ n_i [x(z)_{ii}]^{\frac{\sigma-1}{\sigma}} + n_j [x(z)_{ji}]^{\frac{\sigma-1}{\sigma}} + n_k [x(z)_{ki}]^{\frac{\sigma-1}{\sigma}} \right\}^{\frac{\sigma}{\sigma-1}}$$

---

by Bergstrand (1989), it provides a theoretical background for the gravity model in estimating the volume of sector-specific bilateral trade.

[7] Note that capital is used only for developing new varieties, and not for producing goods. As used in Forslid and Ottaviano (2003) and Baldwin and Krugman (2002), this type of simple formulation of the fixed cost allows us to derive closed form solutions for the endogenous variables.

[8] Some labor is used to abate pollution to reduce the use of environment (see Appendix 1 of Copeland and Taylor 1994).

where $x(z)_{ii}$, $x(z)_{ji}$, and $x(z)_{ki}$ represent the amount of z's individual varieties produced in i, j, and k respectively and consumed in i and; n_i, n_j, and n_k represent the numbers (mass) of varieties produced in each country, i, j, and k. The symbol σ is the elasticity of substitution between varieties in the same sub-sector. As this model characterizes horizontally differentiated varieties for each sub-sector z, we follow Dixit and Stiglitz (1977) in assuming that σ is greater than 1.

Each country imposes an *ad valorem* tariff on products imported from its partner countries. For example, Country i imposes a tariff of rate t_{ij} on products imported from Country j. Each tariff rate is given exogenously in this model. We assume that no tariff is imposed on agricultural products, thus A is the numeraire sector in this model. Perfect competition in the A sector sets the domestic price of agricultural products equal to w_i. Free trade in agricultural products equalizes their price across the three countries, and therefore equalizes w_i internationally. We normalize w_i to 1.

For consumer and producer behavior in the monopolistically competitive manufacturing sub-sectors, we follow the well-known Dixit-Stiglitz derivation. Free entry and exit in the monopolistically competitive sub-sectors make the producers set the prices of their varieties equal to $\sigma/(\sigma-1)$ times their marginal costs. Consumer prices will adjust with the tariff rate for imported varieties. With the wage rate being normalized, sub-sector z's consumer price in Country i will be: $P(z)_{ii} = \tau_i^{\theta(z)}\sigma/(\sigma-1)$ for the domestic varieties and $P(z)_{ij} = (1+t_{ij})\tau_i^{\theta(z)}\sigma/(\sigma-1)$ for the imported varieties from Country j. Due to the full employment of capital, the number (mass) of varieties in z produced in Country i will be equivalent to the total stock of capital specific to that sub-sector in the country so that $n_i = C(z)_i = s(z)K_i$.

The representative consumer maximizes utility based on (2) subject to spending national income I. This yields consumer demand. Thus, for example, individual producers in Country i face the following demand for their varieties in Country j.

$$x(z)_{ij} = \alpha(z)\beta I_j \left[P(z)_{ij}\right]^{-\sigma} / \left[Q(z)_j\right]^{1-\sigma} \tag{4}$$

where $Q(z)_j$ is the price index of sub-sector z varieties in Country j:

$$Q(z)_j = \frac{\sigma}{\sigma-1}\left[s(z)\left(\phi_{ij}\tau_i^{\theta(z)(1-\sigma)}K_i + \tau_j^{\theta(z)(1-\sigma)}K_j + \phi_{jk}\tau_k^{\theta(z)(1-\sigma)}K_k\right)\right]^{\frac{1}{1-\sigma}} \tag{5}$$

The parameter $\phi_{ij} \equiv (1+t_{ij})^{1-\sigma}$ represents the degree of trade openness between Countries i and j. These trade openness parameters are bounded below by 0 (prohibitive tariff) and above by 1 (free trade). The national income I is the sum of all factor revenues (including pollution tax revenue τE) plus total tariff revenue Ω. Both tax and tariff revenues are redistributed lump-sum. Thus, for Country i,

$$(6) \qquad I_i = L_i + K_i \int_0^1 \pi(z)_i s(z) dz + \tau_i E_i + \Omega_i$$

The model is completed by properly determining the endogenous variables. The predetermined factors in the model are the aggregate capital K and its constituents $C(z)$, i.e., capital specific to z. The return to the specific capital $\pi(z)$ can be determined by the fact that it constitutes the fixed cost in manufacturing sub-sector z. Since the total markup is $1/\sigma$ of total revenue, the zero profit condition of monopolistic competition ensures that the fixed cost $\pi(z)$ is the total markup. Using Shepard's Lemma, the optimal level of E can be derived. The revenues from the pollution tax and tariff collection can be readily derived from the equations presented above. The income level for each country is then computed based on the parameter values.

1.2 Pollution Havens in Bilateral Trade

Using the model outlined above, we now analyze bilateral trade flows between Country i and Country j in the three-country setting. We first analyze how they differ for z's produced at different levels of pollution intensity without imposing any assumption on trading bloc formation. In the next sub-section, we will examine how trading bloc formation affects trading patterns. We assume Country i is more abundant in capital ($K_i > K_j$), and has higher level of income than Country j. We also assume that the income difference implies that the pollution tax in Country i is higher ($\tau_i > \tau_j$).

First, we derive the aggregate exports of manufacturing sub-sector z from Country i to Country j. The producer's export revenue from an individual variety of sub-sector z produced in Country i and exported to Country j is the product of the pre-tariff producer price (the domestic price in Country i) and the post-tariff demand for such variety in Country j.

$$(7) \qquad TR(z)_{ij} = x(z)_{ij} P(z)_{ii} = \frac{\alpha(z)\beta I_j \phi_{ij}}{s(z)\left\{\phi_{ij} K_i + K_j \left(\tau_i / \tau_j\right)^{\theta(z)(\sigma-1)} + \phi_{ik} K_k \left(\tau_i / \tau_k\right)^{\theta(z)(\sigma-1)}\right\}}$$

We multiply (7) by the total number of z's varieties produced in Country i to get the expression for the aggregate exports of z's varieties from Country i to Country j.

$$(8) \qquad EX(z)_{ij} = n(z)_i TR(z)_{ij} = \frac{\alpha(z)\beta I_j \phi_{ij} K_i}{\phi_{ij} K_i + K_j \left(\tau_i / \tau_j\right)^{\theta(z)(\sigma-1)} + \phi_{ik} K_k \left(\tau_i / \tau_k\right)^{\theta(z)(\sigma-1)}}$$

Equation (8) provides a gravity-model-like interpretation of bilateral trade flows, where the magnitude of bilateral trade is positively related to I_j, that represents the economic size of the importing country in terms of purchasing power, and to K_i, that represents the economic size of exporting country in terms of production capacity. At the same time, it is negatively related to the trade friction between the two countries (the inverse of ϕ_r). In addition to the traditional gravity model features, equation (8) indicates that an increase in the exporting country's pollution tax reduces the sub-sectoral bilateral trade flow.[9]

Now, we consider how bilateral trade between Country i and Country j varies with pollution intensity in industries (sub-sectors) before the two countries form a trading bloc. To see how the sub-sector-specific volume of exports from i to j changes with pollution intensity, we differentiate the log of (8) with respect to the pollution intensity parameter $\theta(z)$.

$$\frac{\partial \ln EX(z)_{ij}}{\partial \theta(z)} = \frac{(\sigma - 1)\Delta}{\Psi} \tag{9}$$

where

$$\Delta \equiv \left[(K_j/K_i)(\tau_i/\tau_j)^{\theta(z)(\sigma-1)} \ln(\tau_j/\tau_i)\right] + \left[\phi_{ik}(K_k/K_i)(\tau_i/\tau_k)^{\theta(z)(\sigma-1)} \ln(\tau_k/\tau_i)\right] \tag{10}$$

and

$$\Psi \equiv \phi_{ij} + (K_j/K_i)(\tau_i/\tau_j)^{\theta(z)(\sigma-1)} + \phi_{ik}(K_k/K_i)(\tau_i/\tau_k)^{\theta(z)(\sigma-1)} \tag{11}$$

The sign of (9) indicates whether Country i increases or decreases its exports to Country j as pollution intensity of z increases. The sign of the expression is determined by Δ. Note that when the two countries are sufficiently closed to the third country so that ϕ_{ik} is close to 0, the sign of the expression is determined by the first term of Δ. The model virtually becomes a two-country model. Since Country i is richer ($\tau_i > \tau_j$) by assumption, (9) will be negative, indicating that the sub-sectoral bilateral exports from Country i to Country j decreases with more pollution intensity of z. On the other hand, the term Δ would be positive for the exports from Country j to Country i in the equivalent expression to (9). Therefore, the richer trading partner exports fewer "dirty" sub-sector goods as compared to its exports of "clean" sub-sector goods. The poorer trading partner, on the contrary, exports more in "dirty" sub-sectors.

[9] As Copeland and Taylor (2003) discuss, the PHH will not be observed if the negative impact of the pollution tax on the trade flow is offset by the difference in endowment of other factors such as capital.

In a more generic setting with three countries, the results would depend on the size of the third country, or the rest of the world, relative to the two countries. The third country does trade with the two countries separately, which may indirectly influence the pattern of trade between those two countries. When the third country is much smaller than the two countries so that K_k/K_i is sufficiently small, the outcome would be similar to the two-country setting. The richer country (Country i) exports more in "clean" sub-sectors as compared to "dirty" sub-sectors. This scenario is most likely when both i and j are developed countries (North-North trade). Even when the third country is not very small, the same pattern emerges as long as the richer country i is richer than the third country so that $\tau_i > \tau_k$. This may be typical for trade between a developed country and a developing country (North-South trade). Thus in both North-North and North-South bilateral trade relations, the richer nation exports more clean goods while poorer nations export more dirty goods. In this case, the poorer country again becomes the pollution haven.

On the other hand, Δ could become positive for trade between developing countries (South-South trade) where the third country is more capital abundant than the two countries so that K_k/K_i is sufficiently large. This is because the third country k, which could be considered as a group of developed countries, is richer and has a higher level of the pollution tax. Bilateral exports from Country i to Country j increase as z's pollution intensity rises, even though Country i has a higher pollution tax than Country j. In fact both richer and poorer countries export more in the pollution-intensive sub-sectors as compared to the cleaner sub-sectors.[10] In this case, no PHH pattern would be observed in a South-South bilateral trade relation.

1.3 Pollution Havens in Bilateral Trade as Trading Blocs are Formed

Bloc formation leads to a simultaneous reduction in tariff rates between i and j, or increases in ϕ_{ij} and ϕ_{ji}, while keeping the tariff rates toward the third country fixed. Taking the cross derivative of equation (9) with respect to ϕ_{ij} yields the following expression.

$$(12)\qquad \frac{\partial \ln EX(z)_{ij}}{\partial\theta(z)\partial\phi_{ij}} = -\frac{(\sigma-1)\Delta}{\Psi^2}$$

The sign of (12) is the opposite of the sign of Δ. Bloc formation decreases the tariff between Countries i and j. This causes the volume of sub-sector-specific exports to

[10] The intuition behind this outcome is the following. As the pollution intensity increases, the third country's varieties reduce their market share in Country j since the higher pollution tax in the third country make its varieties more expensive than the varieties produced in i and j, thus reducing the demand it faces in Country j's market compared to the varieties produced in i and j. This increases the market share for Country i's varieties in Country j's market.

vary less with pollution intensity as compared to the case before the bloc was formed. For the case of North-North and North-South bilateral trade relations, where we see the PHH pattern before trading blocs are formed, this implies that the richer members' dirty exports expand more as the bloc is formed, while the poorer members' clean exports expand. Thus, the formation of a regional trading bloc generally weakens the PHH pattern that exists between the two countries in North-North or North-South trade relation. Furthermore, note that a decrease in the pollution tax differential τ_i/τ_j toward 1 also reduces the expression in (9). If bloc formation causes a convergence in the pollution tax rate between the two countries, then the bloc formation further reduces the impact of pollution intensity on the volume of trade. Both the intra-bloc tariff reduction and the convergence in the pollution tax ratio weaken the PHH patterns since both essentially weaken the cost advantage of the poorer exporter in pollution intensive industries vis-à-vis its richer bloc partner. It is important to note, however, that the bloc formation effect does weaken the cross-sub-sectoral variation of trade volumes due to pollution intensity but does not reverse the direction of how sub-sectoral trade volumes vary with pollution intensity. On the other hand, for South-South trade, where there is no PHH pattern initially, the smaller variation in pollution intensity implies that the two bloc member trading countries both expand their clean goods exports.

The PHH weakening effect is not necessarily observed when the third country's tariff is reduced. Equation (13) shows the cross derivative of equation (9) with respect to ϕ_{ik}.

$$(13) \quad \frac{\partial \ln EX(z)_{ij}}{\partial\theta(z)\partial\phi_{ik}} = \frac{(\sigma-1)}{\Psi^2}\left(\frac{K_k}{K_i}\right)\left(\frac{\tau_i}{\tau_k}\right)^{\theta(z)(\sigma-1)}\left\{\left(\frac{K_j}{K_i}\right)\left(\frac{\tau_i}{\tau_j}\right)^{\theta(z)(\sigma-1)}\ln\left(\frac{\tau_k}{\tau_j}\right)+\phi_{ij}\ln\left(\frac{\tau_k}{\tau_i}\right)\right\}$$

Note that the cross derivative would be negative when both Countries *i* and *j* have higher pollution taxes than Country *k*. Thus, for North-North bilateral trade, the exports of richer and poorer partners both expand more in clean goods as they become more open to the third country. For South-South bilateral trade, where both countries have lower pollution tax than the third country, both countries' exports of dirty goods would increase as they become more open to the third country. Multilateral trade liberalization, such as accession to the WTO, can be thought of as tariff reduction toward the third country. Our model shows that this enhances the PHH pattern between North and South.

In sum, the model suggests that North-North and North-South trading blocs weakens the PHH pattern that exists prior to bloc formation. The bloc-induced expansion of richer members' exports is biased toward dirty goods, while poorer members' export expansion is biased toward clean goods. On the other hand, the bloc formation between two Southern countries, where there is no explicit PHH pattern initially, expands clean exports in both countries. The trade liberalization outside of trading bloc would enhance the PHH pattern between North and South.

2. Data

2.1 Trade Data

Our national bilateral trade data source is Robert Feenstra's World Trade Flows, 1980-1997. For each importing and exporting nation in each year, the data set provides information on the volume of trade in 34 different manufacturing industries. Trade is measured in thousands of nominal dollars. We deflate this by the producer price index to measure trade in real year 1995 dollars.[11]

In our study, the unit of analysis is a trading pair/year/industry. We have data for 128 nations for 18 years across 34 industries. We work with an unbalanced panel data set that has over 1.7 million observations for 11,434 pairs of trading nations. We choose not to balance this panel because this would increase the number of observations to over 8 million and would raise a host of computation challenges in estimating the multivariate regressions we present below. Thus, in our regressions we drop observations if the World Trade Flows data provides no data or indicates zero trade for a particular trading pair/year/industry. Our gravity regressions are based on positive trade flows between trading partners. For a useful discussion on this point, see Dee and Gali (2003).

The set of nations included in our trade data set are listed in the Appendix, together with their RTA membership information.[12] We investigate total 13 RTAs, that include: European Union (EU), European Free Trade Agreement (EFTA), Central American Common Market (CACM), Caribbean Community (CARICOM), North American Free Trade Agreement (NAFTA), Latin America Integration Agreement (LAIA), Common Market of the South (MERCOSUR), Asia-Pacific Economic Cooperation (APEC), Bangkok Agreement, Andean Community (CAN), ASEAN - Association of Southeast Asian Nations- Free Trade Area (AFTA), Common Market for East and Southern Market (COMESA), and Southern African Development Community

[11] Feenstra (2000) provides a detailed explanation for how the data were created. "The main source for bilateral trade data is the United Nations Statistical Office, which collects the trade data from individual countries, providing them with an international trade classification into which they are to fit their own national systems as best they can. The United Nations (UN) then publishes these data in summary form in the Yearbook of International Trade Statistics and in the full reported detail in Commodity Trade Statistics, but does not attempt to supplement the reported data or make them consistent across countries or years. Statistics Canada has recompiled the UN trade data on a consistent basis, using a version of the Standard International Trade Classification (SITC) Revision 2, modified to match Canadian trade and industry classifications and to accommodate some of the variation in the way countries report their data (Feenstra 2000 page 2)." Feenstra's data sums the Statistics Canada's data according to a 34 manufacturing industry basis used by the U.S. Bureau of Economic Analysis.

[12] The membership data for RTAs incorporates the changes in membership within the RTAs. We have combined the Canada-US Free Trade Agreement (CUSFTA) and NAFTA as a single RTA unit.

(SADC).[13] By 1997, 65% of nations in the trade database are participating in at least one RTA.

For the set of nations in our sample, Figure 1 shows the time trend in the share of world trade taking place within these 13 RTAs. The 1989 formation of APEC leads to a jump in the graph. Growth is observed in the 1990s such that by 1997 almost 65% of world trade takes place within these blocs.

Figure 1: Time Trends in the Share of World Trade within 13 Regional Blocs

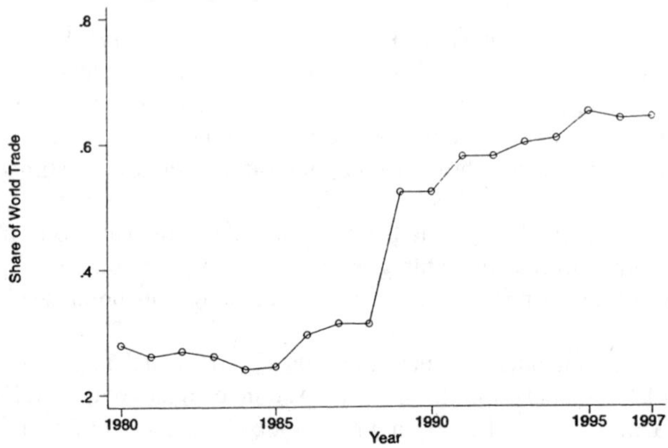

2.2 Measuring Industry Pollution Content

The World Trade Flows database identifies 34 separate industries. These industries are listed in Table 1. For each industry, we create a measure of its energy consumption per dollar of value added. This index is constructed using data from the NBER Manufacturing Industry Productivity Database (Bartelsman and Gray 1997). We take the NBER data set for industries in the United States and construct using 1990 data real energy consumption per dollar of value added for each 4 digit SIC industry. In the tables below this is called the "energy index". Across the 459 4 digit SIC industries, the correlation of this energy index and the industry's capital to labor ratio was 0.75 in 1980 and 0.57 in 1990.

[13] The 13 RTAs are composed of customs unions (EU, CACM, CARICOM, MERCOSUR), free trade agreements (EFTA, NAFTA), preferential agreements (LAIA, CAN, AFTA, COMESA), and other regional integration agreements (SADC, APEC). APEC and SADC are included in our list of RTAs although they are not filed with the WTO as regional trade agreements. While they are not legally binding agreements for tariff reduction, they are active policy forums for the member countries to coordinate trade and investment policies and facilitate trade and investment relations among members.

Table 1: Industry Differences in Energy Intensity

Name	Energy Index	Dirty Industry
Grain, Mill and Bakery Products	0.039	0
Beverages	0.021	0
Tobacco Products	0.008	0
Other Food and Kindred Products	0.041	0
Apparel and Other Textile Products	0.043	0
Leather and Leather Products	0.02	0
Pulp, Paper and Board Mills	0.178	1
Other Paper and Allied Products	0.037	0
Printing and Publishing	0.014	0
Drugs	0.014	0
Soaps, Cleaners and Toliet Goods	0.012	0
Agricultural Chemicals	0.107	1
Industrial Chemicals and Synthetics	0.125	1
Other Chemicals	0.032	0
Rubber Products	0.043	0
Miscellaneous Plastic Products	0.053	0
Primary Metal Industries, Ferrous	0.161	1
Primary Metal Industries, Nonferrous	0.179	1
Fabricated Metal Products	0.039	0
Farm and Garden Machinery	0.018	0
Construction, Mining, etc	0.023	0
Computer and Office Equipment	0.009	0
Other Nonelectric Machinery	0.024	0
Household Appliances	0.023	0
Household Audio and Video	0.011	0
Electronic Components	0.022	0
Other Electrical Machinery	0.025	0
Motor Vehicles and Equipment	0.023	0
Other Transportation Equipment	0.016	0
Lumber, Wood, Furniture	0.042	0
Glass Products	0.102	1
Stone, Clay Concrete, Gypsum	0.104	1
Instruments and Apparatus	0.014	0
Other Manufacturing	0.022	0
Unweighted Mean and s.d. for 34 Industries	0.048	0.049

Note: The Energy Index is a percentage and is defined in the text.

The World Trade Flows database lists the three digit SIC industries that comprise each of the listed 34 industries. We use this information to calculate the energy index for each of the 34 industries. Table 1 presents each industry's energy index. Industries with high energy indices (over 0.10) include; pulp and paper, agricultural chemicals, industrial chemicals, primary metals, glass products and stone and clay.

It is important to note that our index of industry environmental impact is based on data from the United States. Due to capital vintage differences or nation specific regulatory differences, emissions per unit of economic activity may differ across nations. But, it is reasonable to posit that the pollution intensity rankings of industries will be similar across nations. Industries that are high polluters in the United States are likely to be high polluters in other nations.

Table 2: "Dirty" Industries are Highly Polluting

Environmental Indicator	Mean for Dirty Industries	Mean for Non-Dirty Industries
Toxic Air Releases	5265.017	1132.000
Toxic Land Releases	12412.570	853.195
Total Toxic Releases	18812.480	2014.658
Toxic Water Releases	1134.883	29.458
Metal Air Releases	189.700	4.441
Metal Land Releases	5340.283	186.835
Metal Water Releases	20.400	0.624
Total Suspended Solids	108065.700	1542.568
Nitrogen Dioxide	11819.930	1129.647
Particulate Matter (PM10)	2580.467	151.479
Sulfur Dioxide	36582.180	1328.768
Carbon Monoxide	31873.850	1067.179
Volatile Organic Compounds	5921.717	2291.168

Note: The definition of Dirty Industries is given in Table 1. Non-dirty industries represent all industries not classified as "Dirty" in Table 1. The units are pollution in pounds per $ million dollars of value in 1987 U.S dollars. The mean for each set of industries is the unweighted average across the industries in that set.

We recognize that many possible indices may be used for measuring an industry's pollution intensity. To demonstrate that our energy index is positively correlated with other indicators of industry emissions, we use data from the World Bank's World Bank/PRDEI Industrial Pollution Projections Project created by Mala Hettige, David Shaman, David Wheeler and David Witzel.[14] This research team has

[14] See http://wb.forumone.com/niprcd/ippsdata.htm

constructed emissions intensity indicators for 25 of the 34 industries in the World Trade Flows data base. In Table 2, we report the unweighted means for each of the World Bank's pollution indicator for industries that we assign to "dirty", the subset whose energy index is greater than 0.10, versus all other industries. For every World Bank indicator, with the exception of volatile organic compounds, the dirty industry set's average emissions factor is over ten times larger than the average emissions factor for the clean industries. This huge differential raises our confidence that the energy index is highly positively correlated with other measures of pollution.

To provide some facts concerning trends in the pollution intensity of world trade, Figure 2 reports time trends in the average energy intensity of national exports over the years 1980 to 1997.[15] The figure shows that between 1980 and 1997, the pollution content of average world trade has fallen over time. Over this time period, the average energy index declined by 25%. Again, we note that this figure is based on constant industry index weights, thus any reduction in average energy intensity across years reflects a composition shift toward the cleaner industries listed in Table 1.

Figure 2: The Energy Intensity of World Trade From 1980 to 1997

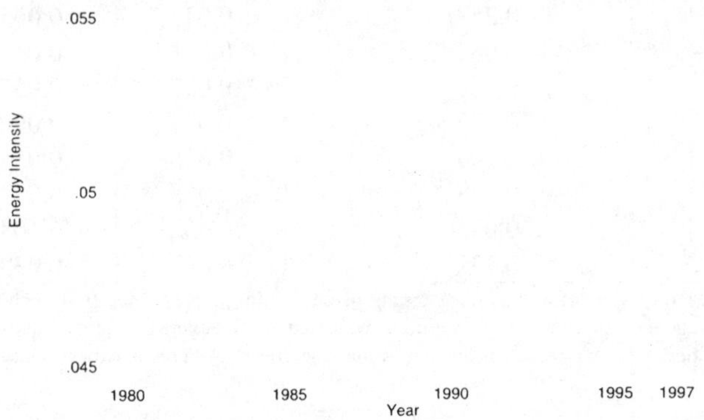

We suspect that this average trend masks much heterogeneity. In Table 3, for each of the thirteen RTA blocs in our sample, we report the 1997 average energy intensity of within-bloc exports and outside-bloc exports by bloc members. For example, the NAFTA bloc consists of the United States, Canada and Mexico. For this bloc, outside-bloc exports represent any exports from these three nations to nations that are not in NAFTA. We seek to establish whether the average energy intensity of trade is higher inside or outside of the bloc. For five of the thirteen blocs, the within-bloc

[15] Define E_j as industry j's energy intensity in the year 1990 where j covers all 34 industries in the trade data base. Define $Exports_{jt}$ as industry j's share in the total exports in year t where the shares sum to one within each year. We construct $\sum_j Exports_{jt} * E_j$. This is graphed in Figure 2.

average energy intensity of exports is greater than the outside-bloc average energy intensity of exports. In the Bangkok bloc, for example, the average energy intensity of within-bloc exports is 0.071, while for Bangkok members the average energy intensity of exports to non-Bangkok members is 0.042. For other blocs such as the EU, NAFTA, and APEC, the two energy intensity averages are quite similar.

Table 3: Trade Facts within RTA Blocs

RTA Bloc	Within-bloc Exports as a Fraction of Total Bloc Member Exports Using 1997 Trade Data	Average Energy Intensity	
		Within-Bloc Exports	Outside of Bloc Exports
All 13 RTAs	0.647	0.046	0.043
EU	0.638	0.049	0.042
EFTA	0.141	0.059	0.063
NAFTA	0.486	0.047	0.042
CACM	0.181	0.057	0.040
CARICOM	0.092	0.048	0.095
MERCOSUR	0.258	0.047	0.065
Bangkok	0.082	0.071	0.042
LAIA	0.183	0.057	0.057
CAN	0.122	0.070	0.090
AFTA	0.204	0.036	0.038
COMESA	0.048	0.058	0.074
SADC	0.076	0.062	0.108
APEC	0.732	0.043	0.039

Note: Within-bloc Exports is defined as trade taking place within the RTA bloc. For such trade, we calculate the average energy intensity. In addition, we calculate the average energy intensity of exports of RTA members who export to nations not in that specific RTA. This is called "Outside of Bloc Exports".

2.3 Additional National Data

The Penn World Tables Version 6.1 provides annual information on each nation's real GDP per-capita and on its population level (Heston, Summers and Aten 2002). Throughout this paper, we proxy for regulatory stringency using national annual per-capita income. We do not have national yearly data on the intensity of environmental regulation. Instead, our approach implicitly adopts a regulatory "J" curve argument, that poorer nations impose less stringent regulation (Selden and Song 1995). From Andrew Rose's database, we use information on whether a bilateral trading pair was ever in a colonial relationship, whether they speak a common language, and their distance from each other.

3. Results

3.1 Bilateral Trade Regressions and Measures of the Aggregate RTA Effect

We begin our econometric analysis by estimating a simple bilateral trade gravity model that resembles equation (8) in Section 1. Corresponding to equation (9), we seek to estimate how the composition of trade varies as a function of the exporting nation's income and the industry's energy index. We estimate this for both trading pairs that are and are not in the same RTA bloc in order to consider the effect of bloc formation. Equation (14) represents our first econometric specification.

$$\begin{aligned} \ln(Trade_{ijmt}) &= c + B_1 \ln(GDP_{it}) + B_2 \ln(GDP_{jt}) + B_3 X_{ij} + B_4[\ln(GDP_{it}) \cdot E_m] \\ &+ B_5[\ln(GDP_{it}) \cdot E_m \cdot RTA_{ijt}] + B_6[\ln(GDP_{it}) \cdot RTA_{ijt}] + B_7 E_m \\ &+ B_8 \ln(Pop_{it}) + B_9 \ln(Pop_{jt}) + \propto_{ij} + U_{ijmt} \end{aligned} \tag{14}$$

In equation (14), the dependent variable is the log of the volume of trade between exporter nation i and importer nation j in industry m in year t. In estimating equation (14), the omitted category is trade between two nations that are not in the same RTA bloc. In equation (14), E_m is industry m's energy index as shown in Table 1. In equation (14), GDP_{it} and GDP_{jt} represent per-capita GDP of the exporting nation i and of the importing nation j respectively in year t. Likewise, Pop_{it} and Pop_{jt} represent population counts of the exporting nation i and of the importing nation j respectively in year t. If any two nation trading pairs are members of the same Regional Trade Agreement, then the RTA dummy equals one. This dummy equals zero if two nations are not members of the same RTA. This specification imposes the assumption of a homogenous RTA membership effect on the composition of within-bloc trade.[16] In one regression, we include $\propto_{ij}$, which is a pair-specific fixed effect between exporter nation i and importer nation j. This fixed effect controls for the pair's geography, historical relationship, and any non-RTA related factors. When we do not include pair fixed effects, we include pair-specific time invariant variables, X_{ij}, such as the log of distance between the nations, and dummies indicating whether both countries in the trading pair speak the same language and ever were in a colonial relationship.

Table 4 presents five OLS estimates of equation (14). The first two columns report the results based on the full sample. Column (1) does not include trading pair

[16] We must note that the specification reported in equation (14) does not include RTA membership as a separate explanatory variable and then include a full set of interactions of this variable with the energy index, and the exporter's GDP per-capita. To make the estimates of equation (14) comparable to the results we present in the next section, where we disaggregate the RTA effect into 13 separate blocs, we sought a more parsimonious specification.

fixed effects while column (2) does. The negative and statistically significant coefficient estimate of B_4 implies that outside of RTA blocs, nations have a greater income elasticity of exports of clean goods relative to dirty goods. We find a small positive interaction term for B_5. All else equal, RTA bloc participation increases an exporter's income elasticity of dirty trade but this coefficient's economic magnitude is quite small. Note that both regressions have very large positive coefficients for the energy index coefficient (B_7). Since our trade variable is measured in value, this indicates that energy intensive goods on average generate more export revenue.
In columns (3), (4), and (5) we split exporters into three groups based on real GDP per capita of the exporters. We classify them as "low income" (less than \$2,000), "middle income" (between \$2,000 and \$10,000 per-capita), and "high income" (more than \$10,000). Interesting, statistically significant results are found both inside and outside of RTA blocs as shown by the sign changes for the coefficients on B_4 and B_5.

Outside of RTA blocs (B_4), both low- and high-income groups have negative coefficients (-5.55 and -4.32 respectively), whereas middle-income group has a positive coefficient (0.93).[17] Thus, outside of RTA blocs, economic growth for low or high-income exporters reduces their exports of dirty goods, while middle-income exporters expand their exports of dirty goods as they grow richer. The coefficient B_5 measures how RTA participation affects the exporter's dirty trade income elasticity. Note that for all three income groups, the sign for the B_5 coefficient is the opposite of the sign for the B_4 coefficient and smaller in absolute value. Thus, the effect inside an RTA ($B_4 + B_5$) is the same sign but smaller in absolute value than the effect outside an RTA (B_4). This indicates that forming a RTA between two countries works to weaken the exporter's economic growth effect on the pollution content of its bilateral exports regardless of the direction of such an effect. However, in absolute value, the coefficient estimates of B_5 are much smaller than the estimates of B_4.

We use these estimated coefficients to construct income elasticities for dirty goods and clean goods exports for different types of nations. We present such income elasticity estimates for trading pairs that are and are not members of the same RTA bloc. Recall from Table 1 that the cleanest industries had an energy index of roughly 0 while the dirtiest industries had an energy index equal to 0.15. For a trading pair that is not in a RTA together, the exporter's income elasticity of dirty goods exports equals:

[17] If we drop China, India and Pakistan from the sample, this removes 22,000 observations from the results reported in Table 4's column (4). In this case, we estimate a negative B_4 coefficient that is borderline significant. We have also re-estimated columns (3-5) of Table 4 where we include 34 industry fixed effects and find quite similar findings. Given the importance of China, India and Pakistan as trading partners, we have re-estimated equation (14) only including observations where one of these three nations is the exporter. We find that the volume of trade with richer importing nations increases as the energy intensity of exports increases. This is consistent with the pollution haven hypothesis.

Table 4: Bilateral Gravity Trade Regressions and the Average RTA Effect

	Exporter Set				
	(1) All	(2) All	(3) Low-Income	(4) Middle-Income	(5) High Income
B_4, on ln(Exporter GDP per-capita) · Energy Index	-2.077 (0.137)	-1.975 (0.036)	-5.550 (0.818)	0.929 (0.471)	-4.316 (0.499)
B_5, on RTA · Energy Index · ln(exporter GDP per-capita)	0.092 (0.031)	0.115 (0.010)	0.302 (0.126)	-0.097 (0.057)	0.136 (0.036)
B_6, on RTA · ln(exporter GDP per-capita)	0.085 (0.006)	0.033 (0.001)	-0.016 (0.021)	0.101 (0.010)	0.073 (0.007)
B_7, on Energy Index	22.623 (1.288)	21.668 (0.337)	43.692 (5.943)	-0.902 (4.025)	44.025 (4.895)
ln(Exporter GDP per-capita)	1.398 (0.018)	1.082 (0.015)	1.011 (0.070)	0.892 (0.053)	1.897 (0.086)
ln(Importer GDP per-capita)	0.854 (0.014)	0.940 (0.013)	0.458 (0.043)	0.619 (0.022)	1.012 (0.018)
ln(Exporter population)	0.636 (0.010)	-0.016 (0.032)	0.386 (0.023)	0.522 (0.016)	0.706 (0.014)
ln(Importer population)	0.487 (0.01)	-0.883 (0.029)	0.316 (0.023)	0.375 (0.016)	0.560 (0.011)
ln(distance)	-0.822 (0.018)		-0.639 (0.068)	-0.596 (0.025)	-0.877 (0.023)
Same language	0.382 (0.038)		0.202 (0.086)	0.173 (0.057)	0.460 (0.052)
Colony	0.890 (0.087)		0.004 (0.142)	0.394 (0.160)	1.144 (0.098)
Time Trend	-0.012 (0.001)	0.037 (0.001)	-0.022 (0.003)	-0.005 (0.002)	-0.018 (0.002)
Constant	-18.648 (0.351)	-3.758 (0.497)	-8.606 (0.721)	- 12.099 (0.631)	-25.704 (0.909)
Observations	1737800	1737800	130396	517443	1089961
Adjusted R^2	0.416	0.533	0.177	0.230	0.490
Pair Fixed Effect	No	Yes	No	No	No

Note: Each column reports a separate OLS estimate of equation (14) in the text. The dependent variable is trade between country *i* and importing nation *j* in year *t* for industry *m*. The dependent variable is the log of trade measured in \$1,000s of 1995 U.S dollars. RTA equals one if the trading pair are both members of one of the thirteen RTAs listed in Table 3. The omitted category is a trading pair that is not in a RTA bloc together and has not been in a colonial relationship and do not speak a common language. In column (3), only trading pairs where the exporter's real GDP per-capita is less than or equal to \$2,000 are included in the estimation. In column (4), only trading pairs where the exporter's real GDP per-capita is greater than or equal to \$2,000 and less than \$10,000 are included in the estimation. In column (5), only trading pairs where the exporter's real GDP per-capita is greater than or equal to \$10,000 are included in the estimation. Robust standard errors are reported in parentheses.

$B_1 + B_4 \cdot 0.15$ and the income elasticity for the cleanest goods exports is B_1. For a trading pair in the same RTA, the exporting nation's income elasticity of dirty goods exports equals: $B_1 + (B_4 + B_5) \cdot 0.15 + B_6$ while the exporting nation's income elasticity of the clean good exports equals $B_1 + B_6$. Based on these formulas, we report trade income elasticities in Table 5. All of the elasticities reported in the table are positive. With the exceptions of clean goods between low- and middle-income countries, we see higher elasticities as the level of income increases. With the exception of clean exports by low-income exporters, RTA participation generally increases the income elasticity.

Table 5: Estimated Dirty and Clean Export Income Elasticities

	(1) Full Sample	(2) Low-Income	(3) Middle-Income	(4) High-Income
Elasticities				
(a) Dirty Industry and RTA Pair	1.185	0.208	1.117	1.344
(b) Dirty Industry and non-RTA Pair	1.086	0.179	1.031	1.250
(c) Clean Industry and RTA Pair	1.483	0.995	0.993	1.971
(d) Clean Industry and non-RTA Pair	1.398	1.011	0.892	1.897
Elasticity Differentials				
(e) Dirty-Clean Differential for RTA Pair (a)-(c)	-0.298	-0.787	0.125	-0.627
(f) Dirty-Clean Differential for non-RTA Pair (b)-(d)	-0.312	-0.833	0.139	-0.647
(g) RTA Effect on Dirty-Clean Differential|(e)|-|(f)|	-0.014	-0.045	-0.015	-0.020

Note: This table reports income elasticites based on the regression results reported in Table 4. Dirty industry indicates that the energy index (presented in Table 1) equals 0.15 and clean industry indicates that the energy index equals 0. The column labeled "Full Sample" is generated using the output reported in Table 4's column (2). The column labeled "Low-Income" is generated using the output reported in Table 4's column (3). The column labeled "Middle-Income" is generated using the output reported in Table 4's column (4). The Column labeled "High-Income" is generated using the output reported in Table 4's column (5).

As shown in the first column of Table 5, overall a 10% increase in per-capita income increases a RTA bloc member's clean exports by 14.83% and its dirty exports by 11.85%. For non-RTA members, these figures are 13.98% and 10.86%. Regardless of whether a trading pair is in a RTA, the income elasticity of dirty exports is less than the income elasticity of clean exports. This finding is consistent with an overall PHH ignoring the RTA effect. Moreover, if we compare RTA and non-RTA pairs, the former elasticity differential is smaller than the latter's. RTA formation shrinks the PHH effect observed outside of RTA blocs. We must acknowledge that this estimated effect, which is based on statistically significant coefficient estimates, is quite small.[18] One plausible explanation for this weak RTA effect is that RTAs are formed between nations with similar capital endowments. Increased mobility of capital within these blocs could further reduce the endowment difference. Another relevant factor is the increasing level of intra-industry trade that is observed particularly among developed economies in recent years regardless of RTA participation.

To summarize these results, regardless of RTA formation, both high- and low-income exporters have greater clean good export elasticities relative to their dirty goods export elasticity. For middle-income exporters, we find the opposite; this group has a greater income elasticity of dirty exports relative to clean exports. For all three income groups, the formation of RTAs slightly shrinks the differences in elasticities between clean and dirty exports.

3.2 The Composition of RTA Bloc Trade Disaggregated by Bloc

In our second bilateral trade specification, we disaggregate the RTA effects by allowing each of the thirteen different RTAs listed in Table 3 to have its own coefficient on $\ln(GDP_{it}) \cdot E_m$ and on E_m. This rich set of interaction terms allows us to test how the composition of energy intensive trade varies by RTA bloc. In which blocs are richer nations the pollution havens and in which blocs are poorer nations the pollution havens? How do the income elasticities of trade content differ for non-RTA members and RTA members? To answer these questions, we estimate the regression presented in (15).

$$
\begin{aligned}
\ln(Trade_{ijmt}) &= c + B_1 \ln(GDP_{it}) + B_2 \ln(GDP_{jt}) + B_3 X_{ij} + B_4[\ln(GDP_{it}) \cdot E_m] \\
(15) \qquad &+ \sum_l B_{5,l}[\ln(GDP_{it}) \cdot E_m \cdot RTA_{ijlt}] + \sum_l B_{6,l}[\ln(GDP_{it}) \cdot RTA_{ijlt}] + B_7 E_m \\
&+ B_8 \ln(Pop_{it}) + B_9 \ln(Pop_{jt}) + \alpha_{ij} + U_{ijmt}
\end{aligned}
$$

In equation (15), RTA_{ijlt} is a dummy variable that equals one if two nations i and j in the year t are in the lth regional trading bloc.[19] With the exception of the RTA-specific interaction terms, equation (15) is identical to equation (14). In this specification we include 26 RTA interaction terms, with two interactions (B_5 and B_6) for each of the thirteen RTAs. The omitted category is a trading pair that is not a member of any of the thirteen RTA blocs listed in Table 3.

---

[18] Since (14.83-11.85) – (13.98-10.86) equals -0.14, the clean/dirty elasticity differential for a RTA pair is 0.014 percentage points smaller than for a non-RTA pair.

[19] We have not attempted to collect tariff data by bilateral trading pair/industry/year. Even if we could collect this huge amount of information, this tariff rate would represent an endogenous variable. We view the RTA dummy as proxying for the tariff levels between trading partners.

Table 6: Bilateral Trade Regressions and Individual RTA Effects

	(1) no pair fixed effect		(2) pair fixed effect	
	Coefficient	(SE)	Coefficient	(SE)
EU				
B_5, on Energy Index · ln(Exporter GDP per-capita)	0.057	(0.045)	0.057	(0.016)
B_6, on ln(Exporter GDP per-capita)	0.106	(0.009)	0.040	(0.002)
EFTA				
B_5, on Energy Index · ln(Exporter GDP per-capita)	0.328	(0.104)	0.329	(0.034)
B_6, on ln(Exporter GDP per-capita)	0.035	(0.014)	-0.003	(0.004)
NAFTA				
B_5, on Energy Index · ln(Exporter GDP per-capita)	0.160	(0.146)	0.153	(0.140)
B_6, on ln(Exporter GDP per-capita)	0.115	(0.034)	0.077	(0.010)
CACM				
B_5, on Energy Index · ln(Exporter GDP per-capita)	-0.149	(0.120)	-0.119	(0.049)
B_6, on ln(Exporter GDP per-capita)	0.123	(0.018)	1.029	(0.195)
CARICOM				
B_5, on Energy Index · ln(Exporter GDP per-capita)	-0.424	(0.271)	-0.345	(0.088)
B_6, on ln(Exporter GDP per-capita)	0.233	(0.033)	-2.901	(0.430)
MERCOSUR				
B_5, on Energy Index · ln(Exporter GDP per-capita)	-0.255	(0.125)	-0.237	(0.090)
B_6, on ln(Exporter GDP per-capita)	0.148	(0.033)	0.100	(0.007)
Bangkok				
B_5, on Energy Index · ln(Exporter GDP per-capita)	0.335	(0.142)	0.337	(0.061)
B_6, on ln(Exporter GDP per-capita)	-0.029	(0.026)	0.321	(0.092)
LAIA				
B_5, on Energy Index · ln(Exporter GDP per-capita)	0.449	(0.077)	0.472	(0.025)
B_6, on ln(Exporter GDP per-capita)	-0.056	(0.010)	-0.021	(0.005)
CAN				
B_5, on Energy Index · ln(Exporter GDP per-capita)	0.063	(0.232)	0.026	(0.090)
B_6, on ln(Exporter GDP per-capita)	0.039	(0.024)	0.089	(0.008)
AFTA				
B_5, on Energy Index · ln(Exporter GDP per-capita)	0.025	(0.126)	0.099	(0.044)
B_6, on ln(Exporter GDP per-capita)	0.087	(0.036)	1.804	(0.083)
COMESA				
B_5, on Energy Index · ln(Exporter GDP per-capita)	-0.233	(0.152)	-0.328	(0.113)
B_6, on ln(Exporter GDP per-capita)	0.066	(0.021)	0.026	(0.009)
SADC				
B_5, on Energy Index · ln(Exporter GDP per-capita)	0.047	(0.146)	0.008	(0.070)
B_6, on ln(Exporter GDP per-capita)	0.170	(0.022)	0.043	(0.008)
APEC				
B_5, on Energy Index · ln(Exporter GDP per-capita)	-0.043	(0.065)	-0.042	(0.020)
B_6, on ln(Exporter GDP per-capita)	0.137	(0.009)	0.044	(0.002)

(continue to the next page)

Table 6, cont'd: Bilateral Trade Regressions and Individual RTA Effects

Energy Index	22.507	(1.337)	21.325	(0.347)
Ln(Exporter GDP per-capita) · Energy Index	-2.065	(0.142)	-1.939	(0.037)
Ln(Exporter GDP per-capita)	1.381	(0.018)	1.028	(0.015)
Ln(Importer GDP per-capita)	0.843	(0.014)	0.898	(0.013)
Ln(Exporter population)	0.639	(0.010)	-0.103	(0.033)
Ln(Importer population)	0.487	(0.009)	-0.932	(0.029)
Ln(distance)	-0.823	(0.019)		
Same Language	0.416	(0.038)		
Colony	0.864	(0.087)		
Time Trend	-0.015	(0.001)	0.039	(0.001)
Constant	-18.407	(0.361)	-1.673	(0.522)
Adjusted R^2	0.421		0.536	

Note: This Table reports two OLS estimates of equation (15) in the text. The unit of analysis is trade between country *i* and importing nation *j* in year *t* for industry *m*. The dependent variable is the log of trade measured in $1,000s of 1995 U.S dollars. The omitted category is a trading pair that is not a member of the 13 RTAs listed in Table 3 and has not been in a colonial relationship and the two nations do not speak a common language. There are 11,434 trading bloc pairs in this data set with 1,737,800 observations.

Table 6 reports two OLS regression estimates. We focus on the right columns estimates in which trading pair fixed effects are included. If the estimate of B_5 is negative and large in magnitude then poorer bloc-member nations are specializing in dirty goods exports relative to richer nations in the same RTA bloc. We observe heterogeneity across the RTA blocs. For NAFTA, CAN and SADC the estimates of B_5 are statistically insignificant. For five RTA blocs, the estimate of B_5 is positive and statistically significant. These five blocs are EU, EFTA, Bangkok, LAIA, and AFTA. On the other hand, we estimate a negative B_5 coefficient for CACM, CARICOM, COMESA and APEC. We must acknowledge that these differences in elasticities are small. Following the same approach we used to generate Table 5, the income elasticity of dirty goods for nations in the LAIA bloc is 0.787 while the income elasticity of dirty goods for trading pairs not in any of the thirteen blocs is 0.737.

In interpreting our regression results reported in Tables 4 and 6, we must acknowledge a few caveats. We estimated these regressions using OLS that implicitly treats RTA participation as exogenous. Within RTA blocs, our focus has been on estimating how the composition of trade varies as a function of bloc member per-capita income. In results available on request, we have used Magee's (2003) instrument set to instrument for RTA participation. These include time invariant variables such as nation pair access to waterways and other transportation infrastructure. Several of the regressions we report include pair-specific fixed effects, and these fixed effects would control for many time invariant trade cost shifters. We recognize that our approach estimates a "total effect" of income differences within-blocs on trade composition. Given our data, we cannot conduct counter-factuals where we estimate how the composition of within-bloc trade is affected when one nation unilaterally increases its regulatory stringency. Recent work by Levinson and Taylor (2004) measures such effects while explicitly controlling for endogeneity of regulatory intensity.

4. Discussion

Most global trade takes place between developed nations (North-North trade) or between developed and developing nations (North-South trade).[20] As shown in Table 7, poorer areas such as Africa are not major global traders. Aggregating across continents, Table 8 shows that very little trade occurs between developing nations (South-South trade). Our theoretical model predicts that pollution havens emerge in poorer nations in both North-North and North-South trade relations. Our empirical findings based on the estimates of equation (14) as reported in Table 4 generally support this prediction. We observe that nations, as their income rises, expand their clean exports more than their dirty exports.

Table 7: Composition of Global Trade

	Destination									
Origin	N.A.	W.E.	Japan	Aust /NZ	L.A.	C.E.E./ Bal/CIS	Africa	M.E.	Asia	World
North America	397	184	64	15	164	7	12	20	135	998
Western Europe	261	1734	45	18	57	149	63	63	139	2559
Japan	136	73	-	9	12	3	4	11	180	433
Australia/NZ	9	10	14	7	1	0	2	4	29	78
Latin America	216	44	7	1	59	3	4	4	15	353
CEE/Baltic/CIS	14	159	4	0	7	76	3	6	18	291
Africa	24	72	4	1	5	1	11	3	18	142
Middle East	40	44	45	2	3	2	10	16	72	252
Asia	252	181	137	24	28	15	18	31	374	1080
World	1351	2501	320	76	335	256	128	158	980	6186

Note: Figures are annual averages for 2000-2002 in US$ billions. "World" in destination includes unspecified destinations. Source: WTO, International Trade Statistics 2003.

While our empirical analysis does not explicitly identify the trade specialization effects of capital endowment, comparing low-income, middle-income and high-income exporters yields some evidence that is consistent with both the factor endowment hypothesis and the pollution haven hypothesis. Relative to low-income and high-income nations, middle-income nations have the greatest income elasticity of exports of

[20] "North" basically corresponds to high-income nations, while "South" includes both middle- and low-income nations.

dirty goods. Low-income nations may have lax environmental regulation, but they do not have the necessary capital to specialize in dirty goods exports. Middle-income nations have greater access to capital than poor nations, and relative to rich-nations they have less stringent environmental regulations.

How does the formation of trading blocs affect these observed trading patterns? RTAs are mostly formed among nations with similar income levels. This expands South-South trade as well as North-North trade. Our theoretical framework predicts that bloc formation reduces the pollution haven effect. The results reported in Table 4 support this prediction. The income elasticity differentials between clean and dirty goods for RTA trading pairs are all smaller in absolute values compared to non-RTA trading pairs (see Table 5). We note that the economic magnitude of the RTA estimated effects are quite small.

Table 8: North and South in Global Trade

	Destination	
Origin	North	South
North	48.1%	17.1%
South	20.5%	14.3%

Note: Figures are based on Table 7. North includes North America, Western Europe, Japan, Australia and New Zealand. South includes all other regions, including unspecified destinations.
Source: WTO, International Trade Statistics 2003.

Disaggregating RTA effects by bloc yields additional insights. Trading blocs differ along a number of dimensions. One source of such heterogeneity is the environmental institutional arrangements built into each RTA. In Section 2, we discussed that the pollution haven effect within a RTA bloc would be mitigated if bloc members harmonize their environmental policies. Some regional trading blocs are taking pro-active steps to minimize any environmental impacts of bloc trade. The chapter on environment in the OECD Report's on regionalism (OECD 2003) provides information on explicit environmental considerations by RTA. For 9 of the 13 RTAs in our sample, Table 9 reports how these RTAs differ with respect to linking environment with trade issues. An interesting pattern emerges. RTAs that have made a greater commitment to harmonizing environmental policy within the bloc (as shown by answering "yes" for the categories listed in Table 9) are more likely to have a positive coefficient estimate of B_5 as reported in Table 6. Since the total effect inside of a RTA is $B_4 + B_5$, this suggests that within-bloc regulatory convergence works to mitigate the pollution haven effect.[21]

[21] For each of the 13 RTA blocs, we calculated its coefficient of variation for per-capita income. The correlation between this coefficient and the OLS estimate of B_5 from Table 6 (the interaction term

Table 9: Linkage with Environment and Selected RTAs

RTA	RTA includes general exception clause for environmental measures	RTA aims to prevent lowering of environmental standards & regulation	Coeffient estimate of B_5
AFTA	Yes	No	0.025
APEC	No	No	-0.043
COMESA	No	No	-0.233
EU/EFTA	Yes	Yes	0.057
MERCOSUR	No	No	-0.0255
NAFTA	Yes	Yes	0.160
SADC	Yes	Yes	0.047

Note: Coefficient estimates of B_5 are from Table 6 Column (1).
Source: OECD Report Table 9.1

5. Conclusion

A growing share of world trade takes place between nations that are members of a RTA bloc. Despite growing public interest in whether the formation and expansion of regional trade agreements such as NAFTA or EU affect environmental quality, the topic is still under-researched. This paper has presented new theoretical and empirical results on the pollution content of bilateral manufacturing trade for different types of nations that are and are not members of regional trading blocs.

Our theoretical framework highlighted the conditions such that the less capital-abundant poorer nation with a lower pollution tax becomes the pollution haven. In a three-country setting, we predict that a pollution haven effect will prevail in North-North and North-South trade relations. In South-South trade, both richer and poorer nations export more in pollution-intensive industries. The formation of trading blocs generally is predicted to reduce the pollution haven effect because, in the presence of product differentiation and intra-industry trade, a reduction in the *ad valorem* tariff within the bloc mitigates cost differential due to asymmetric environmental policies. Moreover, bloc formation could be associated with a convergence in environmental policies among bloc members that further reduces the cost differential. In the presence of regulatory convergence, the factor endowment difference dominates the pollution haven effect.

Outside of RTA blocs, our regression results document evidence of a pollution haven effect. As national per-capita income increases, nations increase their clean exports by a greater percentage than their dirty exports. Nations do increase their exports of dirty goods as they grow richer. When we split nations into low-, middle-,

---

$\ln(GDP_{it})^* E_m{}^*RTA_{ijlt})$ equals -0.13. This is weakly supportive evidence of the hypothesis that the PHH is more likely to be seen in RTA blocs featuring a greater level of within-bloc income inequality.

and high-income nations, we find that middle-income nations have the greatest dirty export income elasticities relative to their clean export income elasticities. Middle-income nations have the capital needed to engage in heavy, polluting manufacturing, and they are more likely to have less stringent environmental regulations than richer nations. This finding is consistent with the Environmental Kuznets Curve hypothesis, but we have presented no evidence that significant environmental degradation has taken place in the middle-income nations.

Our empirical findings with regards to RTA effects generally support out theoretical claims but the magnitudes of these effects are surprisingly small. Regional trading bloc participation slightly weakens the pollution haven effect observed outside of regional trading blocs.[22] We have documented heterogeneity in the composition of trade across RTA blocs. In some blocs, richer nations have a larger dirty exports income elasticity relative to their clean exports income elasticity, while in other blocs the opposite is found. Understanding the causes of this cross-bloc heterogeneity remains a topic for future research.

Future research also could test the hypothesis that regionalized free trade is good for the environment. Expansion is an ongoing trend in many of these RTAs, and the typical new entrant is poorer in the North (for example Eastern Bloc nations joining the EU). An increasing number of North-South RTAs are being negotiated and formed on a bilateral basis that often include relatively pro-active environmental cooperation (e.g., Canada-Chile Free Trade Agreement). If comparable environmental indicators could be collected, such "Southern" expansion of RTAs would offer the opportunity to test whether objective measures of pollution are rising for poorer new entrant bloc nations.

References

Baier, S. and J. Bergstrand, "Economic determinants of free trade agreements," *Journal of International Economics*, forthcoming, available online (2003).

Baldwin R. E. and P. Krugman, "Agglomeration, integration and tax harmonization," NBER Working Paper No. 9290 (2002).

Bartelsman, E. and W. Gray, The NBER Manufacturing Productivity Database (NBER Working Paper No. t0205, Cambridge MA (October 1996).

Bergstrand, J. H., "The generalized gravity equation, monopolistic competition, and the factor-proportions theory in international trade," *Review* of *Economics* and *Statistics* ***71***, 143-153 (1989).

[22] The causal relationship between the size of bilateral trade volumes and the trade bloc formation has been a major research topic in the empirical trade literature. Freund and McLaren (1999) estimate the effects of trade adjustment in four RTAs and find adjustment lags in trade reorientations in some RTAs. For recent empirical estimates of the scale effects of RTAs see Ghosh and Yamarik (2004).

Birdsall N. and D. Wheeler, "Trade policy and industrial pollution in Latin America: Where are the pollution havens?" *Journal of Environment and Development* **2**, 137-149 (1993).

Cole, M. A. and A. J. Rayner, "The Uruguay round and air pollution: Estimating the composition, scale and technique effects of trade liberalisation," *Journal of International Trade and Economic Development* **9**, 343-358 (2000).

Cole, M. A. and R. J. R. Elliot, "Factor endowments or environmental regulations? Determining the trade-environment composition effect," *Journal of Environmental Economics and Management* **46**, 363-383 (2003).

Copeland, B. and M. S. Taylor, "North-south trade and the environment," *Quarterly Journal of Economics* **109**, 755-787 (1994).

Copeland, B. and M. S. Taylor, *Trade and the Environment: Theory and Evidence.* Princeton: Princeton University Press (2003).

Copeland, B. and M. S. Taylor, "Trade, growth and the environment," *Journal of Economic Literature* **42**, 7-71 (2004).

Dee, P. and J. Gali, "The trade and investment effects of preferential trading agreements," NBER Working Paper No. 10160, Cambridge MA (December 2003).

Dixit A. and J. Stiglitz, "Monopolistic competition and optimum product diversity," *American Economic Review* **67**, 297-308 (1977).

Ederington, J. and J. Minier, "Is environmental policy a secondary trade barrier? An empirical analysis." *Canadian Journal of Economics*, **36**, 137-154 (2003).

Endoh, M., "Trade creation and trade diversion in the EEC, the LAFTA and the CMEA: 1960 – 1994," *Applied Economics* **31** 207-216 (1999).

Feenstra, R., "World trade flows, 1980-1997," U.C Davis Working Paper (2000).

Forslid R. and G. Ottaviano, "An analytically solvable core-periphery model," *Journal of Economic Geography* **3**, 229-240 (2003).

Frankel, J. and A. Rose, "An estimate of the effects of common currencies on trade and income," *Quarterly Journal of Economics* **117**, 437 – 466 (2002).

Freund C. L. and J. E. McLaren, "On the dynamics of trade diversion: Evidence from four trade blocs," Board of Governors of Federal Reserve System, International Finance Discussion Paper No. 637, Washington, DC (June 1999).

Ghosh, S., and S. Yamarik, "Does trade creation measure up? A reexamination of the effects of regional trading agreements," *Economic Letters* **82**, 213-219 (2004).

Grether, J-M and J. de Melo, "Globalization and dirty industries: Do pollution havens matter?" NBER Working Paper No. 9776, (June 2003).

Helpman, E. and P. Krugman, *Market Structure and Foreign Trade*, Cambridge MA: MIT Press (1985).

Heston, A., R. Summers, and B. Aten, Penn World Tables Version 6.1 Center for International Comparisons at the University of Pennsylvania (CICUP), October 2002.

Kahn, M.E., "The Geography of U.S pollution intensive trade: Evidence from 1958-1994," *Regional Science and Urban Economics* **33**, 483-500 (2003).

Levinson, A. and M. S. Taylor, "Trade and the environment: Unmasking the pollution haven effect," NBER Working Paper No. 10629 (2004).

Magee, C., "Endogenous preferential trade agreements: An empirical analysis," *Contributions to Economic Analysis & Policy* **2**, Issue 1 Article 15 (2003).

Markusen, J. R., "Explaining the volume of trade: An eclectic approach," *American Economic Review* **76**, 1002-1011 (1986).

Maskus K., "Trade and competitiveness aspects of environmental and labor standards in East Asia," in the World Bank, *East Asia Integrates.* Washington, DC: The World Bank (2003).

OECD (Organisation of Economic Cooperation and Development), *Regionalism and the Multilateral Trading System.* Paris: OECD (2003).

Panagariya A. and S. Suthiwart-Narueput, "Regional integration and the environment," *Environmental and Development Economics* **2,** 5-18 (1997).

Selden, T. M. and D. Song, "Neoclassical growth, the J curve for abatement, and the inverted U curve for pollution," *Journal of Environmental Economics and Management* **29**, 162-168 (1995).

Schiff, M. and L. A. Winters, *Regional Integration and Development.* Washington, DC: The World Bank (2003).

Tobey, J., "Effects of domestic environmental policies on patterns of world trade: An empirical test," *Kyklos* **43**, 191-209 (1990).

WTF (World Trade Flows) 2000, *World Trade Flows 1980-1997 Database,* Institute of Governmental Affairs, University of California, Davis.

WTO (World Trade Organisation) 2002, 'Work of the Committee on Regional Trade Agreements', http://www.wto.org/english/tmtop e/region e/rexom e.htm (accessed 18 June 2002).

WTO (World Trade Organisation) 2003, *International Trade Statistics*, Geneva.

van Beers, C. and J. C. J. M. van den Bergh, "Impact of environmental policy on foreign trade," Timbergen Institute Discussion Paper 069-3, Amsterdam (2000).

Xu X. and L. Song, "Regional cooperation and the environment: Do "dirty" industries migrate?" *Weltwirtschaftliches Archiv* **136**, 137-157 (2000)

Appendix: Countries in Sample and Their RTA Membership

Nation	RTA	Nation	RTA	Nation	RTA
Algeria		Gambia		Nigeria	
Angola	Comesa	Germany	Eu	Norway	Efta
Argentina	Laia, Mercosur	Ghana		Oman	
Australia	Apec	Greece	Eu	Pakistan	
Austria	Eu, Efta	Guatemala	Cacm	Panama	
Bahamas	Caricom	Guinea		Papua N Guinea	Apec
Bahrain		Guinea-Bissau		Paraguay	Laia, Mercosur
Bangladesh	Bangkok	Guyana	Caricom	Peru	Laia, Can
Barbados		Haiti		Philippines	Afta
Belgium-Lux	Eu	Honduras	Cacm	Poland	
Belize	Caricom	Hong kong	Apec	Portugal	Eu, Efta
Benin		Hungary		Qatar	
Bermuda		Iceland	Efta	Romania	
Bhutan		India	Bangkok	Rwanda	Comesa
Bolivia	Laia, Can	Indonesia	Apec, Afta	Sadi Arabia	Gcc
Brazil	Laia, Mercosur	Iran		Senegal	
Bulgaria		Ireland	Eu	Seychelles	Sadc
Burkina Faso		Israel		Sierra Leone	
Burundi	Comesa	Italy	Eu	Singapore	Apec, Afta
Cambodia		Jamaica	Caricom	Somalia	
Cameroon		Japan	Apec	South Africa	Sadc
Canada	Apec, Nafta	Jordan		Spain	Eu
Central Afr Rep		Kenya	Comesa	Sri Lanka	Bangkok
Chad		Korea Rp	Apec, Bangkok	Sudan	Comesa
Chile	Laia, Apec	Kuwait		Sweden	Eu, Efta
China	Bangkok, Apec	Laos P Dem R	Bangkok, Afta	Switzerland	Efta
Colombia	Laia, Can	Lebanon		Syrn Arab Rp	
Comoros	Comesa	Madagascar	Comesa	Taiwan	Apec
Congo		Malawi	Sadc, Comesa	Tanzania	Sadc
Congo Dem Rep		Malaysia	Apec, Afta	Thailand	Apec, Afta
Costa Rica	Cacm	Mali		Togo	
Cyprus		Malta		Trinidad Tbg	Caricom
Denmark	Eu	Mauritania		Tunisia	
Djibouti	Comesa	Mauritius	Sadc, Comesa	Turkey	
Dominican Rp		Mexico	Laia, Nafta,	UK	Eu
Ecuador	Laia, Can	Mongolia		USA	Apec, Nafta
Egypt	Comesa	Morocco		Uganda	Comesa
El Salvador	Cacm	Mozambique	Sadc	Uruguay	Laia, Mercosur
Eq Guinea		Nepal		Venezuela	Laia, Can
Ethiopia	Comesa	Netherlands	Eu	Yemen	
Finland	Eu, Efta	New Zealand	Apec	Zambia	Sadc, Comesa
France	Eu	Nicaragua	Cacm	Zimbabwe	Sadc, Comesa
Gabon		Niger			

Note: Blank after nation means it is not a member of any of 13 RTAs considered in this study.

[11]

Environmental Regulation and International Trade: Empirical Results for Germany, the Netherlands and the US, 1977-1992*

Abay Mulatu, Raymond J.G.M. Florax, and Cees Withagen

Abstract

We empirically investigate the responsiveness of international trade to the stringency of environmental regulation. Stringent environmental regulation may impair the export competitiveness of 'dirty' domestic industries, and as a result, 'pollution havens' emerge in countries where environmental regulation is relatively less stringent. We examine the impact of pollution abatement and control costs on net exports in order to grasp this phenomenon. Theoretically, our analysis is related to a general equilibrium model of trade and pollution nesting the pollution haven motive for trade with the factor endowment motive. We analyze data on two-digit ISIC manufacturing industries during the period 1977–1992 in Germany, the Netherlands and the US, and show that trade patterns in 'dirty' commodities are jointly determined by relative factor endowments and environmental stringency differentials.

KEYWORDS: Trade, Competitiveness, Pollution Haven

*This research is partly funded by the Netherlands Organisation for Scientific Research (NWO) in the Environment and Economy Program. We would like to thank Don Fullerton and two anonymous reviewers for very insightful and helpful comments. Jeroen van den Bergh, Hendrik Folmer, Shelby Gerking, Michael Rauscher, and Aart de Zeeuw provided useful comments and suggestions on earlier versions of this article. The usual disclaimer applies. Abay Mulatu is at the Department of Economic History, London School of Economics, London, UK. Raymond J.G.M. Florax is at the Department of Spatial Economics, Free University, Amsterdam, The Netherlands, and the Department of Agricultural Economics, Purdue University, West Lafayette, USA. Cees A. Withagen is at the Department of Spatial Economics, Free University, Amsterdam, and the Department of Economics, Tilburg University, Tilburg, The Netherlands. He is corresponding author: Department of Spatial Economics, Faculty of Economics, Free University Amsterdam, De Boelelaan 1105, 1081 HV Amsterdam, The Netherlands.

Environmental policymaking, be it unilateral or multilateral, is riddled with competitiveness concerns. Labor unions and industry worry that uneven environmental standards across jurisdictions result in distortions of the 'level playing field' in the market. For instance, in the Netherlands, competitiveness concerns are raised in relation to the unilateral introduction of a carbon tax. In North America, particularly in the US, there has been a lively debate about the economic implications of the North American Free Trade Agreement. Opponents argue that differential environmental standards in Canada, Mexico and the US will result in massive capital flight and loss of jobs in the US. Similarly, in a multilateral setting, a proposal to implement a uniform cross-country CO_2 tax in the European Union has been marred by competitiveness concerns, and it is still under discussion (see Jeppesen 2002). In addition to worries about competition, environmentalists are apprehensive of the emergence of what could be described as 'pollution havens' in the South. They observe the potential of a 'downward spiral' in setting environmental standards. Finally, a 'race-to-the-bottom' can arise from competitiveness concerns inducing governments to undercut each other's environmental standards. The worry about a sequential process of undercutting standards was in fact one of the main reasons for the debacle of the Seattle trade liberalization round of the World Trade Organization in 1999.

The pivotal issue at stake is whether domestic environmental regulation impairs the competitiveness of domestic industries, especially the pollution-intensive industries. The general conclusion emerging from the literature on this topic seems to be rather uniform. Initially, in their early comprehensive account of the empirical literature, Jaffe et al. (1995, p. 157) conclude that "[o]verall, there is relatively little evidence to support the hypothesis that environmental regulation has a large adverse effect on competitiveness … ." Evidence from recent studies suggests this conclusion should be taken with care. Recent studies do occasionally find a negative correlation between trade and environmental stringency, although the findings do not seem to be particularly robust. Recent reviews therefore continue to be cautious: "… [T]he costs imposed by tighter pollution regulation may not be a major determinant of trade patterns" (Copeland and Taylor, 2003, p. 220; see also Mulatu, Florax and Withagen 2003).

Concisely, the literature includes two types of studies. One group of studies investigates cross-country trade flows on the basis of country characteristics and a measure of relative stringency of environmental regulation.[1] The evidence for a negative impact of environmental stringency on international trade flows in these studies is at best mixed. Moreover, their differential results cannot be satisfactorily explained by corresponding differences in theoretical and methodological design, such as a Heckscher-Ohlin versus a gravity-type setup, bilateral versus multilateral trade flows, or the use of cross-section versus panel data (see Mulatu 2004, for details). Another group of studies employs some variant of the cross-commodity equation of the standard Heckscher-Ohlin factor endowment model, with factor intensities and a

[1] See, for instance, Tobey (1990), Diakosavvas (1994), van Beers and van den Bergh (1997), Xu (2000), and Harris, Kònyà and Màtyàs (2002).

measure of environmental compliance costs explaining net exports.[2] These studies by and large document corroborating evidence for the hypothesized negative correlation between compliance costs of environmental regulation and export performance of 'dirty' industries, but they are typically confined to cross-industry trade flows pertaining to the US. Two of the recent studies also account for endogeneity of environmental policy, and show that endogeneity bias implies an increase in the magnitude of the negative effect (Levinson and Taylor 2004; Ederington and Minier 2003).

This article adds to the above literature in a number of respects. First, although we employ a variant of the standard Heckscher-Ohlin based empirical model, in our specification of the baseline empirical model as well as in our readings of the emerging empirical results we pay due attention to recent theoretical insights in the literature (see, e.g., Copeland and Taylor 2003; Kahn and Yoshino 2004). Specifically, we emphasize the joint role of relative factor endowment conditions, pollution-intensity and the geographical 'footlooseness' of industries. Second, although we by and large stick to the traditional exogeneity assumptions in linking environmental regulation to export competitiveness, our analysis provides several empirical novelties. We extend the analysis that is typically confined to the US to a broader set of countries by including data on Germany, and the Netherlands.[3] This multi-country setting lends itself particularly well to cross-country comparisons of the issue under discussion. We also put considerable emphasis on building a database that covers a sufficiently large time-period at a reasonably low level of sectoral aggregation. The data we use stretch from 1977 to 1992, and they are based on a two-digit level classification of the manufacturing industry. Finally, despite the data characteristics, we use the fixed and random effects models typically employed in this literature. However, we also provide results for the so-called time-series cross-section estimator, which allows for heteroskedasticity and contemporaneous autocorrelation. This estimator is typically underutilized in this area of research, although it fits the data characteristics quite well.

We find a general support for the hypothesis that trade patterns in 'dirty' commodities are jointly determined by relative factor endowments and environmental stringency differentials.

The remainder of this article is organized as follows. Section 1 outlines the theoretical Heckscher-Ohlin basis for empirical research on the interrelation between the environment and international trade. This section also discusses the setup of our empirical model. In Section 2, we present the data and some initial exploratory analyses. Section 3 presents and discusses the estimation results, and Section 4 provides conclusions and ramifications.

---

[2] See, Kalt (1988), Grossman and Krueger (1993), Osang and Nandy (2000), Levinson and Taylor (2004), Ederington and Minier (2003), and Ederington, Levinson and Minier (forthcoming).

[3] Bouman (1998) investigates Germany and the US, but his research is concerned with the related issue of the link between stringency of environmental regulation and foreign direct investment.

1. Theory and Empirical Model

All cross-industry empirical studies of environmental policy and trade are rooted in standard international trade theory, using the general empirical trade equation:

(1) $NX = A'\beta + \gamma E + \varepsilon,$

where *NX* is a vector of a country's net exports of commodities, A' is a matrix of factor intensities, *E* is a measure of environmental compliance costs, β and γ are vectors of coefficients to be estimated, and ε is a vector of error terms. Such a specification combines the comparative advantage arguments of the standard Heckscher-Ohlin models (see, e.g., Harkness 1983; Deardorff 1984) and the pollution-haven hypothesis emphasizing the role of cross-country differences in environmental stringency in trade patterns.[4] Antweiler, Copeland and Taylor (2001) and Copeland and Taylor (2003) present a unified framework where the two hypotheses are nested. They underline the joint role of relative factor endowment conditions and differential environmental stringency in the determination of trade patterns in pollution-intensive goods, which are supposedly capital-intensive as well. This implies that a rich capital-abundant country with relatively stringent environmental regulation can maintain its comparative advantage in the pollution-intensive good if the stringency differential is 'small.'

While the above-mentioned empirical trade model is arguably consistent with the central message of Copeland and Taylor (2003), there are further insights that one needs to take into account in interpreting empirical results and in exploring some variants of the baseline empirical model. First, an industry's characteristics do matter in its responsiveness to environmental policy. Second, the fact that trade patterns are jointly determined by relative factor abundance and differences in environmental stringency implies that environmental policy should be sufficiently strict for it to have a distinguishable effect. Third, if a good is relatively labor-intensive and at the same time relatively pollution-intensive, even a moderate stringency of environmental regulation tends to result in a loss of export competitiveness. The converse of course also holds true.[5]

With these insights from theory in mind, we start out with an empirical specification that is similar to the standard cross-commodity equation discussed above. The empirical model relates net exports of an industry to its factor uses, as well as to capital expenditures on pollution abatement and control, and it reads as follows:

(2) $NX^c_{it} = f^c(YEAR_t, MK^c_{it}, SL^c_{it}, UL^c_{it}, RD^c_{it}, TARIFF^c_{it}, PACE^c_{it}) + \varepsilon^c_{it},$

[4] See, for instance, Siebert (1974, 1987), Pethig (1976), and Copeland and Taylor (1994, 1995).

[5] The intersection of a cross-country setting with a cross-commodity model is well established in the standard empirical literature. Under certain assumptions "... [T]he rankings of commodities by net exports and by intensity of use of the relatively abundant factor will be identical" (Harkness 1981, p. 1044).

where i denotes an industry, c a country, t time, and ε is an error term with properties discussed below. We estimate (2) for each country separately, but we drop the index c for notational convenience hereafter.

The dependent variable *NX* measures net exports, scaled by the value of production. Scaling is customary and appropriate because of scale differences between industry groups (see, e.g., Leamer 1988).[6] The explanatory variables are typical of those found in the empirical literature. The variable *YEAR* is meant to capture the year-to-year fluctuation in trade flows. The variable *MK* represents gross fixed new capital formation and proxies the flow of capital services. *SL* and *UL* denote the flow of the services of skilled and unskilled labor, respectively. The labor flow variables are constructed by means of a standard technique from the human capital literature based on production functions, following others like Branson and Monoyios (1977), Kalt (1988), and Stern and Markus (1988).[7] *RD* denotes expenditures for research and development, capturing technological differences across countries, and *TARIFF* represents *ad valorem* tariffs and is obtained as the ratio of duties paid to the custom value of imports. Unfortunately, for Germany and the Netherlands we could not obtain data on *TARIFF* at the appropriate level of aggregation. This may, however, not be detrimental because most Dutch and German exports are within the European Union, and a considerable part of Dutch exports go to Germany. And they are therefore not subject to tariffs. The inclusion of *TARIFF* is hence limited to the US regression. Finally, *PACE* represents capital expenditures for pollution abatement and control as a share of gross fixed new capital formation.

In a strict sense, theory requires the explanatory variables to be the technology

---

[6] Large (small) industry groups use more (fewer) factor inputs, which are included as explanatory variables. The error terms and the explanatory variables are then correlated, violating the Gauss-Markov conditions. The issue of the right scaling variable is a contentious topic in the trade literature. Some consider 'trade size' appropriate (e.g., Deardorff 1984); others favor 'value of production' (e.g., Anderson 1988). We use value of production because it is the most widely used (see Grossman and Krueger 1993; Levinson and Taylor 2004).

[7] Skilled labor is calculated as:

$$SL_{it} = (\overline{W}_{it} - \widetilde{W}_t)L_{it} / 0.10,$$

where *SL* denotes skilled labor services, $\overline{W}$ the average wage in each industry, $\widetilde{W}$ the average wage in the least-paying manufacturing industry, L the number of workers in each industry, and 0.10 is a discount rate. This method rests on the assumption of skilled labor services (i.e., human capital) in an industry being reflected as industry wage differentials, and hence the wages in the least-paying industry reflecting unskilled labor. Similarly, unskilled labor (*UL*) in an industry can be calculated as the product of the number of workers employed in that industry and the average wage in the least-paying industry. This type of calculation goes back to Griliches (1970, p. 85), who proposed a measure of human capital based on "a capitalization of the wage differentials over and above the returns to 'raw', unskilled, or uneducated labor." The discount rate transforms the stock variable into a stream, which is similar to its use in the capitalization of a capital asset with a perpetual yield (see Chiang 1984, p. 464). Note that the actual value of the discount rate does not affect the regression results, apart from changing the scale of the variable.

coefficients, defined as the amount of input per unit of output. In practice, some use factor shares in total value added as a proxy for factor intensity (e.g., Harkness 1981, 1983). Others use factor use, i.e., the total value of each factor input (e.g., Branson and Monoyios 1977; Stern and Markus 1981). The latter is a questionable proxy of relative factor intensity because it does not take account of differences in size across industries. However, in a multiple regression context, where the procedure takes into account the magnitudes of the rest of the variables in comparing the magnitude of one variable across industries, this is not necessarily problematic.

The environmental variable measures capital expenditures rather than recurrent operating costs, although the latter may more suitably reflect the year-to-year burden of compliance with pollution regulation. Because capital expenditures are spread over a number of years, the expenditure at time t may not instantaneously influence the profit function. Hence no instantaneous effect on supply of a specific industry at home as well as abroad would be observed. Levinson and Taylor (2004) contend that pollution abatement capital expenditures may even show a spurious positive correlation with net exports, because industries with thriving exports are more likely to make new investments, which typically have a larger abatement capital component (see Osang and Nandy 2000, and Mulatu 2004, for a more detailed discussion). With these caveats in mind, we nevertheless opt for capital expenditures as a measure of stringency in order to be able to construct a comparable data series for the three countries we consider.

2. Data and Exploratory Analysis

Our dataset pertains to nine two-digit ISIC manufacturing industries, for Germany, the Netherlands, and the US, over the time-period 1977–1992.[8] Table 1 provides the definitions of the industries for various industrial classification systems. Details about how we handle the incompleteness of the panel are described in the notes to Table 2. Monetary values are converted into 1996 US dollars using the US GDP deflator, and Purchasing Power Parity exchange rates to convert the German and Dutch currencies. The data are taken from various OECD publications, except for the data on tariffs and on pollution abatement capital expenditures. The US tariff data are obtained from the website of the National Bureau of Economic Research. The data on the environmental variable are taken from Bouman (1998), who collected the data from the respective national statistics agencies. Bouman (1998, pp. 12-15) discusses the PACE data in the three countries extensively. In all cases, the PACE estimates exclude labor costs related to environmental provisions, costs of relatively more expensive inputs, R&D expenditures and other cost components that cannot be considered capital expenditures. In Germany, the PACE indicator is defined as any new investment directed at the protection of environmental damage arising from production processes

[8] It is standard practice in the literature to use a classification at the four-digit level. Admittedly, the two-digit level can be too high and may potentially conceal the hypothesized correlation between environmental policies and export competitiveness (Levinson and Taylor 2004).

Table 1: Two-digit ISIC classification of industries[a]

ISIC	Industry	RB	NRB	RB & NRB	PI	NPI	PI & NPI
31	Food, Beverages and Tobacco	×				×	
32	Textiles, Apparel and Leather			×		×	
33	Wood products and Furniture	×				×	
34	Paper, Paper products and Printing	×			×		
35	Chemical products			×			×
36	Non-metallic mineral products		×			×	
37	Basic metal products		×		×		
38	Fabricated metal products			×			×
39	Other manufacturing		×			×	

[a] The meaning of the abbreviations is: resource-based (RB), non-resource-based (NRB), pollution-intensive (PI), and non-pollution-intensive (NPI). 'RB & NRB' implies that some sectors within the two-digit category are RB while others are NRB. A similar interpretation holds for the category 'PI & NPI'. See the Appendix for details.

and at producing environmentally benign products. In the Netherlands it is defined as any investment aimed at the reduction of emissions that arise from production processes and end up outside the premises of the firm, and in the US it is defined as any investment aimed at reducing emissions of pollutants, except soil pollutants, arising from production activities. In each of the three countries the PACE data exclude investments directed at protection of workers' health. Consequently, while the definitions of PACE are slightly different in the three countries, they are fairly comparable.

For the industries described in Table 1, a common classification in the environment/trade literature uses the criteria of the industry being pollution-intensive (PI) or non-pollution-intensive (NPI), as well as being resource-based (RB) or non-resource-based (NRB), i.e., not natural-resource-intensive. The rationale behind this categorization is to account for industry heterogeneity according to the expected effect of environmental regulation on industry competitiveness. We expect the effect to be most pronounced for PI and NRB industries (van Beers and van den Bergh 1997).

The categorization of industries as PI is usually based on the emission-intensity per dollar of output, or pollution abatement and control expenditures as percentage of sales or value added. Our classification in Table 1 is based on Low and Yeats (1992), who identify PI industries as industries incurring pollution abatement and control costs of a minimum of 1% of the sales value in 1988. The resource base classification reflects the natural-resource intensity of industries. Non-resource based industries are generally geographically 'footloose.' The resource base classification in Table 1 is based on UNIDO (1982).[9]

[9] A complication is that Low and Yeats (1992), and UNIDO (1982) use the three-digit Standard International Trade Classification (SITC). Details about the conversion to two-digit ISIC industries are presented in the Appendix.

Table 2 presents descriptive statistics of the data for the two-digit ISIC manufacturing industries for each of the three countries. Pollution abatement cost as a share of gross new fixed capital formation for the average two-digit ISIC manufacturing industry is highest in the US and lowest in the Netherlands. For trade of the manufacturing industry, we observe that while both Germany and the Netherlands are net exporters, the US is a net importer.

Table 2: Descriptive statistics for two-digit ISIC industries, 1972–1992[a]

Variable	Mean and standard deviation (in parentheses)		
	Germany	Netherlands	US
Exports	29.949	10.434	29.006
	(47.669)	(11.280)	(50.263)
Imports	21.418	21.230	36.689
	(24.541)	(24.711)	(55.147)
Value of production	105.548	17.622	435.430
	(120.479)	(16.159)	(396.568)
Net exports[b]	0.0002	−0.0764	−0.0604
	(0.0761)	(0.1961)	(0.0851)
Pollution abatement capital expenditure	0.193	0.031	0.676
	(0.246)	(0.052)	(0.817)
New gross fixed capital formation	4.704	0.989	13.839
	(5.841)	(0.893)	(15.037)
Pollution abatement capital expenditure[c]	0.04	0.03	0.05
	(0.03)	(0.04)	(0.04)
Research and development expenditure	2.197	0.286	10.077
	(4.572)	(0.475)	(22.408)
Skilled-labor services	92.725	8.122	329.682
	(154.038)	(9.038)	(484.524)
Tariff	—	—	0.01
			(0.031)
Unskilled labor services	16.865	2.616	49.354
	(20.975)	(2.363)	(52.030)

[a] Data availability across countries varies slightly according to time periods. For the US, data are available for 1973–1991, for Germany from 1975–1992, and for the Netherlands from 1972–1992. For every country, there are some missing observations for specific variables. Years prior to 1977 are not employed because data for research and development are lacking. The database contains 9 two-digit ISIC industries, but for the Netherlands, observations for ISIC 39 are missing. Since one of the econometric techniques we use requires a balanced panel, the years 1987 and 1988 are dropped for the Netherlands because of missing PACE data for ISIC 35 and ISIC 36. Similarly, the year 1987 is dropped for the US data because PACE data are missing. All monetary values are in billions of 1996 US dollars.
[b] Divided by total value of production of the same industry.
[c] Divided by total fixed capital formation of the same industry.

The sectoral variation of PACE and net exports across countries is depicted in Figure 1. It shows that there is no clear-cut negative association between PACE and net exports. We find bivariate correlations of 0.19, 0.49 and –0.11, for Germany, the Netherlands and the US, respectively. Examples of contrary 'evidence' are therefore numerous. For instance, each of the three countries is a significant importer in ISIC 32 and 33, despite the relatively low PACE in these sectors. The PACE levels in ISIC 35 and 37 are relatively high in all countries considered. Nevertheless, Germany is a significant net exporter in both sectors. This result is in line with the results of cross-country comparisons with respect to shares in 'dirty' industry production. For instance, Copeland and Taylor (2003) present evidence that many of the most environmentally stringent countries are among the most prominent producers and exporters of dirty goods. Germany actually has the highest share of world trade in dirty goods in 1988, followed by the US, while the Netherlands comes in sixth.

Does the above inference change if we take variations in factor intensity across sectors into account? Figure 2 provides a comparison of factor use intensity across sectors.[10] From Figure 1 we know that ISIC 34-37 in Germany perform relatively well despite the relatively high PACE level. Figure 2 shows that these sectors are also relatively more intensive in their use of skilled labor. In the US, ISIC 34-35 and 37-38 perform relatively well in trade, despite their relatively higher PACE. These sectors intensively use both man-made capital and skilled labor. Sector 32, however, does poorly in exports, despite its relatively low PACE level. This sector, however, is relatively intensive in unskilled labor. Similar, although slightly less clear-cut, examples can be drawn from Figures 1-2 for the Netherlands. Without attaching undue importance to this result of our exploring the data, these analyses hint at relative factor intensity potentially being an important factor in explaining the effect of PACE on net exports. We will return to this issue in the explanatory regression analyses discussed in the next section.

3. Estimation Results

Prior to discussing the empirical results of the general model in Section 3.2 and some variants in Sections 3.3 and 3.4, we discuss pivotal estimation issues in Section 3.1.

3.1 Estimation Issues

For the estimation of a linear version of the empirical model given in (2), we have panel data referring to nine two-digit ISIC manufacturing industries, in Germany, the Netherlands, and the US, over the time-period 1977–1992. Initially we provide results for three common estimators: ordinary least squares (OLS), and the estimators for fixed and random effects models (FEM and REM, respectively). The assumptions underlying

[10] In order to correct for size differences across industries and countries factor use is expressed as a share in Figure 2, as opposed to the value figures in Table 2. Value figures will also be used in the regressions.

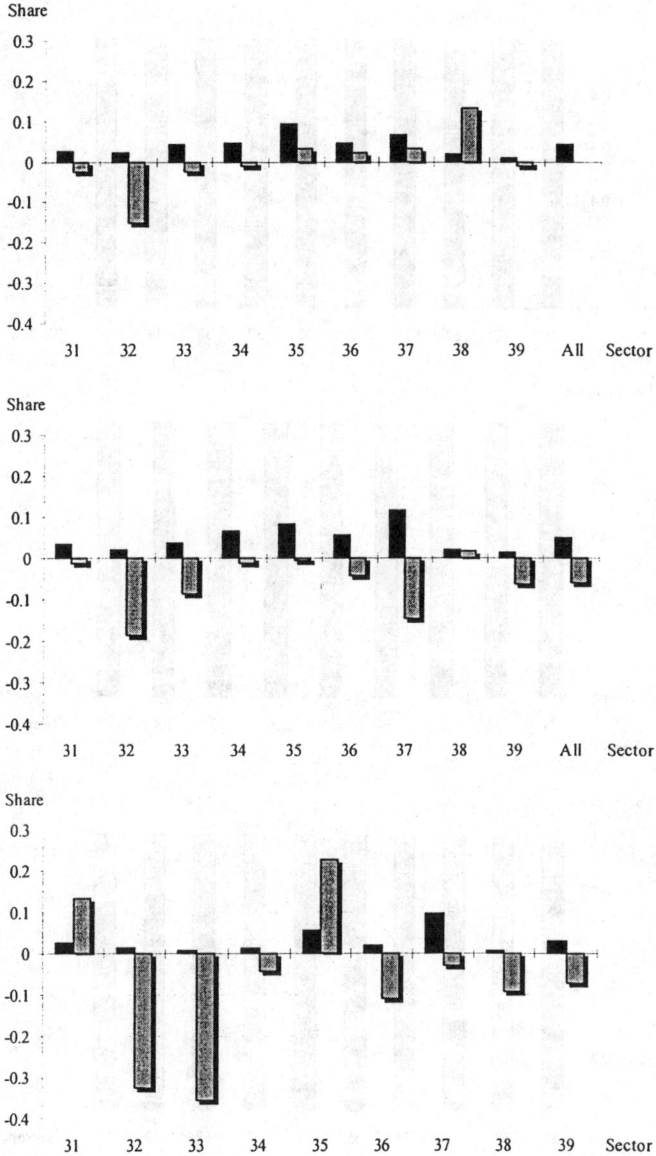

Figure 1: PACE share of capital formation (dark blue) and net export share of production (light green) for different sectors in Germany (top), the Netherlands (middle) and the US (bottom), 1971–1992.

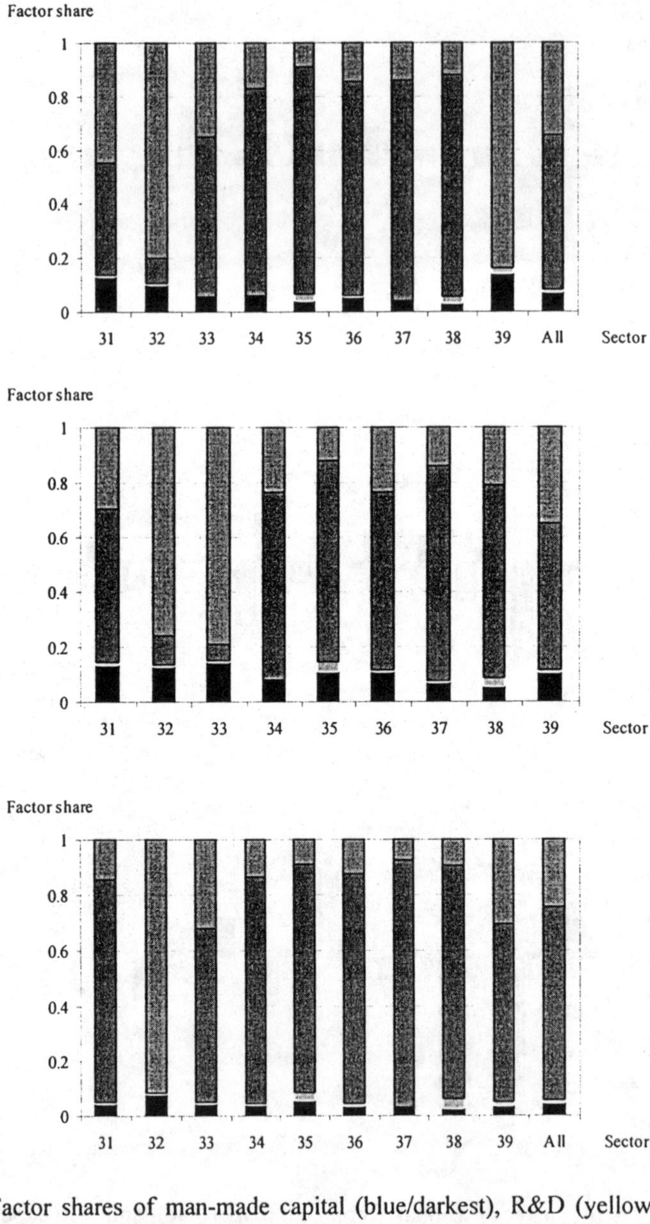

Figure 2: Factor shares of man-made capital (blue/darkest), R&D (yellow/ lightest), skilled labor (orange/dark), and unskilled labor (green/light), for different sectors in Germany (top), the Netherlands (middle) and the US (bottom), 1971–1992.

these estimators are rather restrictive. For OLS it is assumed that $E(\varepsilon_{it}) = 0$ for all i and t, $E(\varepsilon_{it}^2) = \sigma^2$ for all i and t, and $E(\varepsilon_{it}\varepsilon_{js}) = 0$ for all $s \neq t$ or all $i \neq j$. This set of assumptions makes OLS identical to the cross-sectional data analyses predominantly employed in this literature. Simple as well as pooled cross-sectional data analyses using OLS ignore heterogeneity across units with respect to unobservable characteristics, either for lack of variation or as a deliberate modeling choice. This may cause the OLS estimator to be biased.[11] The results for the OLS estimator are reported in the first three columns of Table 3. The Lagrange Multiplier (LM) test indicates that the hypothesis of homoskedasticity is rejected, and OLS is therefore inefficient.

The most common technique accounting for heterogeneity in panel data are the FEM and REM estimators. They account for heterogeneity across units, by means of decomposing the effects of unobservable factors into effects specific to cross-sectional units, to time-periods, and to both cross-sectional units and time-periods (Hsiao 1986, p. 97). These models are used in some of the more recent contributions to the environment/trade literature.[12] Formally, the fixed and random effects variants of equation (2) are given by:

$$(3) \quad NX_{it} = \beta_{0i} + \beta_1 YEAR_t + \beta_2 MK_{it} + \beta_3 SL_{it} + \beta_4 UL_{it} + \beta_5 RD_{it} + \beta_6 TARIFF_{it} + \beta_7 PACE_{it} + \varepsilon_{it}$$

where β_{0i} is a time-invariant group specific attribute, and ε_{it} a classical disturbance with $E(\varepsilon_{it}) = 0$, and $E(\varepsilon_{it}^2) = \sigma^2$. Whereas in the fixed effects model the coefficient β_{0i} is taken as fixed, it is treated as a random variable in the random effects model. So the random effects model splits this attribute into a 'mean intercept' β_0, and a group-specific disturbance μ_i that forms part of the error $\varepsilon_{it} + \mu_i$, where $E(\mu_i) = 0$, $E(\mu_i^2) = \sigma_\mu^2$, $E(\mu_i\mu_j) = 0$ for $i \neq j$, and $E(\varepsilon_i\mu_j) = 0$.[13]

While the fixed and random effects models provide considerable advantage in handling heterogeneity inherent in cross-sectional units, each has its own caveats and they can produce quite different results. The issue of preference of one over the other is highly debated. It is clear though that allowing the unit-specific effect (β_{0i}) to be correlated with the other regressors, as in the fixed effects model, may be attractive in comparison to the assumption that the random effects are uncorrelated with the

[11] Hsiao (1986, pp. 6-7) demonstrates the occurrence of heterogeneity bias in slope coefficients, when omitting group-specific intercepts to account for heterogeneous cross-sectional units in panel data sets. In a simple cross-section analysis, the problem arising from unspecified heterogeneity is similar. The source of the problem is heterogeneity across units with respect to some relevant but unmeasured characteristic that is correlated with the included explanatory variables.

[12] See, for instance, Ederington and Minier (2003), Levinson and Taylor (2004), and Osang and Nandy (2000).

[13] Fixed and random effects models can also include an additional group-invariant period-specific effect, i.e., the so-called 'two-way' variant. We report the results for the 'one-way' variant only because the results of the two-way variant are virtually identical.

 Vol. 3 [2004], No. 2, Article 5

Table 3: Preliminary cross-commodity regressions[a]

Variable	Pooled cross-section			Model Fixed effects[b]			Random effects		
	Germany	Netherlands	US	Germany	Netherlands	US	Germany	Netherlands	US
Constant	–0.012	–0.333**	0.055**				–0.004	–0.174**	–0.164**
	(0.020)	(0.039)	(0.020)				(0.044)	(0.066)	(0.035)
Year	–0.003*	–0.021**	–0.006**	–0.003**	0.004**	0.001	–0.001	0.003	0.0004
	(0.002)	(0.003)	(0.001)	(0.001)	(0.001)	(0.001)	(0.001)	(0.001)	(0.001)
Man-made capital	1.207**	19.156**	0.469**	0.291	1.967	0.194	–0.290	1.567	0.178*
	(0.303)	(1.635)	(0.091)	(0.420)	(1.403)	(0.101)	(0.314)	(1.278)	(0.089)
Skilled labor	0.272**	4.442**	0.017**	–0.085	–1.058	–0.003	0.067	–0.596	0.004
	(0.027)	(0.759)	(0.006)	(0.074)	(0.547)	(0.007)	(0.049)	(0.532)	(0.005)
Unskilled labor	–1.047**	–5.681**	–0.136**	–0.344	–0.385	0.032	–0.151	2.176	0.068
	(0.155)	(1.593)	(0.037)	(0.373)	(2.587)	(0.062)	(0.313)	(1.921)	(0.047)
R&D	–5.515**	–90.031**	–0.258**	1.735	–7.849	–0.317**	–0.059	–6.972	–0.323
	(0.901)	(9.839)	(0.090)	(1.227)	(6.324)	(0.071)	(1.034)	(5.818)	(0.070)
Tariff			–0.418*			1.894**			1.509**
			(0.201)			(0.335)			(0.312)
PACE	–0.399	1.982**	–0.887**	–0.116	0.110	0.287*	–0.031	0.168	0.225
	(0.244)	(0.389)	(0.157)	(0.194)	(0.163)	(0.139)	(0.191)	(0.160)	(0.133)
Number of observations	144	124	126				144	124	126
Number of cross-section	9	8	9	9	8	9	9	8	9
Number of time periods	16	14	14	16	14	14	16	14	14
Degrees of freedom	137	117	118	129	110	110	136	116	117
Adj. R^2	0.67	0.76	0.43	0.88	0.98	0.82			
F	49.60**	67.12**	15.50**	73.37**	397.22**	155.53**			
LR[c]				149.56**	295.72**	154.49**			
Hausman test[d]							19.58**	42.91**	17.89**
LM[e]	75.26**	17.26**	46.28**						

[a] The dependent variable is net exports as a share of the value of production. Standard errors are given in parentheses, and * and ** indicate significance at the 5 and 1% level, respectively. All variables except for Tariff are divided by 10,000.

[b] The estimated coefficients of the group-specific effects are omitted. They are rather similar across data sets, and only few are statistically different from zero.

[c] The LR statistic refers to a test of the null hypothesis of the pooled cross-section model against the fixed effects model. The statistic has a χ^2 distribution with (n–1) degrees of freedom, where n is the number of cross-section units. Note that the estimates of the fixed effects model include coefficients for group-specific effects.

[d] The Hausman test is a test of the null hypothesis of the random effects model against the fixed effects model. The statistic has a χ^2 distribution with 5 degrees of freedom.

[e] The LM statistic refers to a test of the null hypothesis of homoskedasticity against groupwise heteroskedasticity. The statistic has a χ^2 distribution with (n–1) degrees of freedom, where n is the number of cross-section units.

explanatory variables. The fixed effects model is, however, inferior to the random effects model in terms of degrees of freedom, which is a crucial issue in practice (see, e.g., Hsiao 1986; Greene 2000).

The estimation results using the FEM and the REM estimator are reported in the last six columns of Table 3. There are important differences between the two models in terms of the sign and significance of the estimated coefficients. The specification tests show that the fixed effects model is to be preferred over the pooled cross-section and the random effects models. In the fixed effects specification most of the coefficients are not significant. This is not uncommon, and most likely due to a high correlation between the fixed effects and the continuous explanatory variables.

The FEM and REM estimators are, however, criticized for not having appropriate asymptotic properties, because we have a dataset comprising a relatively long time-series for a relatively small number of cross-sections. The so-called time-series/cross-section (TSCS) estimator is an attractive alternative, also because it has greater flexibility in modeling stochastic error processes (Greene 2000, p. 593). We pool the time series of the different cross-sectional units (i.e., nine sectors in the case of Germany and the US, and eight in the case of the Netherlands), but we estimate separate regressions for each country. Formally, the model looks as:

$$(4) \quad NX_{it} = \beta_0 + \beta_1 YEAR_t + \beta_2 MK_{it} + \beta_3 SL_{it} + \beta_4 UL_{it} + \beta_5 RD_{it} + \beta_6 TARIFF_{it} + \beta_7 PACE_{it} + \varepsilon_{it}$$

Several variants can be operationalized based on differing assumptions for the stochastic error component. Because the hypothesis of homoskedasticity is rejected (see Table 3), and correlation across sectors is likely, we have used two variants. The first variant allows for heteroskedasticity across sectors, $E(\varepsilon_{it}^2) = \sigma_{ii}$, and the second variant combines groupwise heteroskedasticity with contemporaneous correlation among sectors, $E(\varepsilon_{it}^2) = \sigma_{ii}$ and $E(\varepsilon_{it}\varepsilon_{jt}) = \sigma_{ij}$. The two variants provide fairly similar results, but the Likelihood Ratio tests show that the second model, which is less restrictive, is to be preferred (see note d to Table 4 for details). Hence we present and discuss the results of this preferred model, using a two-step Generalized Least Squares (GLS) estimator that produces maximum likelihood estimates (see Greene 2000, Ch. 15).[14]

---

[14] An obvious third variant would be to combine heteroskedasticity and cross-correlation with autocorrelation among the disturbances. We have experimented with this variant as well, but it does not converge well, possibly due to the incomplete nature of the panel and lacking degrees of freedom. Note that the inclusion of a time trend variable among the exogenous variables may make autocorrelation less of a problem. If autocorrelation is nevertheless present, the GLS estimator remains unbiased, but it is not necessarily the most efficient estimator.

3.2 General Model and Cross-Country Differences

The estimation results for the preferred variant of model (4) are presented in Table 4. The sign and statistical significance of the coefficient estimates are satisfactory, except for the PACE variable. The coefficient of PACE has the expected sign for Germany and the US, and it is highly significant in the case of the US. For the Netherlands, however, PACE has a counterintuitive positive effect and it is significantly different from zero.

It is plausible to assume that the three countries are relatively abundant in both man-made and skilled labor as well as in high technology services, proxied by R&D, whereas unskilled labor is relatively scarce. The factor-intensity variables, except for the unskilled-labor variable and the tariff variable, are then expected to have a positive sign. Table 4 shows that this is exactly what occurs, except for the R&D variable, which has the 'wrong' sign and is significantly different from zero in all cases. This finding is similar to the result documented by Kalt (1988), and the aberration may be caused by omitted variables, endogeneity and/or multicollinearity.[15] Endogeneity is potentially a crucial issue in our empirical analysis as well, and we discuss it briefly below. Levinson and Taylor (2004), Ederington and Minier (2003) and Ederington, Levinson and Minier (forthcoming) also find the expected signs for skilled labor, man-made capital and tariffs. Our positive coefficient estimate for man-made capital is similar to the results of Kalt (1988), and Grossman and Krueger (1993), but it is different from the findings of Branson and Monoyios (1977), and Stern and Markus (1981).

Stringency of environmental regulation, relative factor abundance of a country, and sectoral differences in factor intensity jointly determine trade patterns. Copeland and Taylor (2003, p. 214) cogently argue along these lines:

> "... for some types of industries factor endowment effects can reinforce pollution policy effects. Polluting industries which are intensive in unskilled labor or natural resources that are abundant in low income countries will be attracted to those low income countries both by factor abundance and by less stringent policy. We would therefore expect to see heterogeneity across industries as well as countries in the effects of the interaction between policy differences and factor endowments in determining trade patterns."

There is of course potentially an issue of heterogeneity across sectors regarding the effect of PACE on net exports. Different sectors may react differently to stringency differentials. We will turn to this issue in the next subsection, but we first note that the magnitude of the estimated effect of environmental stringency in the different countries

[15] In general, this type of research suffers from high correlations among explanatory variables. We have tried to mitigate this problem by specifying the model in first-order differences, which would simultaneously reduce the potential inefficiency due to serial correlation. This specification does however not lead to plausible results, and it is again complicated by the incomplete nature of the panel.

Table 4: Baseline cross-commodity regressions[a]

Variable	Country		
	Germany	Netherlands	US
Constant	−0.031**	−0.280**	−0.004
	(0.005)	(0.017)	(0.006)
Year	0.0008	−0.0161**	−0.003**
	(0.0005)	(0.002)	(0.0004)
Man-made capital	0.021	18.070**	0.282**
	(0.075)	(0.917)	(0.023)
Skilled labor	0.168**	2.092**	0.015**
	(0.005)	(0.490)	(0.001)
Unskilled labor	−0.420**	−2.937**	−0.069**
	(0.017)	(0.966)	(0.009)
R&D	−2.376**	−64.044**	−0.287**
	(0.276)	(5.709)	(0.020)
Tariff			0.008
			(0.042)
PACE	−0.008	1.458**	−0.351**
	(0.044)	(0.216)	(0.051)
Log-likelihood	355.39	120.35	368.68
No. of observations	144	112	126
No. of cross-section units	9	8	9
No. of time periods	16	14	14
No. of estimated parameters[b]	52	43	53
Homoskedasticity (LR)[c]	149.91**	73.32**	91.27**
Cross-correlation (LR)[d]	195.54**	122.49**	202.47**

[a] With heteroskedasticity and cross-correlation. The dependent variable is net exports as a share of the value of production. Standard errors are given in parentheses, and * and ** indicate significance at the 5 and 1% level, respectively. All variables except for Tariff are divided by 10,000.

[b] The number of estimated parameters corresponds to the loss of degrees of freedom. The total loss corresponds to the number of coefficient estimates plus the estimated parameters of the variance-covariance structure.

[c] The LR statistic refers to a test of the null-hypothesis of homoskedasticity against groupwise heteroskedasticity, and it has a χ^2 distribution with $(n - 1)$ degrees of freedom, where n is the number of cross-section units.

[d] The LR statistic refers to a test of the null-hypothesis of no cross-correlation, and the test has a χ^2 distribution with $n(n - 1)/2$ degrees of freedom, where n is the number of cross-section units. To test the hypothesis of both homoskedasticity and no cross-correlated disturbances, the test statistic is the sum of the LR statistic of homoskedasticity and the LR statistic of no cross-correlation. The combined test has a χ^2 distribution with $n(n - 1)/2$ degrees of freedom, where n is the number of cross-section units. The joint test is rejected in all cases.

is in line with the stringency of their environmental policy. As indicated by the PACE variable, the US is the most stringent, while the Netherlands is the most lax. Pollution abatement control and capital expenditures (PACE) as a percentage of total investment for the manufacturing industry amount to 5% in the US, 4% in Germany, and 3% in the Netherlands. The importance of relative factor endowments in determining trade patterns implies that PACE outlays should be substantial enough to have a measurable effect on net exports. The relatively low level of PACE in the Netherlands may therefore go a long way towards explaining the positive coefficient. Of course one should be cautious not to infer strong generalizations from a comparison of just three countries. The positive coefficient can, however, only be taken as clear evidence for the rejection of the hypothesis that stringency of environmental regulation can be a source of comparative disadvantage in dirty goods, if one is willing to buy into the appropriateness of a one-sided statistical test.

Alternatively, it has been suggested that the effect may be ironed out by the provision of government subsidies to industries that are particularly hard hit by regulation (see, e.g., van Beers and van den Bergh 1997). Arguably, the Dutch so-called 'polder model', which typically adheres to consensus among stakeholders and elaborate negotiations in public decision-making, may also contribute to flattening out the most severe impacts of policy measures.[16] We do, however, not find it reasonable to take the positive sign of the PACE variable as a corroboration of the Porter hypothesis, which maintains that environmental stringency and competitiveness are positively correlated because it enhances innovations and the elimination of X-inefficiency. One can agree with a weaker interpretation of the Porter hypothesis that firms react to well-designed environmental policies by investing in innovation, thereby curtailing the burden of compliance with regulation. However, our model has not been designed in such a way that it could test the Porter hypothesis.

3.3 Variation According to Pollution and Natural Resource Intensity

It is common practice in the environment/trade literature to split samples of observations on cross-country and cross-commodity data for separate analyses, among other things, to account for heterogeneity across sectors. Van Beers and van den Bergh (1997) and Xu (2000), for example, consider separate analyses for pollution intensive (PI), and for pollution intensive and non-resource-based (NRB) industries. A recent study by Ederington, Levinson and Minier (forthcoming) presents a more elaborate treatment of the issue of different characteristics of industries and trading partners in the analysis of the responsiveness of trade flows to environmental policy. We also utilize the distinction between the characteristics of industries and estimate model (4) allowing the estimated coefficient of PACE to vary according to pollution intensity (see the first

[16] The word 'polder' signifies the classic story of the Dutch's extensive reclamation of land from water. Such arduous undertaking presumably required a unique institution of consensus-building and collaboration, that to date still characterizes Dutch policymaking.

three columns of Table 5), and to pollution intensity and non-resource-base features (see the last three columns of Table 5).

Levinson and Taylor (2004) point out that one of the disquieting aspects of the few studies in the literature documenting a negative effect of pollution policy on competitiveness is that these effects are found to be weaker when the data are restricted to PI industries. Obviously, the theoretical expectation is that the effect of stringency of environmental policy is stronger for pollution intensive industries. Limited data availability and the relatively high level of sectoral aggregation make that our categorization of industries as PI or NPI is not very strict. In fact, NPI comprises a mix of both PI and NPI industries, whereas the PI category exclusively concerns pollution intensive industries, specifically the paper and basic metal industries (ISIC 34 and 37; see Table A1 in the Appendix for details).

The estimation results are presented in the first three columns of Table 5. The earlier results for the factor intensity variables remain intact. The results for PACE are similar as well. For the Netherlands, the counterintuitive positive effect shows up again, now in both PI and NPI industries. For Germany, the effect on PI industries is negative, but the effect in NPI industries is significant and positive. In the US, the greatest negative effect occurs for PI industries, and although the effect in NPI industries is positive, it is not significantly different from zero. A slightly different specification, containing the interaction between PACE and PI in addition to PACE, would show whether the estimate of PACE for PI industries is larger in absolute value than the estimate of PACE for all industries, as we would expect. The results indicate that this is only the case for Germany.[17]

In the last three columns of Table 5, we show the results for a somewhat stricter categorization of industries. In effect, we separate ISIC 37, the basic metal industry, which is the only industry that is both PI and NRB, from all other industries. The theoretical expectation is that the effect of stringency of environmental policy on net exports will be the most pronounced in the PI/NRB sector.

The results show that the estimates for the factor intensity variables are virtually identical. For the Netherlands, the PACE results are similar to our earlier findings. They are persistently significant and positive, and do not vary over the different industry categorizations. With respect to Germany, both coefficient estimates are now negative, as expected, but neither is significantly different from zero. For the US, we once again find the most negative effect on the sector that is most vulnerable in terms of environmental regulation. However, the effect on all other industries is now positive and significant.

Before continuing to summarize the results of our successive empirical investigation, we address the issue of endogeneity pointed out earlier. In economic terms endogeneity means that environmental stringency is determined by endogenous variables, such as capital input, if capital is polluting, or output, if output is polluting. In statistical terms the problem of endogeneity will manifest itself in the violation of the

---

[17] Results are not shown here, but are available upon request.

Table 5: Cross-commodity regressions of net exports (pollution/natural-resource intensity)[a]

Variable	Specification					
	with a distinction between PI and NPI sectors[b]			with a distinction between PI/NRB and NPI/RB sectors[c]		
	Germany	Netherlands	US	Germany	Netherlands	US
Constant	–0.035**	–0.263**	–0.030**	–0.027**	–0.280**	–0.034**
	(0.006)	(0.018)	(0.007)	(0.005)	(0.017)	(0.006)
Year	0.0005	–0.015**	–0.003**	0.0007	–0.016**	–0.002**
	(0.0005)	(0.003)	(0.0004)	(0.0005)	(0.002)	(0.0004)
Man-made capital	0.162	16.558**	0.429**	0.014	17.894**	0.153**
	(0.109)	(1.014)	(0.030)	(0.076)	(0.927)	(0.020)
Skilled labor	0.204**	2.282**	0.019**	0.160**	2.063**	0.019**
	(0.007)	(0.587)	(0.002)	(0.006)	(0.512)	(0.001)
Unskilled labor	–0.567**	–3.649**	–0.065**	–0.424**	–2.892**	–0.068**
	(0.039)	(1.094)	(0.008)	(0.021)	(0.971)	(0.006)
R&D	–3.131**	–60.654**	–0.414**	–2.092**	–63.562**	–0.291**
	(0.330)	(6.326)	(0.027)	(0.299)	(5.891)	(0.022)
Tariff			–0.079			0.096
			(0.060)			(0.057)
PACE–NPI[d]	0.631**	0.672	0.673**			
	(0.055)	(0.514)	(0.062)			
PACE–PI[d]	–0.036*	0.401**	–0.134**			
	(0.019)	(0.072)	(0.020)			
PACE–NPI/RB[e]				–0.024	1.432**	0.141**
				(0.049)	(0.330)	(0.037)
PACE–PI/NRB[e]				0.093	0.052	–0.793**
				(0.061)	(0.375)	(0.082)
Log-likelihood	342.40	123.85	349.52	355.45	120.73	365.69
No. of observations	144	112	126	144	112	126
No. of cross-section units	9	8	9	9	8	9
No. of time periods	16	14	14	16	14	14
No. of estimated parameters	53	44	54	53	44	54
Homoskedasticity (LR)	143.51**	17.90**	99.13**	151.50**	16.95**	110.70**
Cross-correlation (LR)	167.01**	126.45**	143.30**	193.71**	123.30**	154.85**

[a] See the notes to Table 4.
[b] PI and NPI represent pollution-intensive and non-pollution-intensive sectors, respectively.
[c] PI/NRB and NPI/RB are pollution-intensive non-resource-based and non-pollution-intensive resource-based sectors, respectively.
[d] Effectively, PACE–PI concerns ISIC 34 and 37, and PACE–NPI all other industries.
[e] Effectively, PACE–PI/NRB concerns ISIC 37, and PACE–NPI/RB all other industries.

condition that the covariance between the explanatory variables and the error terms is zero, a condition that is to be met for the validity of our regressions (see, e.g., Levinson and Taylor 2004). Ideally, instrumental variable estimation should be used, but we refrained from that option after discovering that adequate data for Germany and the Netherlands are lacking. As a second best solution we therefore supplement the above results with a set of regressions where the explanatory variables from the base-year T are used to explain net exports from $T \rightarrow T + n$, where T is 1977 and n is 16 for

Germany and the Netherlands and 15 for the US (see, e.g., Levinson 1996; List and Co 2000). While this specification likely reduces the impact of endogeneity, it focuses on explaining cross-sectional variation only, and therefore obviously has downsides as well. With such caveats in mind, estimation results are presented in Table 6.

Overall, the results remain largely intact. The only noticeable exception is the result for Germany in the last set of regressions. The effect of PACE for PI and NRB industries is now as expected and significantly different from zero. These additional results are reassuring and can be taken as an indication for our results being reasonably robust to various misspecification biases.

In sum, there are variations in the effect of PACE both across countries and across industries following their pollution intensity and degree of being resource-based. For Germany and the US, there is some evidence of a negative association between PACE and net exports. This evidence is stronger for the US than for Germany. We also see signs that the effect is stronger for pollution intensive and non-resource-based industries. For the Netherlands, however, there is no evidence whatsoever for the hypothesized negative link. We also observe that the magnitude of the effect of environmental policy on net exports is generally greater for countries where the environmental regulation is stricter. However, there is no unequivocal support for the theoretically expected negative correlation between strictness of environmental regulation and international competitiveness of industry, not even for the intensive polluters and 'footloose' industries, although our results for the US indicate that the situation comes very close to the hypothesized effect. This is also the case for Germany when allowing for differences in terms of pollution intensity and being footloose.

These conclusions are of course dependent on our limited ability to clearly distinguish pollution-intensive and non-resource-based industries. Following Copeland and Taylor's (2003) observation that relative factor-abundance conditions and stringency of regulation interact in a variety of ways, resulting in heterogeneous effects across industries, we investigate sectoral variation *per se* in the next subsection.

3.4 Sectoral Variation *Per Se*

We explicitly address whether or not there is sectoral variation, and whether this variation can be explained by concurrent variations in pollution, natural resource, and factor intensities. Hence, we allow for sectoral variation in the PACE coefficient of equation (4), and then heuristically interpret the different characteristics of a sector in terms of the relative vulnerability of a sector vis-à-vis the strictness of environmental policy.

The estimation results allowing for sectoral variation, presented in Table 7, show that the outcomes for the factor-intensity variables are slightly different from those reported above. The most significant changes are the sign reversals of man-made capital for Germany, and unskilled labor for the US.

With respect to the industry-by-industry PACE effects clear sectoral variation is apparent across sectors as well as countries. For Germany, ISIC 32-34 (textiles, wood,

Vol. 3 [2004], No. 2, Article 5

Table 6: Cross-commodity regressions with explanatory variables from 1977[a]

Variable	Baseline regression			With a distinction between PI and NPI sectors[b]			With a distinction between PI/NRB and NPI/RB sectors[b]		
	Germany	Netherlands	US	Germany	Netherlands	US	Germany	Netherlands	US
Constant	−0.030*	−0.593**	0.046**	−0.039**	−0.529**	0.035*	−0.023*	−0.597**	0.026
	(0.017)	(0.025)	(0.019)	(0.016)	(0.014)	(0.021)	(0.014)	(0.015)	(0.019)
Year	−0.002*	0.002	−0.004**	−0.002**	0.002**	−0.004**	−0.002**	0.002*	−0.004**
	(0.001)	(0.002)	(0.001)	(0.001)	(0.001)	(0.001)	(0.001)	(0.001)	(0.001)
Man-made capital	1.825**	31.649**	1.057**	2.243**	11.308**	1.118**	2.860**	66.382**	0.435*
	(0.385)	(2.168)	(0.183)	(0.379)	(1.724)	(0.188)	(0.359)	(2.746)	(0.248)
Skilled labor	0.337**	7.581**	0.051**	0.371**	9.144**	0.050**	0.512**	13.685**	0.053**
	(0.026)	(0.882)	(0.010)	(0.026)	(0.499)	(0.010)	(0.033)	(0.682)	(0.010)
Unskilled labor	−1.345**	−6.257**	−0.281**	−1.559**	−14.432**	−0.272**	−1.798**	−11.874**	−0.258**
	(0.118)	(1.196)	(0.045)	(0.124)	(0.829)	(0.045)	(0.119)	(0.823)	(0.04)
R&D	−7.544**	−183.57**	−0.915**	−8.315**	−111.62**	−0.93**	−15.107**	−369.993**	−0.85**
	(1.015)	(19.624)	(0.233)	(0.981)	(11.729)	(0.232)	(1.364)	(17.575)	(0.224)
Tariff			−0.043			−0.056			−0.010
			(0.219			(0.219)			(0.210)
PACE	−0.102	3.624**	−1.267**						
	(0.087)	(0.524)	(0.145)						
PACE–NPI[b]				0.480**	9.093**	0.254			
				(0.118)	(0.552)	(0.193)			
PACE–PI[b]				−0.077**	1.324**	−0.394**			
				(0.028)	(0.093)	(0.047)			
PACE–NPI/RB[b]							−0.237**	−5.054**	−0.623**
							(0.076)	(0.681)	(0.229)
PACE–PI/NRB[b]							−2.171**	7.233**	−0.533**
							(0.302)	(0.502)	(0.150)
Number of observations	144	127	135	144	127	135	144	127	135
Adj. R^2	0.77	0.92	0.55	0.79	0.98	0.56	0.79	0.97	0.59
F	80.54**	242.49**	135.24**	216.15**	714.35**	22.10**	101.91**	595.38**	25.31**

[a] See note a of Table 4.

[b] See the notes of Table 5 for the meaning of the abbreviations.

Table 7: Cross-commodity regressions with a sector specific PACE coefficient[a]

Variable	Country		
	Germany	Netherlands[b]	US
Constant	0.010	–0.239**	–0.075**
	(0.006)	(0.022)	(0.007)
Year	0.0004	–0.0110**	–0.0014**
	(0.0003)	(0.002)	(0.0004)
Man-made capital	–0.536**	8.275**	0.078**
	(0.075)	(1.030)	(0.024)
Skilled labor	0.094**	4.598**	0.011**
	(0.007)	(0.669)	(0.002)
Unskilled labor	–0.171**	–8.855**	0.067**
	(0.055)	(1.342)	(0.013)
R&D	–0.371	–50.181**	–0.308**
	(0.302)	(5.940)	(0.023)
Tariff			0.209*
			(0.098)
PACE–ISIC 31	0.808**	9.088**	0.610**
	(0.228)	(1.260)	(0.168)
PACE–ISIC 32	–8.822**	–1.278	–4.458**
	(0.520)	(0.984)	(1.064)
PACE–ISIC 33	–0.304**	–8.700**	0.007
	(0.078)	(3.133)	(0.118)
PACE–ISIC 34	–0.016**	0.640*	0.115*
	(0.059)	(0.666)	(0.052)
PACE–ISIC 35	0.297**	0.911**	0.260**
	(0.092)	(0.421)	(0.102)
PACE–ISIC 36	0.877**	1.681**	0.496**
	(0.078)	(0.630)	(0.067)
PACE–ISIC 37	0.484**	0.932**	–0.423**
	(0.090)	(0.199)	(0.081)
PACE–ISIC 38	5.180**	–12.103**	–3.264**
	(1.492)	(3.940)	(0.710)
PACE–ISIC 39	–0.664**		3.603**
	(1.403)		(0.538)
Log-likelihood	396.51	139.40	388.04
No. of observations	143	112	126
No. of cross-section units	9	8	9
No. of time periods	16	14	14
Parameters	60	50	61
Homoskedasticity (LR)	115.50**	18.37**	131.07**
Cross-correlation (LR)	173.57**	127.41**	137.36**

[a] See the notes to Table 4.

[b] Data for ISIC 39 in the Netherlands are missing; see also footnote a to Table 2.

and paper industries) as well as ISIC 39 (other manufacturing) show declining net exports resulting from environmental regulation. For the Netherlands, ISIC 32, although not statistically significant, and 33 show similar effects. ISIC 38 (metal products) shows the greatest environmental policy impact. Similarly, in the US, ISIC 32 shows a negative correlation, as well as ISIC 37 (basic metal products) and ISIC 38.

Can we heuristically interpret these findings in terms of pollution, resource base, and factor intensities? Going back to the results allowing for variation in PI and NRB industries (see Tables 5 and 6) shows that the current results are not at odds with those obtained earlier. Sectors 34 and 37 are susceptible to environmental regulation, although the current results reveal that this conclusion holds only for ISIC 34 in Germany, and ISIC 37 in the US. We now also find that ISIC 32, 33 (except for the US), and 38 (except for Germany) are prone to be affected by environmental policy. Hence, while pollution and natural-resource intensity can be seen as necessary criteria for the categorization of heterogeneity of industries in view of environmental policy-making, they are not sufficient. Arguably, other sectoral characteristics, specifically factor intensity, are most likely relevant as well. For instance, ISIC 32 and 33 (except in the US) are affected by environmental policy, but they are at the same time the most unskilled-labor-intensive industries (see also Figure 2). Consequently, while these industries are most likely not subject to the most stringent environmental regulation, the relative factor abundance condition (i.e., the relative scarcity of unskilled labor) implies that even a moderately stringent regulation evokes a decline in net exports. We take this as plain support for Copeland and Taylor's (2003) view that trade patterns are determined jointly by relative factor abundance and relative stringency of environmental regulation. Even more so, the non-susceptibility of ISIC 31 and 35-36 that we find, can be explained by the high intensity as well as abundance of man-made capital and skilled labor in these sectors in each of the countries considered. Admittedly, the case is less clear-cut for ISIC 38, for which we find rather counterintuitive results for Germany, and also for 39, for which the result is mixed. Finally, although the results for ISIC 34 and 37 may seem erratic, we would like to point out that these sectors are relatively intensive in terms of skilled labor and man-made capital. Since the three countries are abundant in these factors, the factor intensity of these industries neutralizes the eroding effect of stringent environmental regulation on export competitiveness. Overall it has to be stressed that, while mixed results are obtained, there is support for the general tendency of the interplay of factor intensity and pollution intensity in determining export competitiveness.

Summing up, the evidence provided by the estimation results allowing for sectoral variation does not lend support to the existence of an unconditional negative link between stringency and net exports for the manufacturing industry in general. There is apparent variation across industries. *Ceteris paribus*, pollution intensive industries are more susceptible to stringency of environmental regulation. Moreover, the impact of environmental policy is conditional upon sectoral variation in factor intensity and a country's characteristics in terms of factor abundance. Since the countries considered are relatively scarce in unskilled-labor services, even a moderately stringent environmental regulation is prone to lead to a decline in net exports of those industries that are intensive in the use of unskilled-labor services. Conversely, industries that are sufficiently intensive in skilled labor and man-made capital, which are the abundant factors in each of the three countries, have a more resilient export performance, even under conditions of strict environmental regulation.

This result sheds new light on the results obtained in the literature so far. As mentioned above, the results have not been very consistent and clear (see also Mulatu, Florax and Withagen 2003, for a quantitative assessment). The cross-country studies, based either on the Heckscher-Ohlin or the gravity modeling approach, show mixed results. While Diakosavvas (1994) and van Beers and van den Bergh (1997) find evidence of the hypothesized negative link; Tobey (1990), Xu (2000), and Harris, Kònyà and Màtyàs (2002) do not. A recent cross-country study by Kahn and Yoshino (2004) deserves special mention. Although their study differs from ours in terms of the focus of the research question, the methodology and the data, both their study and ours emphasize the joint role of factor abundance and stringency of regulation in determining trade patterns in dirty goods. Kahn and Yoshino (2004) present general evidence that middle-income countries, relative to both richer and poorer countries, have a higher income elasticity of dirty goods exports. This is because they have less stringent regulation, like poorer economies, but they also tend to be relatively capital-abundant, which makes them specialize in comparatively dirty capital-intensive goods.

The cross-commodity studies, which are concerned with the US manufacturing industry and use the HO model, also show varying results, but on average they find evidence for the proclaimed negative link between compliance costs of environmental regulation and export performance of 'dirty' industries. The novelty of our results is the revelation of the link between environmental regulation and trade flows not merely being an issue of pollution intensity, but the conditioning of this link to differentials in stringency and relative factor abundance.

4. Conclusions

In this article, we obtain new empirical evidence on the effect of environmental regulation compliance costs on the export performance of the manufacturing industry. We specify empirical models that in general terms are based on a recently developed general equilibrium model of trade and pollution (Copeland and Taylor 2003). This theoretical model asserts that trade patterns in dirty goods are determined by the interaction of the stringency of environmental regulation and relative factor abundance. Consequently, concurrent empirical analyses should find heterogeneity across sectors and countries with respect to the impact of stringency of regulation on net exports. The observed heterogeneity should be attributable to stringency differentials, sectoral factor intensity, and country-by-country relative factor abundance.

We estimate empirical models for a newly developed dataset concerned with nine two-digit ISIC manufacturing industries in Germany, the Netherlands, and the US. The analysis involves a series of cross-commodity regressions of net exports on suggested determinants of commodity trade, and pollution abatement capital expenditures. We pay due attention to variations across sectors and countries, and utilize an econometric technique referred to as the time-series cross-section estimator that fits our data characteristics well.

The empirical results show that the estimated effects of stringency of environmental regulation on export competitiveness differ across the three countries. For the US, we consistently find evidence for the hypothesis that stringency of environmental regulation is a source of comparative disadvantage in dirty industries. For Germany, the evidence of this negative link is born out for pollution-intensive industries. For the Netherlands, however, neither the aggregate analysis, nor the analysis allowing for variation across pollution intensity and footloose industries, shows the hypothesized effect. Only when allowing for sectoral variation *per se*, we find evidence for the negative link in the wood and fabricated metal industries.

We show that sectoral variation in pollution intensity and the extent to which industries are resource-based are necessary but not sufficient criteria to explain the industry-by-industry and country-by-country variation in the observed effects of environmental policy on international trade. The observed empirical regularities are, however, consistent with an interpretation that does not only account for stringency differentials in environmental regulation and the dirtiness of different industries, but also explicitly considers the factor intensity of the different industries and the relative factor abundance of countries. For a country in which a specific production factor is relatively scarce and at the same time an industry intensively uses this factor, even a modestly stringent environmental regulation will evoke a decline in net exports. This situation occurs for the textiles, wood, and fabricated metal industries in all three countries, except for the latter in Germany. Conversely, industries that are intensive in production factors that are abundant in the countries considered show resilient export performances even though environmental regulation is relatively strict. We show that this situation prevails in, for instance, the food, chemical, and non-metallic mineral industries in the three countries considered.

Given the lack of robustness of research results in this area, and simultaneously the importance attached to environmental policy-making in the context of international trade, further empirical research is necessary. We recommend such research be focused on the triad of environmental regulation, factor intensity of industries, and relative factor abundance of countries, rather than the consideration of environmental regulation in isolation.

Appendix: Data Sources

Detailed data sources are as follows:

- *pollution abatement cost*, Bouman (1998)
- *GDP deflator*, OECD (Organization for Economic Co-operation and Development), 199 OECD main economic indicators (1980–1997), OECD, Paris
- *exchange rate*, OECD, Purchasing-power-parity, http://www.oecd.org/

- *US GDP deflator*, US Bureau of Economic Analysis, http://www.bea.doc.gov/
- *value added*, *capital*, *labor*, *import* and *export*, and *R&D expenditures*, The OECD STANDATABASE for industrial analysis (1974–1993, 1975–1994, 1976–1995, and 1978–1992), R&D expenditures in industry, 1977–1998 (OECD 2000 edition), 1974–1995 (OECD 1997 edition), Foreign trade by commodities (OECD 1989–1996, vol. 3), Industrial structure statistics core data (1974–1993, 1975–1994, 1976–1995, and 1978–1992), OECD, Paris
- *tariffs* (for the US), The US National Bureau of Economic Research

Table A1 provides the classification of industries with respect to pollution intensity (Low and Yeats 1992), resource-base (UNIDO 1982), and industry classifications (Industrial Statistics Yearbook, 1995, Volume ii, Commodity production statistics, United Nations).

Table A1: Standard industrial classifications, including pollution intensity and resource-base classification

Standard Industrial Classification (SIC)	International Standard Industrial Classification (ISIC)	Standard International Trade Classification (SITC)	Resource-Based (RB) or Non-Resource-Based (NRB)	Pollution-Intensive (PI) or Non-Pollution-Intensive (NPI)
20	31	0111	RB	NPI
21	31	1210	RB	NPI
22	32	6512	NRB	NPI
24	33	2433	RB	NPI
25	33	2433	RB	NPI
26	34	6411	RB	PI
27	34	6411	RB	PI
28	35	599	NRB	PI
29	35	2741	RB	NPI
30	35	2741	RB	NPI
31	32	6113	RB	NPI
32	36	6651	NRB	NPI
33	37	6711–6714	NRB	PI
34	381	6960–6980	NRB	PI
35	382	7115	NRB	NPI
36	383	7221	NRB	NPI
37	384	7353	NRB	NPI
38	385	8641–8642	RB	NPI
39	39	8914–8918	NRB	NPI

References

Anderson, J. (1988). Comment on "cross-section estimation". In R. Feenstra (ed.), *Empirical Methods for International Trade*. Cambridge MA: MIT Press (83–87).

Antweiler, W., Copeland, B., and Taylor, S. (2001). Is free trade good for the environment? *American Economic Review*, **91**, 877–907.

Beers, C. v., and Bergh, J. van den (1997). An empirical multi-country analysis of the impact of environmental regulation on foreign trade flows. *Kyklos*, **50**, 29–46.

Bouman, M. (1998). *Environmental Costs and Capital Flight*. PhD thesis, Tinbergen Institute Research Series, University of Amsterdam, Amsterdam.

Branson, W., and Monoyios, N. (1977). Factor inputs in U.S. trade. *Journal of International Economics*, **7**, 111–131.

Chiang, A. (1984). *Fundamental Methods of Mathematical Economics* (third ed.). Singapore: McGraw-Hill.

Copeland, B., and Taylor, S. (1994). North-South trade and the environment. *Quarterly Journal of Economics*, **109**, 755–787.

Copeland, B., and Taylor, S. (1995). Trade and transboundary pollution. *American Economic Review*, **85**, 716–737.

Copeland, B., and Taylor, S. (2003). *International Trade and the Environment: Theory* and *Evidence*. Princeton: Princeton University Press.

Deardorff, A. (1984). Testing trade theories and predicting trade flows. In R. Jones and P. Kenen (eds.), *Handbook of International Economics*. Amsterdam: North-Holland (**1**, 468–517)

Diakosavvas, D. (1994). The impact of environmental policies on agricultural trade. *Journal of International Development*, **6**, 207–218.

Ederington, J., and Minier, J. (2003). Is environmental policy a secondary trade barrier? *Canadian Journal of Economics*, **36**, 137–154.

Ederington, J., Levinson, L., and Minier, J. (2004). Footloose and pollution-free, *Review of Economics and Statistics* (forthcoming).

Greene, W. (2000). *Econometric Analysis* (fourth ed.). Upper Saddle River NJ: Prentice Hall International.

Griliches, Z. (1970). Notes on the role of education in production functions and growth. In W. Hansen (ed.), *Education, Income, and Human Capital* USA: National Bureau of Economic Research (71–115).

Grossman, G., and Krueger, A. (1993). Environmental impacts of a North American free trade agreement. In P. Garber (ed.), *The U.S.-Mexican Free Trade Agreement* Cambridge MA: MIT Press (13–56).

Harkness, J. (1981). Cross-section test of the Heckscher-Ohlin Theorem: Reply. *American Economic Review*, **71**, 1044–1048.

Harkness, J. (1983). The factors-proportions model with many nations, goods and factors: theory and evidence. *The Review of Economics and Statistics*, **65**, 298–305.

Harris, M., Kònyà, L., and Màtyàs, L. (2002). Modeling the impact of environmental regulation on bilateral trade flows: OECD, 1990–1996. *The World Economy*, **25**, 387–405.

Hsiao, C. (1986). *Analysis of Panel Data*. New York: Cambridge University Press.

Jaffe, A., Peterson, S., Portney, P., and Stavins, R. (1995). Environmental regulation and the competitiveness of US manufacturing: what does the evidence tell us? *Journal Economic Literature*, **33**, 132–163.

Jeppesen, T. (2002). *Environmental Regulation in a Federal Framework: Framing Environmental Policy in the European Union*. Cheltenham: Edward Elgar.

Kahn, M. and Yoshino, Y. (2004). Testing for pollution havens inside and outside of regional trading blocs. *Advances in Economic Analysis & Policy*, **4**, no. 2, Article 4. http://www.bepress.com/bejeap/advances/vol4/iss2/art4

Kalt, J. (1988). The impact of domestic environmental regulatory policies on U.S. international competitiveness. In M. Spence and H. Hazard (eds.), *International* competitiveness. Ballinger: Harper and Row (221–262).

Leamer, E. (1988). Cross-section estimation of the effects of trade barriers. In R. Feenstra (ed.), *Empirical Methods for International Trade* Cambridge MA: MIT Press (51–82).

Levinson, A. (1996). Environmental regulations and manufacturers' location choices: evidence from the Census of Manufactures. *Journal of Public Economics*, **62**, 5-29.

Levinson, A., and Taylor, S. (2004). Trade and the environment. Unmasking the pollution haven Effect. NBER Working Paper No. 10629.

List, J. and Co. C. (2000). The effects of environmental regulations on foreign direct investment. *Journal of Environmental Economics and Management*, **40**, 1– 20.

Low, P., and Yeats, A. (1992). Do 'dirty' industries migrate? In P. Low (ed.), *International Trade and the Environment*. Washington DC: World Bank (89–103).

Mulatu, A. (2004). *Relative Stringency of Environmental Regulation and International Competitiveness*. PhD thesis, Tinbergen Institute Research Series, Vrije Universiteit Amsterdam.

Mulatu, A., Florax, R., and Withagen, C. (2003). Environmental regulation and competitiveness: an exploratory meta-analysis. In C. Boehringer and A. Loeschel (eds.), *Empirical Modeling of the Economy and the Environment* (ZEW Economic Studies). Heidelberg: Physica Verlag (**20**, 23–54).

Osang, T., and Nandy, A. (2000). Impact of U.S. environmental regulation on the competitiveness of manufacturing industries. Working paper, Southern Methodist University; Public Service Electric and Gas, Newark.

Pethig, R. (1976). Pollution, welfare, and environmental policy in the theory of comparative advantage. *Journal of Environmental Economics and Management*, **2**, 160–169.

Siebert, H. (1974). Environmental protection and international specialization. *Weltwirtschaftliches Archiv*, **110**, 494–508.

Siebert, H. (1987). Environmental quality and the gains from trade. *Kyklos*, **30**, 657–673.

Stern, R., and Markus, K. (1981). Determinants of the structure of U.S. foreign trade, 1958–76. *Journal of International Economics*, **11**, 207–224.

Tobey, J. (1990). The impact of domestic environmental policy on international trade: an empirical test. *Kyklos*, **43**, 191–209.

Xu, X. (2000). International trade and environmental regulation: time series evidence and cross-section test. *Journal of Environmental, Agricultural and Resource Economics*, **17**, 233–257.

[12]

Pollution Havens and Foreign Direct Investment: Dirty Secret or Popular Myth?*

Beata Smarzynska Javorcik and Shang-Jin Wei

Abstract

The "pollution haven" hypothesis refers to the possibility that multinational firms, particularly those engaged in highly polluting activities, relocate to countries with weaker environmental standards. Despite the plausibility and popularity of this hypothesis, the existing literature has found only limited evidence to support it. To enhance our ability to detect the possible "dirty secret," this study makes improvements in four areas. First, we focus on investment flows from multiple countries to 25 economies in Eastern Europe and the former Soviet Union. Transition countries are a suitable region for studying this question, as they offer a large variation in terms of environmental standards. Second, we take into explicit account the effect of host country corruption. Third, we include information on both the polluting-intensity of the potential investor and the environmental stringency in the potential host country, which allows us to test whether dirty industries are relatively more attracted to locations with weak standards. And fourth, we rely on firm-level rather than industry-level data. Despite these improvements, we find no support for the "pollution haven" hypothesis. If anything, firms in less polluting industries are more likely to invest in the region. We find no systematic evidence that FDI from "dirtier" industries is more likely to go to countries with weak environmental regulations.

KEYWORDS: Pollution, corruption, foreign direct investment

*Javorcik is at: The World Bank, Development Economics Research Group, MSN MC3-303, 1818 H St NW, Washington, DC 20433. Tel: (202) 458-8485. Fax: (202) 522-1159. Wei is at: The International Monetary Fund, Research Department, Room 10-700, 700 19th Street NW, Washington, DC 20431. Tel: (202) 623-5980. Fax: (202) 623-7271. We wish to thank Hans Peter Lankes, Susan Bartolo and Matthias Busse for kindly providing some of the data and Rachel Rubinfeld for superb research assistance. We are also grateful to Simon Evenett, Don Fullerton, Ann Harrison, Michael Nicholson, Nina Pavcnik, participants of the EIIT conference, World Bank trade seminar and Yale School of Management seminar as well as two anonymous referees for their comments. The views expressed in the paper are those of the authors and should not be attributed to the World Bank, the IMF or any other organization with which the authors are affiliated.

The possibility that pollution-intensive multinational firms relocate to developing countries with less stringent environmental standards has been labeled the "pollution haven" hypothesis. The logic sounds plausible: if it costs money to conform to more stringent environmental requirements in developed countries, profit-maximizing firms would want to relocate their production activities. The believers abound. For example, the Sierra Club states that that "in our global economy, corporations move operations freely around the world, escaping tough pollution control laws, labor standards, and even the taxes that pay for social and environmental needs."[1]

The evidence on the "pollution haven" hypothesis as related to foreign direct investment (FDI) is, however, mixed. The existing literature can be divided into three strands: studies of within-country plant location choice, examination of industrial composition of foreign direct investment inflows within a single country, and analysis of inter-country FDI flows. The first strand encompasses work by Levinson (1996), who finds little evidence that inter-state differences in environmental regulations influence the geographic location of U.S. plants, as well as several studies reaching the opposite conclusion. For instance, Becker and Henderson (2000) demonstrate that the annual designation of air quality attainment status, which triggers specific equipment requirements at the county level in the U.S., reduces the number of firm establishments in polluting industries in nonattainment areas. Similarly, Keller and Levinson (2002) and Fredriksson et al. (2003) show that differences in abatement costs between U.S. states have a moderate deterrent effect on FDI. Further, Dean et al. (2003) examine FDI inflows into Chinese provinces and find that investment from OECD countries is attracted to regions with high pollution levies (contradicting the predictions of the "pollution haven hypothesis") while the opposite is true of FDI originating Hong Kong, Macao and Taiwan.

The second approach has been followed by Eskeland and Harrison (2003) who examine the distribution of FDI across industries in Mexico, Venezuela, Morocco and Cote d'Ivoire and find little evidence in support of the "pollution haven" hypothesis. The third approach has been adopted by only one study, Xing and Kolstad (1998), which reports a positive association between the amount of sulfur emissions in a host country and inflows of U.S. FDI in heavily polluting industries. This evidence is based on a fairly small sample (no more than 22 observations in each regression), so its robustness is subject to debate.

Given that the possibility of moving polluting activities to developing countries has been at the center of the globalization debate, there is a further need to examine this question in the context of developing economies. This is the objective of this study, which follows the third approach outlined above.[2] Our contribution to the literature lies

[1] See "A Fair Trade Bill of Rights" at the Sierra Club website (http://www.sierraclub.org/trade/ftaa/rights.asp).

[2] It is beyond the scope of this study to make any normative statements about the "pollution haven" hypothesis, our objective is only to produce new evidence shedding light on its validity. While environmental activists would strongly condemn relocation of polluting activities to developing countries, a theoretical contribution of Oates and Schwab (1988) demonstrates that if pollution is local and

in four areas. First, we examine the location decision pertaining to FDI flows from multiple source countries to multiple developing economies, namely to 25 economies in Eastern Europe and the former Soviet Union. Transition countries are a suitable region for studying the "pollution haven" hypothesis. While all of them share the common heritage of more than half a century of central planning, possess relatively high human capital endowments and opened up to the world at the same time, they also offer a reasonably large variation in terms of environmental standards. Thus our sample includes countries with relatively high income and relatively high environmental standards (in terms of conformity with the European Union standards) such as Poland and Czech Republic, and at the same time also covers economies that are substantially poorer and are perceived to have much weaker environmental regimes, such as Azerbaijan and Uzbekistan. Moreover, there is a significant variation in the sample in terms of the environmental sustainability rankings. For instance, in 2002 Latvia and Hungary were ranked 10th and 11th among 142 countries while Turkmenistan and Ukraine occupied the 131st and 136th positions respectively. Therefore, while similar in some respects, the countries included in our data set allow us to compare investment destinations quite different in terms of environmental attainment and stringency of laws.

In addition to an intrinsic interest in transition, a focus on Eastern Europe and the former Soviet Union can offer insights into the broader question of the role of FDI in economic development throughout the world. While investment in other developing regions has been studied extensively, one finding of that research has been the importance of previous investment experience as a determinant of current FDI flows (see Hallward-Driemeier, 1996). Thus, the impact of current policy variables may be obscured and overcome by a long history of past policies, for which it is difficult to control. Transition economies offer almost a natural control since FDI in the region was negligible prior to 1989.

Second, we postulate there may be features of developing countries that deter FDI but at the same time are correlated with laxity of environmental protection. A leading example of such a feature is a host country's weakness in public institutions, particularly the prevalence of bureaucratic corruption. Host country A may have less stringent environmental protection than country B, which might make country A more attractive than country B to foreign direct investment, particularly from the "dirty" industries. On the other hand, country A may also have a more severe corruption problem, which tends to discourage inward foreign investment, including those from the "dirty" industries. Indeed, it seems reasonable to expect corruption and laxity of environmental protection to go together, so that statistical analysis on the effect of environmental policy on FDI that omits local corruption might fail to detect an effect.[3] To the best of our knowledge, the only study examining the impact of both corruption and environmental standards on FDI inflows is that of Fredriksson et al. (2003)

---

preferences differ, allowing jurisdictions to choose their own level of environmental standards may be welfare increasing.

[3] Several studies have demonstrated that corruption in a host country is a significant deterrent to inward FDI (Hines, 1995; Wei, 1997, 2000a and 2000b; and Smarzynska and Wei, 2000).

mentioned earlier. A similar argument could be made with respect to democracy. Authoritarian governments are less likely to be concerned with protecting the environment. At the same time, multinationals may be less likely to invest in countries with authoritarian governments, perhaps out of concern for unpredictable policy changes. Indeed Busse (2004) shows that in the 1990s more democratic countries received more FDI.

Third, unlike the existing studies which *either* ask whether FDI is attracted to locations with laxer standards *or* examine whether polluting industries are more likely to undertake FDI, we take into account both the polluting-intensity of the potential investor and the environmental stringency in the potential host country. This allows us to test whether dirty industries are *relatively* more attracted to locations with weak standards.[4] To capture the strength of environmental protection in host countries, we adopt several different measures that complement one another. Specifically, we employ four types of proxies: (i) a degree of participation (ratification, signature but no ratification, or neither) in three different international environmental protection treaties, covering transboundary aspects of environmental impact, air pollution and industrial accidents; (ii) an index of the strength of the air and water ambient and emission standards system as rated by European Bank for Reconstruction and Development (EBRD, 1997); (iii) Environmental Sustainability Index (ESI), which is a measure of a country's overall progress towards environmental sustainability developed jointly by the World Economic Forum, the Yale Center for Environmental Law and Policy, and the Center for International Earth Science Information Network; (iv) the actual reduction in emissions of carbon dioxide and lead during the period in question (scaled by the GDP growth). Reductions in emissions may be viewed as proxies for a host country's effective enforcement of environmental policies. For the first two measures, effectiveness of enforcement is adjusted by using information on the number of environmental NGOs in a country relative to its population size.

We compute pollution intensity for all multinational firms in our sample, based on the data on actual pollution emissions and abatement cost of U.S. firms filed with the U.S. Environmental Protection Agency (EPA). More precisely, pollution intensity is computed for each four-digit SIC industry; and every firm in the sample is assigned to a four-digit SIC industry. We have the emissions data for 269 and the abatement cost data for 140 four-digit SIC industries in our sample. This is a labor-intensive task, but the payoff is an enhanced precision in assigning pollution intensity to the production activities of multinational firms. This is, however, still not a perfect measure. Note that we do not need to assume that the multinational firms in the source and host countries have identical pollution intensity. Instead, what we need is a weaker assumption: the relative pollution intensity between the overseas activities of the two multinational firms (e.g., Dupont Poland and Nike Poland) is proportional to their pollution intensity at home (e.g. Dupont and Nike in the U.S.).

[4] Only Dean et al. (2003) takes into account both aspects of the issue.

Fourth, as Zarsky (1999, p. 66) stated after surveying the empirical literature, "the quality of the evidence, both statistical and case study, is poor compared to the research needs. In terms of location decisions, most of the statistical studies rely on very aggregated data about 'industry choices' which shed little light on firms or production stages." This suggests the usefulness of employing a firm-level data set. Hence instead of using country- or industry-level figures, we make use of a unique firm-level data set that describes the investment decision by 143 multinational firms in 25 countries in Eastern Europe and the former Soviet Union. This represents potentially 3,575 (=143x25) investment decisions at the firm level. One of the advantages of employing firm level data is our ability to explicitly control for the characteristics of investing firms, which affect their investment decisions, such as a firm's size, R&D intensity and previous experience of trading with the region.

Levinson and Taylor (2004) point out that the existing studies may have failed to detect the pollution haven effect due to biases introduced into the estimation by aggregation, unobserved heterogeneity and endogeneity of environmental standards. Our data set allows us overcome two of these difficulties, while the third one is unlikely to be a serious concern in this case. Our analysis does not suffer from the aggregation bias as we use disaggregated measures of pollution intensity and abatement costs (at the four-digit SIC industry level). We are able to control for unobserved heterogeneity of industries by introducing firm-specific fixed effects. In a separate specification, we control for unobserved heterogeneity of host countries by adding host country fixed effects. Finally, the potential endogeneity problem is mitigated by the fact that a large part of the effort to improve environmental standards in Eastern Europe has been undertaken in the context of the expected accession to the European Union (EU).

Rather than keeping the suspense, we spell out the bottom line right now. We find little support for the "pollution haven" hypothesis. While there is some evidence indicating that countries with less stringent environmental regulations, particularly when proxied by participation in international environmental treaties, may be more attractive to foreign investors, this finding is not robust. Moreover, there is no evidence suggesting that multinationals in pollution-intensive activities are more likely to invest in locations with weaker environmental regulations. On the contrary, we find that FDI inflows to transition economies are more likely to take place in clean industries.

The first section below describes the methodology and the data employed. Section 2 discusses the empirical results. The last section concludes.

1. Methodology and Key Variables

1.1 Empirical Model

Let FDI_{jk} be a dummy that takes the value of one if firm j has established an investment or has concrete plans to invest in host country k, and zero otherwise.

Our strategy is to estimate some variant of the following probit specification:

$$\text{Prob}[FDI_{jk} = 1] = X_j \Gamma_j + H_k \Phi_k + Z_{jk} \Pi + \beta E_k + \gamma D_j E_k + \varphi D_j + e_{jk}$$

where X_j is a vector of variables describing the characteristics of firm j; H_k is a vector of variables describing the characteristics of host country k other than its environmental standards; Z_{jk} is a vector of variables describing the relationship between host country k and the source country where firm j originates; and Γ_j, Φ_k and Π are vectors of parameters with corresponding dimensions. D_j is an index of firm j's pollution intensity or "dirtiness," and E_k is an index of host country k's environmental standards, possibly adjusted by the strength of enforcement. e_{jk} is an iid normally distributed error term.

The parameter β captures a "volume effect:" a negative (or positive) β implies that a stronger environmental protection in a host country tends to discourage (or encourage) inward FDI. The parameter γ captures a "composition effect:" a negative (or positive) γ implies that more (or less) pollution-intensive FDI would go to a host country with relatively weaker environmental standards. In other words, the "pollution haven" hypothesis can be represented by $\beta < 0$ and/or $\gamma < 0$.

As we are also concerned about unobserved country characteristics that may affect FDI inflows, we also estimate a logit specification with host country fixed effects, in which case γ is the coefficient of interest, as the volume effect will be captured by the fixed effects. Similarly, to take into account unobserved investor characteristics we employ a logit specification with fixed effects for investing companies.

Crucial to our empirical strategy is to have plausible measures of pollution intensity D_j by multinational firms and of the strength of environment protection by host countries, E_k. We will discuss the construction of these measures next.

1.2 Measuring Pollution Intensity

We use two measures of pollution intensity of industries: one based on pollution emissions and one on abatement costs.[5] The first measure is compiled using the Toxics Release Inventory (TRI) data collected by the U.S. Environmental Protection Agency. TRI provides a comprehensive overview of toxic chemical pollution from manufacturing facilities in the United States. In 1997, the reportable TRI chemical list included 576 individually listed chemicals and 28 chemical categories. The database contains information on releases of toxic substances into air, water, land and underground injections measured by weight. A median value of emissions for reporting facilities in each 4-digit SIC code is found and then normalized by the average shipping volume in the industry.[6] The information on shipping volumes comes from the 1997 Economic Census CD-ROM. The average shipping volume was found by dividing the 1997 value of sales, shipments and receipts (given in thousands of dollars) by the

[5] The emissions and abatement cost figures are also used in Eskeland and Harrison (2003).

[6] The information on all facilities that reported both an SIC code and emissions was taken into account. The SIC code and the emissions data were matched for each facility. If a facility reported more than one SIC code, each unique facility and SIC pair was treated as an individual observation. The value of emissions for each observation was divided by the number of SIC codes reported for the facility. The data were then regrouped by SIC code and the median of all observations for a particular 4-digit SIC code was calculated for each emissions category (air, land, water, underground).

number of establishments reporting for that SIC code. If data on shipping volumes were not available for a particular SIC code, that code was dropped.

A histogram analysis of the pollution intensity (not reported to save space) indicates that the data are highly skewed. A very small number of observations are more than three standard deviations away from the mean. As we are not sure if this is caused by the outliers in the EPA data or genuine difference in pollution intensity, we adopt a simple transformation that would help us avoid the dominance of outliers in our subsequent statistical analyses. More precisely, the data on pollution were converted into a pollution intensity index taking on the values from 0 to 2 by using the following criteria: the index takes on the value of zero if emissions in all four categories (air, land, water and underground) are in the lowest 33 percentiles; the value of 2 if emissions in any category are in the top 33 percentiles; and finally, the value of 1 in all other cases.

The second measure, the abatement index, is based on the data on total pollution abatement expenditures, as reported in the Manufacturers' Pollution Abatement Capital Expenditures and Operating Costs Survey (Census Bureau, 1994). First, we aggregate operating costs and capital expenditures related to pollution abatement. Then, we follow a similar procedure to that described above to obtain median values (normalized by sales) for each SIC code. The abatement data are also converted into an index ranging from 0 (if the normalized value was below the 33rd percentile) to 2 (if the value was above the 66th percentile).

Both indices are calculated based on two 4-digit SIC codes that describe the operations of each firm in our sample.[7] If the index values for the two industry codes differ, the higher value is used. Note that dropping all firms for which index values differ between the two SIC codes does not change the conclusions of the paper. For illustration, Appendix Table A1 lists the classification of industries at the 3-digit SIC level. The regressions, however, use classification at the (more detailed) 4-digit level.

The main drawback of both indices is that they are based on the U.S. data. Thus we are assuming that the relative ranking of pollution intensity of two facilities set up by American investors overseas (e.g., Dupont Poland and Nike Poland) is the same as the relative ranking of their pollution intensity at home (e.g., Dupont and Nike in the U.S.).[8] Further, it may be argued that the total amount of pollutants emitted does not take into account the differences in toxicity risks associated with different substances. Thus, an industry emitting a large quantity of a relatively harmless substance would rank as a greater polluter than another industry emitting a small quantity of a very potent pollutant. However, Dasgupta and Meisner (1998) have shown that at the aggregate level, there is no significant variation in rankings of industries based on total emissions and toxicity risks.

---

[7] Note that *Worldscope* database, from which we obtained firm SIC codes, reports up to nine 4-digit SIC codes for each firm ranked in order of importance. We used the first two.

[8] Of course, this assumption would be violated if industries can separate various stages of production and move pollution-intensive stages abroad.

1.3 Strength of Environmental Standards

We measure the strength of environmental protection of the host countries in several different ways, recognizing that each of them has its own advantages and limitations. The definitions of all measures are described below.

1.3.1 Participation in International Treaties

Treaties I = participation in international treaties. Five international treaties have been developed by the United Nations Economic Commission for Europe during the past twenty-one years. We take into account three treaties that came into effect before or during the time relevant for our data set (i.e., before 1995): the Convention on Long-range Transboundary Air Pollution, the Convention on Environmental Impact Assessment in a Transboundary Context and the Convention on the Transboundary Effects of Industrial Accidents.[9] The information on treaties comes from the United Nations Economic Commission for Europe. An index is created by awarding each country 1 point for ratifying each treaty prior to 1996 and 0.5 point for signing each treaty before 1996 or ratifying it after that time.[10] Thus, the index can range from 0 to 3.

Treaties II = ratification of international treaties. Since only ratification of a treaty results in a country having a legal obligation to comply, we restrict the index to awarding a point only for ratifying a treaty before 1996 and zero otherwise. Again the index ranges from 0 to 3.

Mindful that participation in treaties and enforcement on the ground are not the same thing, we also construct a separate measure that adjusts for possible strength of enforcement. Since active NGO movement tends to exert pressure on the government to enforce environmental regulation, we adjust for enforcement by making use of information on the number of environmentally oriented NGOs in a host country. Thus,

Enforcement-adjusted treaty index = *Treaties* × number of environmental NGOs per million people in country *k*. The figures on NGOs come from the *Environmental Encyclopedia and Directory* (1998).

1.3.2 Quality of Air and Water Ambient and Emission Standards

EBRD index = index of air and water ambient and emission standards in country *k*, which ranges from 1 denoting the weakest standards to 3 denoting the strongest. The index value of 1 is awarded to countries with a maximum permissible concentration (MPC) system in place and the MPC being broadly based on the former Soviet system. The index equals 2 for countries with a new system being introduced, either as an evolution of MPC or in order to meet EU requirements. The highest index denotes countries having essentially a new standards system is in place, often following EU

[9] We exclude the Convention on the Protection and Use of Transboundary Water-courses and International Lakes as it may not be applicable to all countries.

[10] The 1996 cutoff is determined by the coverage of our sample. We want to give more weight to treaties that were ratified during the period investments included in our sample took place.

requirements. The index reflects laws on the books but not their enforcement. The source is EBRD (1997).

Similar to the previous measure, we also construct an enforcement-adjusted standard index as *Enforcement-adjusted EBRD index* = *EBRD index* × number of environmental NGOs per million people in country *k*.

1.3.3 Environmental Sustainability Index

The Environmental Sustainability Index is a measure of overall progress towards environmental sustainability. It was developed jointly by The World Economic Forum, the Yale Center for Environmental Law and Policy, and the Center for International Earth Science Information Network for 142 countries. The ESI scores are based upon a set of 20 core "indicators," each of which combines two to eight variables for a total of 68 underlying variables. The ESI permits cross-national comparisons of environmental progress in a systematic and quantitative fashion. In this study, we use the overall index as well as three of its components, which we consider the most relevant to our analysis.

ESI overall = the overall index expressed in standard normal percentiles.

ESI env quality = a component of the ESI index capturing quality of air and water, availability of water per capita and severity of human induced soil degradation; expressed in standard normal percentiles.

ESI env stress = reflects anthropogenic stress on the natural environment; based on data capturing, among others, air pollution emissions over populated areas, fertilizer and pesticide use; expressed in standard normal percentiles.

ESI env institutions = reflects the extent to which a country possesses institutions and underlying social attitudes that will enable it to foster effective responses to environmental challenges; based on information including, among others, country's technological achievement, presence of NGOs, control of corruption, subsidies on energy usage; expressed in standard normal percentiles.

Since the ESI was developed only in 2001 there is a regrettable mismatch between the time period it covers and that of our sample. Note, however, that the mismatch is smaller than it may seem as, due to lags in data reporting, the figures used to build the index were a couple years old at the time it was constructed.

1.3.4 Observed Actual Reduction in Various Pollutants

For a number of transition economies, we have collected data on the actual observed percentage reduction of two major pollutants (lead and CO_2). These might be viewed as result-based, enforcement-effort-adjusted, alternative measures of the strength of the environmental standard in the countries. Since these changes in emissions of lead and CO_2 may be largely due to output drop experienced by many transition countries during the early 1990s, we make an adjustment to take this into account. Thus we define

Reduction in lead emissions = percentage reduction in total lead emissions adjusted for percentage change in GDP during 1990-96. Source: OECD (1999, p. 47) for emissions; and *World Bank Global Development Indicators* for GDP change.

Reduction in CO_2 emissions = percentage reduction in CO_2 emissions relative to the growth rate of GDP during 1992-95. Both CO_2 and GDP figures are from the *World Development Indicators* database.

Variables *Treaties* and *EBRD index* reflect environmental standards on the book. *Adjusted Treaties* and *Adjusted EBRD index* are standards on the book adjusted for the strength of enforcement. *ESI* variables capture environmental sustainability of a country as well as its attitude towards environmental protection. The last two variables reflect the actual progress that has been made in lowering pollution emissions, thus they capture the combination of laws and their implementation. Again, each measure is useful in some ways but suffers from some drawbacks in other ways. Our strategy is to be broad-minded and not to rely exclusively on any particular measure.[11]

All measures are listed in Appendix Table A2. Summary statistics are presented in Table 1, while correlations can be found in Appendix Table A3.

1.4 Measuring Corruption in Host Countries

1.4.1 *Global Competitiveness Report/World Development Report* Corruption Index

Corruption, by its very nature, is difficult to measure. Most of the available indices are based on subjective perceptions from surveys of firms or individuals.[12] Many of them do not cover enough transition economies to be useful to us. Thus, we adopt a composite measure based on the *Global Competitiveness Report* (GCR) 1997 and the *World Development Report* (WDR) 1997 corruption indices. The *Global Competitiveness Report* is produced jointly by the Geneva-based World Economic Forum and Harvard Institute for International Development. The survey for the report was conducted in late 1996 on 2,827 firms in 58 countries. The GCR survey asked respondents to rate the level of corruption in their country on a one-to-seven scale, based on the extent of "irregular, additional payments connected with imports and exports permits, business licenses, exchange controls, tax assessments, police protection or loan applications." The GCR Corruption Index is based on the country average of the individual ratings.

---

[11] While it would be interesting to endogenize the environmental standards as has been done in an excellent paper by Fredriksson et al. (2003), the data limitations prevent us from doing so. The potential endogeneity problem is mitigated by the fact that a large part of the effort to improve environmental standards in Eastern Europe has been undertaken in the context of the expected accession to the European Union. We also hope that the inclusion of host country fixed effects would attenuate the problem.

[12] Wei (2000b) discusses the relative merits and drawbacks of four types of corruption measures.

Table 1: Summary Statistics

Variable	No. of obs.	Mean	Std. dev.	Min	Max
Firm size	143	14.2	1.7	8.0	18.7
R&D-intensity	143	2.9	3.2	0.0	17.4
Regional experience	143	.52	0.5	0.0	1.0
Emissions	116	0.8	0.7	0.0	2.0
Abatement	83	1.3	0.9	0.0	2.0
GDP	25	8.6	1.6	6.1	12.1
Relative GDP per capita	5725	3.4	1.1	-0.8	5.7
Corporate tax	23	29.6	6.4	15.0	40.0
Distance	3289	7.9	0.9	4.6	9.7
Corruption I	19	4.5	0.8	2.6	5.5
Corruption II	23	4.3	1.8	0.0	8.0
Treaties I	25	1.5	0.9	0.0	3.0
Treaties II	25	0.9	0.9	0.0	3.0
EBRD index	25	1.5	0.7	1.0	3.0
NGOs/ pop	25	2.9	4.2	0.2	19.3
Reduction in lead emissions	25	13.0	28.1	-42.1*	84.8
Reduction in CO_2 emissions	10	27.1	23.8	-25.9*	63.4
ESI overall	22	50.0	8.0	36.8	63.2
ESI env quality	22	48.6	10.8	25.7	65.4
ESI env stress	22	59.2	12.1	31.0	76.8
ESI env institutions	22	41.5	13.1	20.5	66.2

* a negative value indicates that the decrease in GDP was larger than the reduction in emissions.
Firm size, GDP, relative GDP per capita and distance are expressed in logarithms.

The WDR index comes from a 1996 World Bank survey of 3,866 firms in 73 countries conducted in preparation for the *World Development Report 1997*. Question 14 of that survey asks: "Is it common for firms in my line of business to have to pay some irregular, 'additional' payments to get things done?" The respondents were asked to rate the level of corruption on a one-to-six scale. The WDR corruption index is based on the country average of the individual answers.

For both corruption indices, the original sources are such that a higher number implies lower corruption. To avoid awkwardness in interpretation, they are re-scaled in this paper so that a high number now implies high corruption.

Each measure covers a different subset of countries for which we have investment data, thus we use a composite corruption index derived by Wei (2000b) and call it *Corruption I*. Since both measures come from surveys with similar methodologies and similar questions and are highly correlated (0.83), Wei combined them using the following procedure: (1) use GCR as the benchmark; (2) compute the ratio of GCR to WDR for all countries that are available in both GCR and the WDR; and (3) for those countries that are covered by WDR but not GCR (which is relatively rare), convert the WDR rating into the GCR scale by using the ratio in (2).

1.4.2 Neumann Corruption Index

Additionally, we use a corruption measure based on the information obtained by Peter Neumann (1994), a journalist at a German business publication *Impulse*, from people with business experience in each host country, mainly German exporters (*Corruption II*). Neumann interviewed on average ten individuals (or minimum three) per country with a guarantee of strict confidentiality. The measure indicates the proportion of the transactions that involved corrupt payments. Note that in the year in which Peter Neumann conducted his interviews, it was not against any German law for German firms to offer bribes to foreign government officials. The Neumann measure has two advantages, it is a "harder" and "more objective" than the other index used and it is based on the information collected in 1994 which was just a year before our FDI data were obtained.[13]

1.5 Other Variables

Following the existing literature on determinants of FDI, our regressions also include proxies for market size (GDP), corporate tax rate, openness to trade (sum of exports and imports divided by the GDP), distance between source and host country, and the GDP per capita of the source economy relative to that of the host.[14] GDP and GDP per capita data refer to 1993 and come from EBRD (1994) for transition economies and *IMF International Financial Statistics* for source economies. The use of two different sources is motivated by a more complete coverage of transition countries by the EBRD. Relative GDP per capita is defined as the log difference between the GDP per capita of the source and home economy. Information on corporate tax rates is from PriceWaterhouseCoopers. Tax rates are expressed in percentages. If several rates apply, the highest one is used. Trade figures are from the UN COMTRADE database. Distance between the capital cities is measured in kilometers. The primary source is Rudloff (1981), supplemented by Pearce and Smith (1984).

The model also includes a proxy for the level of democracy in the host country. As mentioned earlier, authoritarian governments may be less concerned with protecting the environment and at the same time, multinationals may be less likely to invest in countries with authoritarian governments, perhaps out of concern for unpredictable policy changes. The proxy for democracy is a composite measure based on two indicators constructed by Freedom House (2002): political rights and civil liberties. Both indicators range from 1 to 7 with higher numbers implying fewer rights and

[13] Neumann's index was first used by Ades and Di Tella (1997).

[14] For a review of the literature on FDI determinants see Caves (1982) and Froot (1993).

liberties. In the analysis, we use a composite measure derived by Helliwell (1994) and recently employed by Busse (2004)

Democracy = [14 – (Political Rights + Civil Liberties)]/12.

The dependent variable comes from a unique firm-level data set based on the EBRD Foreign Investment Survey conducted in January 1995. A brief questionnaire was sent to about 9,500 firms from all over the world asking them about their planned or undertaken investment projects in Eastern Europe and the former Soviet Union.[15] Additional information about the type of the project was requested. The criterion for including a firm in the survey was a firm's listing in a commercial database *Worldscope*, which provides detailed financial statements and business descriptions for public companies located in more than fifty countries. Sending the questionnaire to all firms listed assured that all major public companies in the world were included. Responses were obtained from 1,405 firms.[16] Unfortunately, the survey did not ask about the time when each investment was undertaken. However, since the magnitude of FDI inflows into the region was marginal before 1989, the information collected pertains to the period 1989-94.[17] Thus our dataset is cross-sectional in nature but varies in two dimensions—by firm and by potential destination country.

Our empirical analysis focuses on investment in manufacturing facilities, since investment in service sectors or in distribution alone is not likely to have a significant environmental impact. As the objective of this study is to explore the impact of government policies on the magnitude and nature of FDI inflows, firms in the oil, gas and coal sector, which are likely to be attracted to natural resource endowments, are excluded from the estimations. Moreover, to limit the number of zeros on the left hand side of the regression, we include only firms with at least one manufacturing investment in the region. Thus, our final sample includes 143 firms that undertook 355 investment projects in the region. The potential number of observation in our sample is equal to 3,575 (=143×25) investment decisions at the firm level.[18]

Information on firm characteristics, such as a firm's size and R&D intensity, is from *Worldscope*. From the survey, we get the proxy for regional experience, defined as a dummy variable taking on the value of one if a given firm had had a trading relationship with the region before the transition process began in 1990 and zero otherwise.

---

[15] The source countries in the sample listed in order of importance include: United Kingdom, United States, Japan, Canada, Germany, France, Finland, Switzerland, South Africa, Sweden, Ireland, Australia, Norway, Italy, Malaysia, Netherlands, Belgium, Denmark, New Zealand, Austria, Singapore, Portugal, Argentina, Colombia, Greece, Philippines, South Korea and Hong Kong, China.

[16] The question of response bias is discussed in the Appendix.

[17] Eastern Europe and the Soviet Union were virtually closed to foreign investment before 1989 (see Meyer, 1995; Dunning and Rojec, 1993; Hunya, 1997).

[18] The actual number in the regressions is smaller due to missing values of pollution or environmental standard proxies.

2. Estimation Results

2.1 Basic Regressions

We begin the test of the "pollution haven" hypothesis with regressions employing participation in international treaties as a proxy for environmental standards in a host country. The results, presented in terms of marginal effects, can be found in Table 2. Standard errors, adjusted for clustering by host country, are listed in brackets. As stated in the last section, the coefficient on the environmental regime (labeled *Env Std*) captures a *volume effect* – the (marginal) effect of the strength of environmental protection in a host country on the overall volume of inward FDI. The coefficient on the product of the host country's environmental regime and the investing firm's pollution intensity captures a *composition effect* – whether stronger environmental protection discourages the investment from more polluting industries by a greater amount.

The results provide some support for the volume effect but no evidence of the composition effect. The coefficient on environmental standards is negative and statistically significant in three out of six regressions. Interestingly, in all but one case, it is statistically significant only when environmental standard is accompanied by a corruption proxy as an explanatory variable. This gives support to our argument suggesting the importance of including the corruption variable. The interaction term between environmental regime and the emission index does not reach the conventional significance levels. Probably the strongest finding relevant to the "pollution haven" hypothesis is that firms in less polluting industries (as proxied by the emissions index) appear to be more likely to undertake FDI in transition economies. The last finding has two potential explanations: the region is oversaturated with heavy industry and thus attracts less FDI to these sectors (which tend to be pollution-intensive) and/or foreign investors may be unwilling to acquire existing companies in polluting sectors out of concern about unresolved liability for the past environmental damage.

The other variables have the expected signs. We find that larger and less R&D-intensive firms are more likely to undertake FDI. As far as host country characteristics are concerned, the data indicate that larger and more democratic countries as well as those located close to the investor's home country are more attractive to FDI. High corporate taxes and a greater incidence of corruption tend to discourage foreign investors, but they do not appear to be significant in all regressions. Higher difference in income levels between source and host countries is associated with more FDI, while the opposite is true in the case of host countries which are more open to trade. These effects, however, appear to be significant only in two cases.

Table 3 presents results of regressions in which environmental standards on the books are adjusted for enforcement. The findings suggest the presence of a volume effect. In three out of four regressions we find a negative and significant coefficient on the measures of environmental standards, which indicates that countries with higher standards attract less FDI. There is no evidence in favor of a composition effect. The interaction term between environmental regime and the emission index reaches the conventional significance levels only in two cases, but contrary to our expectations it bears a positive sign. The presence of environmental NGOs does not have a deterring effect on FDI, to the contrary, it seems to be associated with higher inflows.

Table 2: Participation in International Treaties

	Treaties I (ratification and/or signing)			Treaties II (ratification only)		
Firm size	0.020***	0.0195***	0.0173***	0.0195***	0.0195***	0.0172***
	[0.003]	[0.0033]	[0.0035]	[0.0034]	[0.0034]	[0.0034]
R&D intensity	-0.008***	-0.0079***	-0.0071***	-0.0079***	-0.0079***	-0.0071***
	[0.003]	[0.0028]	[0.0027]	[0.0027]	[0.0027]	[0.0026]
Regional experience	0.006	0.0057	0.0077	0.0053	0.0053	0.0074
	[0.011]	[0.0111]	[0.0113]	[0.0112]	[0.0112]	[0.0113]
Distance	-0.031***	-0.0306***	-0.0285**	-0.0303***	-0.0302***	-0.0280***
	[0.009]	[0.0092]	[0.0113]	[0.0087]	[0.0086]	[0.0102]
Relative GDP per capita	0.01	0.0096	0.0194*	0.0078	0.0078	0.0218*
	[0.012]	[0.0122]	[0.0117]	[0.0119]	[0.0120]	[0.0118]
GDP	0.048***	0.0474***	0.0460***	0.0470***	0.0470***	0.0482***
	[0.005]	[0.0054]	[0.0052]	[0.0055]	[0.0055]	[0.0056]
Corporate tax	-0.001	-0.0013	-0.0001	-0.0013*	-0.0014*	-0.0003
	[0.001]	[0.0008]	[0.0011]	[0.0007]	[0.0007]	[0.0012]
Democracy	0.057	0.0557	0.1310***	0.0574*	0.0578*	0.1372***
	[0.043]	[0.0423]	[0.0380]	[0.0327]	[0.0325]	[0.0323]
Openness	-0.037*	-0.0373*	-0.0241	-0.032	-0.0319	-0.0249
	[0.022]	[0.0214]	[0.0192]	[0.0199]	[0.0198]	[0.0193]
Corruption I	**-0.023***	**-0.0227****		-0.0168	-0.0167	
	[0.012]	**[0.0114]**		[0.0111]	[0.0111]	
Env Std	**-0.019***	**-0.0228****	-0.0154	**-0.0179***	-0.0158	-0.0168
	[0.011]	**[0.0116]**	[0.0167]	**[0.0108]**	[0.0118]	[0.0118]
Emissions	**-0.014*****	**-0.0258****	-0.0221	**-0.0143*****	-0.011	-0.0099
	[0.004]	**[0.0128]**	[0.0138]	**[0.0044]**	[0.0070]	[0.0068]
Env Std×Emissions		0.006	0.005		-0.0028	-0.0027
		[0.0067]	[0.0072]		[0.0052]	[0.0054]
Number of obs	2204	2204	2668	2204	2204	2668
Psedo R^2	0.28	0.28	0.24	0.28	0.28	0.25
obs. P	0.12	0.12	0.11	0.12	0.12	0.11
pred. P	0.06	0.06	0.06	0.06	0.06	0.06

All results are presented in terms of marginal effects evaluated at the sample mean.
Standard errors, listed in brackets, have been adjusted for clustering by host country.
***, **, * denotes significance at the 1, 5 and 10 percent level respectively.

2.2 Robustness Checks and Extensions

To check the robustness of our findings, we present results from regressions using alternative measures of environmental regulation (Table 4). No support is found for the "pollution haven" when the raw or enforcement-adjusted EBRD's rating of the air and water standard is used. Furthermore, there is no support for the hypothesis when reductions in emissions are used as a proxy for environmental regime or when the ESI indices are employed. Strikingly, we find again that multinationals in cleaner industries are more likely to engage in FDI. Thus, the evidence from Table 4 is consistent with that found in the previous table.

Table 3: Enforcement Adjusted Treaties

	Treaties I		Treaties II	
Firm size	0.0200***	0.0168***	0.0204***	0.0174***
	[0.0032]	[0.0034]	[0.0036]	[0.0035]
R&D intensity	-0.0077***	-0.0069***	-0.0075***	-0.0067***
	[0.0027]	[0.0025]	[0.0027]	[0.0024]
Regional experience	0.0055	0.0077	0.0048	0.0072
	[0.0113]	[0.0112]	[0.0112]	[0.0113]
Distance	-0.0269***	-0.0248***	-0.0270***	-0.0245***
	[0.0089]	[0.0087]	[0.0086]	[0.0087]
Relative GDP per capita	-0.0059	0.0223*	-0.013	0.0123
	[0.0120]	[0.0123]	[0.0145]	[0.0117]
GDP	0.0411***	0.0491***	0.0375***	0.0436***
	[0.0036]	[0.0052]	[0.0051]	[0.0052]
Corporate tax	-0.0007	0.0005	0.0001	0.0007
	[0.0004]	[0.0011]	[0.0004]	[0.0011]
Democracy	0.1259***	0.1228***	0.0800**	0.1222***
	[0.0388]	[0.0379]	[0.0325]	[0.0281]
Openness	-0.0103	-0.019	-0.0091	-0.0166
	[0.0177]	[0.0181]	[0.0171]	[0.0186]
Corruption I	0.012		0.0073	
	[0.0108]		[0.0113]	
Env Std	**-0.0210*** **	-0.0032	**-0.0092*** **	**-0.0068*** **
	[0.0051]	[0.0064]	**[0.0028]**	**[0.0022]**
Emissions	**-0.0186*** **	**-0.0166*** **	**-0.0134** **	**-0.0106* **
	[0.0061]	**[0.0058]**	**[0.0060]**	**[0.0054]**
Env Std ×Emissions	**0.0010*** **	**0.0008** **	-0.0001	-0.0011
	[0.0003]	**[0.0004]**	[0.0017]	[0.0018]
NGOs/pop	**0.0340*** **	0.0072	**0.0031*** **	**0.0027* **
	[0.0081]	[0.0094]	**[0.0010]**	**[0.0015]**
Number of obs	2204	2668	2204	2668
Psedo R^2	0.29	0.25	0.28	0.25
obs. P	0.12	0.11	0.12	0.11
pred. P	0.06	0.05	0.06	0.05

All results are presented in terms of marginal effects evaluated at the sample mean. Standard errors, listed in brackets, have been adjusted for clustering by host country.
***, **, * denotes significance at the 1, 5 and 10 percent level respectively.

Table 4: Alternative Measures of Environmental Standards in Host Countries

	EBRD index	Adjusted EBRD index	Reduction in emissions		Environmental Sustainability Index			
			CO_2	lead	overall	env quality	env stress	env institutions
Firm size	0.019***	0.0191***	0.0195***	0.0256***	0.0198***	0.0202***	0.0202***	0.0198***
	[0.003]	[0.0034]	[0.0034]	[0.0035]	[0.0038]	[0.0038]	[0.0037]	[0.0036]
R&D	-0.008***	-0.008***	-0.0080***	-0.0094*	-0.008***	-0.008***	-0.008***	-0.008***
Intensity	[0.003]	[0.0028]	[0.0028]	[0.0049]	[0.0030]	[0.0030]	[0.0030]	[0.0030]
Regional	0.006	0.0058	0.0058	0.014	0.005	0.005	0.0049	0.005
Experience	[0.011]	[0.0112]	[0.0112]	[0.0179]	[0.0119]	[0.0119]	[0.0119]	[0.0120]
Distance	-0.030***	-0.0256***	-0.0297***	-0.0268*	-0.031***	-0.031***	-0.032***	-0.031***
	[0.009]	[0.0078]	[0.0093]	[0.0141]	[0.0097]	[0.0095]	[0.0100]	[0.0096]
Relative GDP per capita	0.014	0.0073	0.0104	-0.0094	0.0175	0.0128	0.0113	0.0171
	[0.012]	[0.0114]	[0.0129]	[0.0267]	[0.0131]	[0.0138]	[0.0136]	[0.0134]
GDP	0.046***	0.0373***	0.0485***	0.0631***	0.0499***	0.0480***	0.0504***	0.0497***
	[0.006]	[0.0067]	[0.0053]	[0.0090]	[0.0034]	[0.0037]	[0.0045]	[0.0038]
Corporate	-0.001	-0.0001	-0.0013*	-0.0058*	-0.0006	-0.0004	0.0000	-0.001
Tax	[0.001]	[0.0005]	[0.0007]	[0.0030]	[0.0010]	[0.0010]	[0.0016]	[0.0011]
Democracy	0.007	0.0807**	0.0181	-0.0656	0.0284	0.0577	0.0453	0.0064
	[0.039]	[0.0344]	[0.0407]	[0.0853]	[0.0436]	[0.0447]	[0.0487]	[0.0477]
Openness	-0.033	-0.0107	-0.0313	-0.0418*	-0.0380*	-0.0344	-0.0327	-0.0346
	[0.022]	[0.0182]	[0.0211]	[0.0228]	[0.0222]	[0.0220]	[0.0244]	[0.0240]
Corruption	**-0.025***	0.0232	**-0.0272****	**-0.0445*****	**-0.0275****	**-0.0233****	**-0.0325*****	-0.021
I	**[0.013]**	[0.0202]	**[0.0113]**	**[0.0089]**	**[0.0127]**	**[0.0112]**	**[0.0116]**	[0.0166]
Env Std	0.008	**0.0241****	0.0001	**0.0014*****	0.0003	0.0006	0.0006	0.0007
	[0.011]	**[0.0099]**	[0.0003]	**[0.0004]**	[0.0010]	[0.0004]	[0.0008]	[0.0012]
Emissions	**-0.026****	**-0.018*****	**-0.016*****	-0.0107	**-0.081****	-0.0386	-0.0162	**-0.053*****
	[0.010]	**[0.0052]**	**[0.0050]**	[0.0107]	**[0.0355]**	[0.0287]	[0.0199]	**[0.0201]**
Env Std×	0.006	**0.0007*****	0.0002	0.0000	**0.0012***	0.0005	0.0000	**0.0008***
Emissions	[0.005]	**[0.0002]**	[0.0002]	[0.0003]	**[0.0006]**	[0.0005]	[0.0003]	**[0.0004]**
NGOs/pop		**-0.0443****						
		[0.0186]						
Number of obs	2204	2204	2204	1044	2088	2088	2088	2088
Psedo R^2	0.28	0.28	0.28	0.23	0.27	0.27	0.27	0.27
obs. P	0.12	0.12	0.11	0.12	0.12	0.12	0.12	0.11
pred. P	0.05	0.06	0.07	0.06	0.06	0.06	0.06	0.04

All results are presented in terms of marginal effects evaluated at the sample mean.
Standard errors, listed in brackets, have been adjusted for clustering by host country.
***, **, * denotes significance at the 1, 5 and 10 percent level respectively.

As another robustness test, we re-estimate the model using the abatement cost index to capture a firm's pollution intensity (Table 5). When this change is made, four of the twelve interaction terms are significant, but their signs are negative in two regressions and positive in the other two cases. The same is true of the coefficients on environmental standards. Thus, again the data produce no robust support for the "pollution haven" hypothesis. Note that when abatement costs are used as a proxy for industry pollution intensity, we find no systematic relationship between "dirtiness' of the sector and the firm's propensity to undertake FDI.

Further, we re-estimate our model employing *Corruption II* (Neumann index) as a measure of corruption incidence in host countries. The regression results, presented in Table 6, give no support to the hypothesis.

Since our measures of investor pollution intensity have been calculated using U.S. data, one could argue that this calls for limiting our sample to U.S. firms only. When we re-estimate our model on this sub-sample (see Table 7), in five out of twelve cases we find evidence in support of the volume effect. There is, however, no support for the composition effect. As before, the data suggest that firms in dirty industries are less interested in establishing presence in the region.

One may argue that omitted country specific variables may have affected our results. Thus, we also re-estimate the models replacing country specific variables with fixed effects (see Tables 8 and 9). We find no support at all for the composition effect when either the emissions or the abatement cost index is used. The previous result suggesting weak propensity of firms in dirty industries to invest in the region, is still present, albeit only in a handful of cases. When industry pollution intensity is proxied with abatement cost index the latter result disappears.

Being concerned about unobserved investor characteristics, we also estimate a model with firm fixed effects, which, does not change our earlier conclusions. As before, we find no support whatsoever for the composition effect when either of the pollution-intensity proxies is used. A handful of cases produce results consistent with the volume effect (see Tables 10 and 11).

We have performed further robustness checks that are not reported here to save space. For instance, we narrowed the sample to include only firms in highly polluting industries (emissions or abatement index equal to two) but the results produced gave very little support to the "pollution haven" hypothesis.

As alternative way of adjusting the measures of environmental standards for enforcement, we divided them by the corruption index. This adjustment, however, did not alter the qualitative results since the coefficients of interest remained insignificant in almost all the regressions.

Table 5: Alternative Measure of Pollution Intensity (Abatement Costs)

	Treaties I	Treaties II	Enforcement adjusted		EBRD index	Adjusted EBRD index	Reduction in emissions		Environmental Sustainability Index			
			treaties I	treaties II			CO_2	lead	overall	env quality	env stress	env institutions
Corruption I	-0.017	-0.0085	**0.0253****	0.0117	**-0.0224***	0.0179	**-0.0240****	**-0.036*****	**-0.0231***	-0.0165	**-0.027****	-0.0163
	[0.012]	[0.0101]	**[0.0102]**	[0.0116]	**[0.0121]**	[0.0216]	**[0.0104]**	**[0.0083]**	**[0.0124]**	[0.0112]	**[0.0113]**	[0.0159]
Env Std	-0.026	-0.023	**-0.0228*****	**-0.0099*****	-0.0023	**0.0195***	-0.0003	0.0008	0.0007	**0.0011****	0.0012	0.001
	[0.021]	[0.0145]	**[0.0064]**	**[0.0033]**	[0.0113]	**[0.0109]**	[0.0004]	[0.0007]	[0.0011]	**[0.0005]**	[0.0009]	[0.0013]
Abatement	-0.004	0.0013	0.0068	0.0018	-0.012	0.0054	-0.0005	0.007	-0.0266	0.0034	0.0236	-0.0149
	[0.021]	[0.0105]	[0.0044]	[0.0063]	[0.0097]	[0.0047]	[0.0038]	[0.0222]	[0.0450]	[0.0255]	[0.0176]	[0.0253]
Env Std	0.003	0.0008	**-0.0007****	0.001	**0.0070***	**-0.0005****	**0.0003***	0.0001	0.0005	0.0000	-0.0004	0.0004
×Abatement	[0.009]	[0.0058]	**[0.0003]**	[0.0024]	**[0.0039]**	**[0.0002]**	**[0.0001]**	[0.0006]	[0.0008]	[0.0005]	[0.0003]	[0.0005]
NGOs/pop			**0.0389*****	**0.0025*****		-0.0338						
			[0.0097]	**[0.0010]**		[0.0207]						
Number of obs	1577	1577	1577	1577	1577	1577	1577	747	1494	1494	1494	1494
Psedo R^2	0.34	0.34	0.35	0.34	0.34	0.34	0.34	0.31	0.33	0.33	0.32	0.33
obs. P	0.10	0.10	0.11	0.11	0.11	0.11	0.10	0.12	0.12	0.12	0.12	0.12
pred. P	0.04	0.04	0.04	0.04	0.04	0.04	0.04	0.04	0.04	0.04	0.04	0.06

All results are presented in terms of marginal effects evaluated at the sample mean.
Standard errors, listed in brackets, have been adjusted for clustering by host country.
***, **, * denotes significance at the 1, 5 and 10 percent level respectively.

Table 6: Alternative Measure of Corruption (Neumann Index)

	Treaties I	Treaties II	Enforcement adjusted		EBRD index	Adjusted EBRD index	Reduction in emissions		Environmental Sustainability Index			
			treaties I	treaties II			CO_2	lead	overall	env quality	env stress	env institutions
Corruption II	0.000	0.0015	-0.0023	-0.0001	-0.0009	0.0035	0.0012	**-0.0314***	-0.0011	-0.0002	0.0038	-0.0042
	[0.006]	[0.0064]	[0.0053]	[0.0057]	[0.0052]	[0.0044]	[0.0049]	**[0.0169]**	[0.0052]	[0.0053]	[0.0048]	[0.0057]
Env Std	-0.021	**-0.0240***	-0.0046	**-0.0082*****	**0.0249***	**0.015*****	**0.0009****	**0.002*****	**0.0022****	**0.0014****	0.0011	**0.0022****
	[0.024]	**[0.0145]**	[0.0075]	**[0.0025]**	**[0.0141]**	**[0.0037]**	**[0.0004]**	**[0.0005]**	**[0.0010]**	**[0.0006]**	[0.0008]	**[0.0011]**
Emissions	**-0.031****	**-0.015****	**-0.017*****	**-0.0120***	**-0.024****	**-0.016*****	**-0.016*****	-0.0167	**-0.0751***	-0.0306	-0.0234	**-0.0493****
	[0.014]	**[0.0073]**	**[0.0063]**	**[0.0061]**	**[0.0105]**	**[0.0051]**	**[0.0051]**	[0.0147]	**[0.0391]**	[0.0247]	[0.0197]	**[0.0202]**
Env Std	0.01	0.0013	**0.0009****	-0.0006	0.0059	**0.0006****	0.0003	0.0002	**0.0012***	0.0003	0.0002	**0.0008***
×Emissions	[0.007]	[0.0047]	**[0.0004]**	[0.0019]	[0.0055]	**[0.0002]**	[0.0002]	[0.0004]	**[0.0007]**	[0.0004]	[0.0003]	**[0.0004]**
NGOs/pop			0.0092	**0.0027***		**-0.027*****						
			[0.0111]	**[0.0016]**		**[0.0078]**						
Number of obs	2436	2436	2436	2436	2436	2436	2436	1160	2320	2320	2320	2320
Psedo R^2	0.23	0.23	0.24	0.24	0.24	0.25	0.24	0.21	0.23	0.23	0.23	0.23
obs. P	0.11	0.11	0.11	0.11	0.11	0.11	0.11	0.10	0.12	0.12	0.12	0.12
pred. P	0.06	0.07	0.06	0.06	0.06	0.06	0.06	0.07	0.07	0.07	0.07	0.07

All results are presented in terms of marginal effects evaluated at the sample mean.
Standard errors, listed in brackets, have been adjusted for clustering by host country.
***, **, * denotes significance at the 1, 5 and 10 percent level respectively.

Table 7: The Sub-Sample of U.S. Firms

	Treaties I	Treaties II	Enforcement adjusted		EBRD index	Adjusted EBRD index	Reduction in emissions		Environmental Sustainability Index			
			treaties I	treaties II			CO_2	lead	overall	env quality	env stress	env institutions
Corruption I	-0.001	0.0022	0.015	0.0104	-0.0078	-0.0092	-0.0063	**0.0403*****	-0.0032	-0.0011	-0.0129	0.0099
	[0.010]	[0.0115]	[0.0125]	[0.0147]	[0.0107]	[0.0216]	[0.0118]	**[0.0053]**	[0.0103]	[0.0113]	[0.0126]	[0.0117]
Env Std	**-0.029****	**-0.024****	**-0.018*****	**-0.0111*****	-0.0157	0.0005	0.0001	**-0.002*****	0.0005	0.0009	**0.002*****	0.001
	[0.011]	**[0.0109]**	**[0.0051]**	**[0.0042]**	[0.0111]	[0.0090]	[0.0003]	**[0.0002]**	[0.0011]	[0.0006]	**[0.0008]**	[0.0007]
Emissions	**-0.09*****	**-0.066*****	**-0.047*****	**-0.0550*****	**-0.08*****	**-0.048*****	**-0.05*****	**-0.0407****	**-0.210*****	**-0.083****	0.0068	**-0.144*****
	[0.031]	**[0.0201]**	**[0.0163]**	**[0.0169]**	**[0.0224]**	**[0.0168]**	**[0.0167]**	**[0.0203]**	**[0.0546]**	**[0.0346]**	[0.0212]	**[0.0324]**
Env Std	**0.026****	**0.0178***	0.0003	0.0059	**0.023*****	0.0005	0.0002	0.0000	**0.003*****	0.0007	**-0.0009****	**0.002*****
×Emissions	**[0.013]**	**[0.0100]**	[0.0014]	[0.0046]	**[0.0080]**	[0.0012]	[0.0003]	[0.0005]	**[0.0009]**	[0.0006]	**[0.0005]**	**[0.0006]**
NGOs/pop			**0.027*****	-0.0004		-0.0022						
			[0.0085]	[0.0015]		[0.0168]						
Number of obs	399	399	399	399	399	399	399	189	378	378	378	378
Psedo R^2	0.37	0.37	0.37	0.37	0.37	0.36	0.36	0.36	0.38	0.36	0.36	0.38
obs. P	0.12	0.12	0.12	0.12	0.12	0.12	0.12	0.11	0.12	0.12	0.12	0.12
pred. P	0.04	0.04	0.04	0.04	0.04	0.04	0.04	0.04	0.04	0.05	0.04	0.03

All results are presented in terms of marginal effects evaluated at the sample mean.
Standard errors are listed in brackets.
***, **, * denotes significance at the 1, 5 and 10 percent level respectively.

Table 8: Regressions with Host Country Fixed Effects — Treaties and EBRD index

		Treaties I	Treaties II	Enforcement adjusted treaties I	Enforcement adjusted treaties II	EBRD index	Adjusted EBRD index
Panel A. Specification with Emissions							
Firm size	0.305***	0.305***	0.306***	0.306***	0.305***	0.304***	0.3055***
	[0.052]	[0.052]	[0.0524]	[0.052]	[0.0523]	[0.0523]	[0.0524]
R&D	-0.117***	-0.116***	-0.1167***	-0.117***	-0.116***	-0.1164***	-0.1164***
Intensity	[0.028]	[0.028]	[0.0284]	[0.028]	[0.0284]	[0.0284]	[0.0284]
Regional	0.111	0.11	0.1115	0.112	0.1104	0.11	0.112
experience	[0.158]	[0.158]	[0.1584]	[0.158]	[0.1585]	[0.1585]	[0.1585]
Distance	-0.361***	-0.362***	-0.3595***	-0.355***	-0.359***	-0.3638***	-0.3541***
	[0.083]	[0.083]	[0.0831]	[0.083]	[0.0831]	[0.0832]	[0.0831]
Emissions	**-0.183***	-0.331	-0.1272	**-0.231***	-0.1369	-0.3882	**-0.2361****
	[0.099]	[0.362]	[0.1869]	**[0.119]**	[0.1320]	[0.2557]	**[0.1153]**
Env Std×Emissions		0.076	-0.0473	0.01	-0.0271	0.1006	0.0093
		[0.179]	[0.1347]	[0.014]	[0.0515]	[0.1150]	[0.0102]
Number of obs	2552	2552	2552	2552	2552	2552	2552
LR Chi^2	67.95	68.14	68.08	68.52	68.23	68.72	68.79
Prob > Chi^2	0.00	0.00	0.00	0.00	0.00	0.00	0.00
Pseudo R2	0.05	0.05	0.05	0.05	0.05	0.05	0.05
Panel B. Specification with Abatement							
Abatement	0.063	-0.143	0.048	0.128	-0.028	-0.2085	0.1184
	[0.113]	[0.537]	[0.2382]	[0.134]	[0.1583]	[0.3042]	[0.1313]
Env Std×		0.102	0.0122	-0.013	0.0558	0.125	-0.009
Abatement		[0.260]	[0.1698]	[0.014]	[0.0696]	[0.1308]	[0.0107]
Number of obs	1328	1328	1328	1328	1328	1328	1328
LR Chi^2	55.27	55.43	55.28	56.08	55.95	56.18	55.96
Prob > Chi^2	0.00	0.00	0.00	0.00	0.00	0.00	0.00
Psedo R^2	0.07	0.07	0.07	0.07	0.07	0.07	0.07

Regressions in Panel B also include firm size, R&D intensity, regional experience and distance, which are not reported to save space.

All results are presented in terms of marginal effects evaluated at the sample mean. Standard errors are listed in brackets.

***, **, * denotes significance at the 1, 5 and 10 percent level respectively.

Summing up, despite our best efforts to give a chance for the "pollution haven" hypothesis to show up in the data, we were unable to find more than occasional weak support for the hypothesis. Our data indicate that host country environmental standards have very little impact on FDI inflows both in terms of the volume and in terms of composition. We do not find robust evidence of foreign investment in pollution-intensive industries flocking to countries with weak environmental regimes. On the contrary, the data indicate that firms in cleaner sectors are more likely to undertake FDI.

Table 9: Regressions with Host Country Fixed Effects—Reduction in Emissions and ESI Index

	Reduction in emissions		Environmental Sustainability Index			
	CO_2	lead	overall	env quality	env stress	Env Institutions
Panel A. Specification with Emissions						
Firm size	0.3056***	0.3585***	0.2999***	0.3005***	0.3007***	0.2998***
	[0.0524]	[0.0795]	[0.0525]	[0.0526]	[0.0526]	[0.0525]
R&D intensity	-0.1167***	-0.1345***	-0.1154***	-0.1157***	-0.1158***	-0.1155***
	[0.0284]	[0.0445]	[0.0285]	[0.0285]	[0.0285]	[0.0285]
Regional experience	0.1111	0.2188	0.0965	0.0972	0.0971	0.0964
	[0.1585]	[0.2416]	[0.1592]	[0.1591]	[0.1591]	[0.1591]
Distance	-0.3608***	-0.3859***	-0.3645***	-0.3638***	-0.3644***	-0.3662***
	[0.0830]	[0.1200]	[0.0832]	[0.0833]	[0.0834]	[0.0832]
Emissions	**-0.2348****	-0.2586	-1.0258	-0.4074	-0.2773	-0.6802
	[0.1162]	[0.2900]	[0.7531]	[0.4482]	[0.4312]	[0.4354]
Env Std× Emissions	0.0051	0.004	0.0158	0.0045	0.0018	0.0105
	[0.0059]	[0.0076]	[0.0139]	[0.0086]	[0.0075]	[0.0088]
Number of obs	2552	1160	2436	2436	2436	2436
LR Chi^2	68.72	40.42	67.54	66.51	66.29	67.66
Prob > Chi^2	0.00	0.00	0.00	0.00	0.00	0.00
Pseudo R2	0.05	0.07	0.05	0.05	0.05	0.05
Panel B. Specification with Abatement						
Abatement	-0.0009	0.2851	-0.6205	0.0344	0.5541	-0.3706
	[0.1360]	[0.4743]	[0.9537]	[0.5203]	[0.4790]	[0.5771]
Env Std× Abatement	0.0063	-0.0017	0.0124	0.0006	-0.0088	0.0087
	[0.0076]	[0.0117]	[0.0173]	[0.0098]	[0.0083]	[0.0114]
Number of obs	1328	747	1328	1328	1328	1328
LR Chi^2	55.96	36.52	55.79	55.28	56.4	55.85
Prob > Chi^2	0.00	0.00	0.00	0.00	0.00	0.00
Pseudo R^2	0.07	0.11	0.07	0.07	0.07	0.07

Regressions in Panel B also include firm size, R&D intensity, regional experience and distance, which are not reported to save space. All results are presented in terms of marginal effects evaluated at the sample mean. Standard errors are listed in brackets.
***, **, * denotes significance at the 1, 5 and 10 percent level respectively.

3. Concluding Remarks

This paper tests whether polluting activities move from industrialized countries to developing economies, as claimed by many anti-globalization activists. To enhance our ability to detect the possible "dirty secret" that multinational firms flock to countries with weak environmental protection and that this is particularly the case for more pollution-intensive industries, we examine the location decision pertaining to FDI flows from multiple source countries to 25 economies in Eastern Europe and the former Soviet Union. Transition countries are a suitable region for studying the "pollution haven" hypothesis. While all of them share the common heritage of more than half a century of central planning and opened up to FDI at the same time, they also offer a reasonably large variation in terms of environmental standards.

As a host country's weakness in public institutions, particularly the prevalence of bureaucratic corruption, may deter FDI but at the same time may be correlated with laxity of environmental protection, statistical analysis on the effect of environmental policy on FDI that omits local corruption might fail to detect an effect. Therefore, we explicitly account the effect of host country corruption.

Further, unlike existing studies that examine *either* whether FDI is more likely to flow to locations with laxer standards *or* whether polluting industries are more likely to engage FDI, we take into account both the polluting-intensity of the potential investor and the environmental stringency in the potential host country. Thus we are able to test whether dirty industries are *relatively* more attracted to locations with weak standards.

Finally, we rely on firm-level rather than country- or industry-level data, which allows us to avoid aggregation bias and gives us an opportunity to control for firm characteristics that may affect investment decisions.

Despite these improvements, we find no support for the "pollution haven" hypothesis. On the contrary, our results indicate that firms in less polluting industries are more likely to invest in the region. Our findings are still subject to caveats. The measures of pollution intensity and of the strength of environmental protection are possibly too noisy for us to obtain precise estimates. While we have employed multiple measures of environmental standards, we have not allowed them to be endogenous. Future work will, hopefully, improve on these dimensions.

Appendix: Response Bias

The response rate in the survey was close to fifteen percent. It is likely that firms that perceived the survey as more relevant (for instance, firms that had invested or considered investing in transition economies) were more likely to respond. To check this hypothesis, we examine the list of major foreign investors in Poland compiled by the Polish State Investment Agency (PAIZ, 1995). Poland was chosen for this exercise since it was the most popular destination country in the sample. Out of 329 firms on the list, 118 received the EBRD survey and fifty percent of them responded.

Formal tests on difference in means could not reject the null hypothesis that the means of firm specific variables in the respondent and non-respondent groups were not significantly different from each other. Thus, among the investing firms, the decision to respond to the survey was not systematically related to firm characteristics observable in the data. Unfortunately, it was not possible to identify which among the firms that did not respond to the survey were not interested in undertaking investment in Eastern Europe and the former Soviet Union. We have no reason, however, to suspect that in the case of these firms, the decision to answer the survey was systematically related to their characteristics. Therefore, the data set can be treated as if the investing firms had been over-sampled. This does not affect the results presented in this study, however, since all of regressions focus on firms that invested in at least one country in the region.

Table 10: Regressions with Firm Fixed Effects — Emissions

	Treaties I	Treaties II	Enforcement adjusted		EBRD index	Adjusted EBRD index	Reduction in emissions		Environmental Sustainability Index			
			treaties I	treaties II			CO_2	lead	overall	env quality	env stress	env institutions
Distance	-1.96***	-1.931***	-1.841***	-1.811***	-1.93***	-1.78***	-1.93***	-1.76***	-1.93***	-1.929***	-1.93***	-1.93***
	[0.195]	[0.1928]	[0.201]	[0.1971]	[0.1950]	[0.2016]	[0.1949]	[0.2872]	[0.1967]	[0.1976]	[0.1963]	[0.1969]
Relative GDP	-0.11	-0.2314	-0.399	-0.8448**	0.0451	-0.1453	-0.0501	0.352	-0.0357	-0.0297	-0.0393	-0.028
per capita	[0.259]	[0.2732]	[0.296]	[0.3879]	[0.2601]	[0.2666]	[0.2517]	[0.5633]	[0.2891]	[0.2538]	[0.2513]	[0.2840]
GDP	0.890***	0.8843***	0.758***	0.6582***	0.8712***	0.762***	0.898***	0.909***	0.868***	0.862***	0.872***	0.868***
	[0.085]	[0.0855]	[0.105]	[0.1204]	[0.0824]	[0.1042]	[0.0961]	[0.1985]	[0.0825]	[0.0853]	[0.0839]	[0.0822]
Corporate tax	-0.028**	-0.0333**	-0.017	-0.0046	-0.0188	-0.0071	-0.0222	-0.0306	-0.0148	-0.0138	-0.0079	-0.0148
	[0.014]	[0.0137]	[0.012]	[0.0123]	[0.0120]	[0.0122]	[0.0156]	[0.0663]	[0.0128]	[0.0127]	[0.0218]	[0.0118]
Democracy	0.234	-0.0115	1.058	-0.0205	-0.824	0.2615	-0.5113	-0.6156	-0.4249	-0.3205	-0.194	-0.4235
	[0.769]	[0.7014]	[0.927]	[0.7231]	[0.7471]	[0.7567]	[0.6896]	[1.4110]	[0.7636]	[0.8406]	[0.9181]	[0.6852]
Openness	-0.247	-0.0437	0.215	0.478	-0.2359	0.1295	-0.1791	-0.9223	-0.285	-0.2795	-0.2254	-0.2827
	[0.316]	[0.3432]	[0.376]	[0.4138]	[0.3089]	[0.3659]	[0.3503]	[0.5724]	[0.3103]	[0.3108]	[0.3490]	[0.3111]
Corruption I	0.156	0.2719	0.553*	**0.6645****	0.0169	**0.6321***	-0.0055	-0.3487	-0.0524	-0.036	-0.0631	-0.0447
	[0.224]	[0.2290]	[0.308]	**[0.3073]**	[0.1983]	**[0.3515]**	[0.2049]	[0.2138]	[0.1901]	[0.2001]	[0.1891]	[0.2209]
Env Std	-0.483	**-0.4069****	**-0.309****	**-0.1814****	0.1733	**0.3450****	0.0011	0.0055	-0.0037	0.0025	0.0036	-0.001
	[0.307]	**[0.1956]**	**[0.135]**	**[0.0732]**	[0.1811]	**[0.1641]**	[0.0083]	[0.0113]	[0.0193]	[0.0101]	[0.0142]	[0.0180]
Env Std×	-0.025	-0.0297	0.002	0.0151	0.0369	0.0016	0.004	0.0039	0.0043	-0.0009	0.0014	0.0023
Emissions	[0.181]	[0.1297]	[0.017]	[0.0463]	[0.1040]	[0.0123]	[0.0050]	[0.0082]	[0.0134]	[0.0076]	[0.0069]	[0.0089]
NGOs/pop			**0.511****	**0.0551***		**-0.614****						
			[0.213]	**[0.0307]**		**[0.3055]**						
Number of obs	3078	3078	3078	3078	3078	3078	3078	900	2916	2916	2916	2916
LR Chi^2	675.25	679.18	677.95	680.48	673.43	677.24	672.58	249.88	643.8	643.76	643.88	643.76
Prob > Chi^2	0.00	0.00	0.00	0.00	0.00	0.00	0.00	0.00	0.00	0.00	0.00	0.00
Pseudo R^2	0.44	0.44	0.44	0.44	0.44	0.44	0.44	0.46	0.43	0.43	0.43	0.43

All results are presented in terms of marginal effects evaluated at the sample mean.
Standard errors are listed in brackets.
***, **, * denotes significance at the 1, 5 and 10 percent level respectively.

Table 11: Regressions with Firm Fixed Effects — Abatement

	Treaties I	Treaties II	Enforcement adjusted		EBRD index	Adjusted EBRD index	Reduction in emissions		Environmental Sustainability Index			
			treaties I	treaties II			CO_2	lead	overall	env quality	env stress	env institutions
Distance	-2.16***	-2.13***	-2.064***	-2.012***	-2.13***	-2.009***	-2.14***	-1.631***	-2.119***	-2.105***	-2.144***	-2.0990***
	[0.256]	[0.253]	[0.264]	[0.260]	[0.257]	[0.266]	[0.2570]	[0.3683]	[0.2575]	[0.2572]	[0.2599]	[0.2582]
Relative GDP	0.062	-0.082	-0.465	-0.860*	0.13	-0.074	0.1091	-0.1029	0.2791	0.1498	0.0991	0.3182
Per cap	[0.312]	[0.333]	[0.382]	[0.516]	[0.316]	[0.333]	[0.3083]	[0.8876]	[0.3498]	[0.3061]	[0.3035]	[0.3606]
GDP	0.812***	0.809***	0.683***	0.607***	0.799***	0.728***	0.789***	0.9514***	0.8083***	0.7692***	0.8139***	0.8088***
	[0.104]	[0.105]	[0.133]	[0.155]	[0.101]	[0.130]	[0.1193]	[0.2867]	[0.1030]	[0.1058]	[0.1049]	[0.1026]
Corporate tax	-0.027	-0.034**	-0.019	-0.006	-0.018	-0.008	-0.015	-0.0582	-0.013	-0.0126	0.0035	-0.0183
	[0.016]	[0.016]	[0.014]	[0.015]	[0.014]	[0.015]	[0.0202]	[0.0902]	[0.0154]	[0.0152]	[0.0258]	[0.0140]
Democracy	0.446	0.323	2.184*	0.5	-0.058	0.875	0.0555	0.2474	0.3914	0.6182	0.7178	0.1845
	[0.944]	[0.865]	[1.220]	[0.903]	[0.914]	[0.952]	[0.8582]	[1.7207]	[0.9599]	[1.0428]	[1.1018]	[0.8794]
Openness	-0.545	-0.344	0.298	0.443	-0.542	0.042	-0.5875	-0.3468	-0.5949	-0.5396	-0.3752	-0.5329
	[0.388]	[0.424]	[0.507]	[0.552]	[0.386]	[0.473]	[0.4579]	[1.0266]	[0.3844]	[0.3832]	[0.4289]	[0.3843]
Corruption I	-0.04	0.11	**0.751***	**0.705***	-0.164	0.632	-0.1956	**-0.5456****	-0.1524	-0.0918	-0.2099	-0.0344
	[0.269]	[0.279]	**[0.402]**	**[0.398]**	[0.239]	[0.426]	[0.2548]	**[0.2630]**	[0.2344]	[0.2478]	[0.2280]	[0.2767]
Env Std	-0.156	-0.301	**-0.395****	**-0.207***	-0.059	**0.357***	-0.0093	0.0136	0.0039	0.0082	0.0198	0.0143
	[0.219]	[0.279]	**[0.177]**	**[0.110]**	[0.238]	**[0.195]**	[0.0126]	[0.0199]	[0.0270]	[0.0143]	[0.0180]	[0.0254]
Env Std	-0.15	-0.077	-0.014	0.023	0.084	-0.008	0.0059	-0.0015	0.01	0.0012	-0.0034	0.0064
×Abatement	[0.444]	[0.149]	[0.017]	[0.055]	[0.113]	[0.013]	[0.0059]	[0.0117]	[0.0151]	[0.0084]	[0.0074]	[0.0103]
NGOs/pop			**0.706****	**0.084****		-0.591						
			[0.279]	**[0.037]**		[0.365]						
Number of obs	2166	2166	2166	2166	2166	2166	2166	621	2052	2052	2052	2052
LR Chi^2	486.07	488.77	493.91	492.79	485	491.12	485.41	182.69	460.61	460.54	460.64	460.83
Prob > Chi^2	0.00	0.00	0.00	0.00	0.00	0.00	0.00	0.00	0.00	0.00	0.00	0.00
Pseudo R^2	0.4535	0.456	0.4608	0.4597	0.4525	0.4582	0.45	0.48	0.44	0.44	0.44	0.44

All results are presented in terms of marginal effects evaluated at the sample mean.
Standard errors are listed in brackets.
***, **, * denotes significance at the 1, 5 and 10 percent level respectively.

References

Ades, Alberto, and Rafael Di Tella. 1997. "National Champions and Corruption: Some Unpleasant Interventionist Arithmetic," *Economic Journal,* 107(443): 1023-1042.

Becker, Randy, and Vernon Henderson. 2000. "Effects of Air Quality Regulation on Polluting Industries," *Journal of Political Economy*, 108(2): 379-421.

Busse, Matthias. 2004. "Transnational Corporations and Repression of Political Rights and Civil Liberties: An Empirical Approach," *Kyklos,* 57(1): 45-66.

Caves, Richard. 1982. *Multinational Enterprise and Economic Analysis.* Cambridge University Press: New York.

Census Bureau. 1994. "Pollution Abatement Costs and Expenditures: 1994," *Current Industrial Reports.* U.S. Department of Commerce, ESA.

Dasgupta, Susmita and Craig Meisner. 1998. "Accounting for Toxicity Risks in Pollution Control: Does It Matter? *World Bank Policy Research Working Paper* No. 2002.

Dean, Judith, Mary Lovely and Hua Wang. 2003. "Foreign Direct Investment and Pollution Havens: Evaluating the Evidence from China," U.S. International Trade Commission, mimeo, Washington, DC.

Dunning, John H and Matija Rojec. 1993. *Foreign Privatization in Central & Eastern Europe.* CEEPN: Ljubljana, Slovenia.

Eskeland, Gunnar, and Ann Harrison. 2003. "Moving to Greener Pastures? Multinationals and the Pollution-haven Hypothesis," *Journal of Development Economics*, 70(1): 1-23.

Environmental Encyclopedia and Directory. 1998. Europe Publications Ltd.: London.

European Bank for Reconstruction and Development. 1994. *Transition Report*. London.

______ . 1997. *Transition Report*. London.

Fredriksson, Per, John List and Daniel Millimet. 2003. "Bureaucratic Corruption, Environmental Policy and Inbound US FDI: Theory and Evidence," *Journal of Public Economics*, 87: 1407-1430.

Fredriksson, Per and Jakob Svensson. 2003. "Political Instability, Corruption and Policy Formation: The Case of Environmental Policy," *Journal of Public Economics,* 87: 1383-1405.

Freedom House. 2002. *Freedom in the World: The Annual Survey of Political Rights and Civil Liberties 2000-2001.* New Brunswick: Transaction Publications.

Froot, Kenneth. 1993. *Foreign Direct Investment.* Chicago: University of Chicago Press.

Hallward-Driemeier, Mary. 1996. "Understanding Foreign Direct Investment by Firms: Market Pull, Cost Push and Knowledge Accumulation," mimeo, MIT.

Helliwell, John. 1994. Empirical Linkages Between Democracy and Economic Growth, *British Journal of Political Science,* 24: 225-248.

Hines, James, 1995, "Forbidden payment: Foreign Bribery and American Business after 1977," NBER Working Paper 5266, September.

Hunya, Gabor. 1997. "Large Privatisation, Restructuring and Foreign Direct Investment" in Salvatore Zecchini, ed., *Lessons from the Economic Transition. Central and Eastern Europe in the 1990s.* Kluwer Academic Publishers: Dordrecht, Boston and London. 275-300.

Keller, Wolfgang and Arik Levinson. 2002. "Pollution Abatement Costs and Foreign Direct Investment to U.S. States," *Review of Economics and Statistics*, 84(4): 691-703.

Levinson, Arik. 1996. "Environmental Regulation and Manufacturers' Location Choices: Evidence from the Census of Manufactures," *Journal of Public Economics,* 62(1-2): 5-29.

Levinson, Arik and M. Scott Taylor. 2004. "Unmasking the Pollution Haven Effect," NBER Working Paper 10629, July.

Meyer, Klaus. 1995. "Direct Foreign Investment in Eastern Europe. The Role of Labor Costs," *Comparative Economic Studies,* 37(3): 69-88.

Neumann, Peter. 1994. "Bose: fast alle bestechen," *Impulse,* Gruner + Jahr AG&Co.: Hamburg.

Oates, Wallace E. and Robert M. Schwab. 1988. "Economic Competition among Jurisdictions: Efficiency Enhancing or Distortion Inducing?" *Journal of Public Economics*, 35(3): 333-354.

OECD. 1999. *Environment in the Transition to a Market Economy.* Progress in Central and Eastern Europe and the New Independent States. OECD: Paris.

PAIZ (Polish State Investment Agency). 1995. *Major Investor List.* PAIZ: Warsaw.

Pearce, A. E., and Charles G. Smith. 1984. *The Hutchinson World Weather Guide.* Hutchinson: London.

Rudloff, Willy. 1981. *World Climates, with Tables of Climatic Data and Practical Suggestions.* Wissenschaftliche Verlagsgesellschaft: Stuttgart.

Smarzynska, Beata and Shang-Jin Wei. 2000. "Corruption and Composition of Foreign Direct Investment: Firm Level Evidence," NBER Working Paper 7969, October.

Wei, Shang-Jin. 1997. "Why is Corruption So Much More Taxing Than Taxes? Arbitrariness Kills," NBER Working Paper 6255, November.

______ . 2000a. "How Taxing is Corruption on International Investors?" *Review of Economics and Statistics*, 82(1): 1-11.

______ . 2000b. "Local Corruption and Global Capital Flows," *Brookings Papers on Economic Activity*, No. 2.

Xing, Yuqing, and Charles Kolstad. 1998. "Do Lax Environmental Regulations Attract Foreign Investment?" Working Paper 28-98. University of California Santa Barbara.

Zarsky, Lyuba. 1999. "Havens, Halos and Spaghetti: Untangling the Evidence about Foreign Direct Investment and the Environment" in *Foreign Direct Investment and the Environment.* OECD. 47-74.

Appendix Tables

Table A1: Classification of 3-digit SIC Sectors by Pollution Intensity

High Pollution			
102 ~	Copper ores	276 #	Manifold business forms
109 ~	Miscellaneous metal ores	278 ~	Blank books and bookbinding
131 ~	Crude petroleum and natural gas	282 *	Plastics materials and synthetics
132 ~	Natural gas liquids	283 *	Drugs
172 ~	Painting and paper hanging special trade contractors	284 *	Soaps, cleaners, and toilet goods
175 ~	Carpentry and floor work special trade contractors	287 *	Agricultural chemicals
179 ~	Miscellaneous special trade contractors	308	Miscellaneous plastics products, n.e.c.
205 #	Bakery products	314 ~	Footwear, except rubber
206 #	Sugar and confectionery products	325 ~	Structural clay products
224 ~	Narrow fabric mills	326 *	Pottery and related products
226 *	Textile finishing, except wool	327 *	Concrete, gypsum, and plaster products
235 ~	Hats, caps, and millinery	328 ~	Cut stone and stone products
239 ~	Miscellaneous fabricated textile products	329 *	Miscellaneous nonmetallic mineral products
241 ^	Logging	331 *	Blast furnace and basic steel products
242	Sawmills and planing mills	332 *	Iron and steel foundries
243	Millwork, plywood, and structural members	333 #	Primary nonferrous metals
249	Miscellaneous wood products	341 #	Metal cans and shipping containers
251	Household furniture	344 *	Fabricated structural metal products
252 #	Office furniture	345 *	Screw machine products, bolts, etc.
254	Partitions and fixtures	347 *	Metal services, n.e.c.
259 ~	Miscellaneous furniture and fixtures	354 *	Metalworking machinery
261 #	Pulp mills	367 *	Electronic components and accessories
262 #	Paper mills	373	Ship and boat building and repairing
263 #	Paperboard mills	385 ~	Ophthalmic goods
272 ~	Periodicals	387 ~	Watches, clocks, watchcases, and parts
273 #	Books	393 ~	Musical instruments
275	Commercial printing		
Low Pollution		**Medium Pollution**	
142 ~	Crushed and broken stone, including riprap	201	Meat products
154 ~	General building contractors—nonresidential buildings	204 *	Grain mill products
202 ~	Dairy products	205 *	Bakery products
204 #	Grain mill products	214 #	Tobacco stemming and redrying
206 *	Sugar and confectionery products	221 #	Broad woven fabric mills, cotton
211 ~	Cigarettes	226 #	Textile finishing, except wool
214 *	Tobacco stemming and redrying	245 ~	Wood buildings and mobile homes
221 *	Broad woven fabric mills, cotton	252 *	Office furniture
222	Broad woven fabric mills, manmade fiber and silk	253 ~	Public building and related furniture
223 ~	Broad woven fabric mills, wool	262 *	Paper mills
227 ~	Carpets and rugs	263 *	Paperboard mills
232 ~	Men's and boys' furnishings	265 #	Paperboard containers and boxes

Table A1 continued

Low Pollution		Medium Pollution	
261 *	Pulp mills	271 ^	Newspapers
265 *	Paperboard containers and boxes	282 #	Plastics materials and synthetics
273 *	Books	285	Paints and allied products
274 ~	Miscellaneous publishing	287 #	Agricultural chemicals
276 *	Manifold business forms	291	Petroleum refining
277 ~	Greeting cards	295 *	Asphalt paving and roofing materials
279 ~	Printing trade services	306	Fabricated rubber products, n.e.c.
283 #	Drugs	311	Leather tanning and finishing
284 #	Soaps, cleaners, and toilet goods	321 ~	Flat glass
295 #	Asphalt paving and roofing materials	322	Glass and glassware, pressed or blown
299	Miscellaneous petroleum and coal products	323	Products of purchased glass
301	Tires and inner tubes	324	Cement, hydraulic
302 ~	Rubber and plastics footwear	326 #	Pottery and related products
305	Hose and belting and gaskets and packing	327 #	Concrete, gypsum, and plaster products
317 ~	Handbags and personal leather goods	329 #	Miscellaneous nonmetallic mineral products
331 #	Blast furnace and basic steel products	332 #	Iron and steel foundries
333 *	Primary nonferrous metals	336 *	Nonferrous foundries (castings)
334	Secondary nonferrous metals	339	Miscellaneous primary metal products
335	Nonferrous rolling and drawing	344 #	Fabricated structural metal products
336 #	Nonferrous foundries (castings)	346 *	Metal forgings and stampings
341 *	Metal cans and shipping containers	347 #	Metal services, n.e.c.
343	Plumbing and heating, except electric	348 *	Ordnance and accessories, n.e.c.
345 #	Screw machine products, bolts, etc.	353 #	Construction and related machinery
346 #	Metal forgings and stampings	354 #	Metalworking machinery
348 #	Ordnance and accessories, n.e.c.	356 *	General industrial machinery
353 *	Construction and related machinery	358	Refrigeration and service machinery
356 #	General industrial machinery	359 ~	Industrial machinery, n.e.c.
361	Electric distribution equipment	362	Electrical industrial apparatus
364 #	Electric lighting and wiring equipment	364 *	Electric lighting and wiring equipment
366	Communications equipment	365 ~	Household audio and video equipment
367 #	Electronic components and accessories	372 #	Aircraft and parts
372 *	Aircraft and parts	374	Railroad equipment
376	Guided missiles, space vehicles, parts	382 ~	Measuring and controlling devices
381 ~	Search and navigation equipment	386	Photographic equipment and supplies
391	Jewelry, silverware, and plated ware	395 #	Pens, pencils, office, and art supplies
395 *	Pens, pencils, office, and art supplies		

emission but in another abatement category
* abatement but in another emission category
^ only abatement data were available
~ only emission data were available

Table A2: Measures of Environmental Standards and Corruption

Country	Corruption I	Corruption II	Treaties I	Treaties II	EBRD index	NGOs/ pop	Reduction in emissions		Environmental Sustainability Index			
							Lead	CO_2	overall	env quality	env stress	env institutions
Albania	4.4		2.0	2	1	1.3		59.3	44.2	44.6	65.4	39.6
Armenia	4.5	6	1.5	0	1	2.2		17.2	50.6	50.3	74.2	39.3
Azerbaijan	5.5	6	0.5	0	1	1.6		-9.9	46.4	38.9	65.2	27.8
Belarus	5.0	4	1.5	1	1	1.3	63.4	8.9	48.0	53.6	66.0	28.6
Bulgaria	5.5	4	3.0	3	1	3.1	20.7	-19.4	47.4	25.7	59.2	33.5
Croatia		4	2.0	1	2	5.6	15.8	2.5	54.1	57.0	59.1	49.3
Czech Rep.	3.3	4	2.0	1	3	1.6		22.4	57.2	53.3	31.0	60.0
Estonia	2.6	2	1.5	0	2	19.3	42.6	20.3	57.7	59.1	66.5	54.1
Georgia	5.0	4	0.5	0	1	2.0		84.8				
Hungary	3.9	6	2.5	2	3	1.2		1.5	61.0	50.4	64.1	56.6
Kazakhstan	5.1	4	1.5	0	1	0.9	-25.9	0.9	41.6	48.8	76.8	21.5
Kyrgyzstan	4.9	4	0.5	0	1	1.1		17.1	39.6	42.8	67.8	26.8
Latvia	4.6	4	2.0	1	2	2.7		12.5	56.3	58.3	55.2	50.7
Lithuania	3.9	0	2.0	1	1	2.1	22.7	13.2	60.3	57.9	64.4	49.1
Macedonia FYR		8	1.0	0	1	11.4		-1.9	39.2	38.7	37.8	38.5
Moldova	5.0		3.0	3	1	1.4		10.2	47.4	49.4	68.7	36.0
Poland	4.6	4	2.0	1	3	0.6	45.2	23.3	47.6	34.3	45.5	45.8
Romania		6	1.5	1	2	0.7		18.8	44.1	36.8	62.1	38.4
Russia	5.3	8	2.5	2	1	0.2		-12.4	56.2	65.4	69.8	42.6
Slovak Rep.	3.6	4	1.5	1	2	5.3	36.3	28.1	63.2	60.9	49.5	60.0
Slovenia		2	1.5	1	2	3.5		-0.2	59.9	63.8	43.4	66.2
Tajikistan		4	0.0	0	1	0.9		76.0				
Turkmenistan		4	0.0	0	1	1.2		-42.1				
Ukraine	4.3	4	1.5	1	1	0.2	18.0	-18.9	36.8	32.8	45.7	28.2
Uzbekistan	5.2	4	0.0	0	1	0.4	32.5	11.8	41.6	46.9	64.8	20.5

Table A3: Correlations between Country Specific Variables

	GDP	Relative GDP per capita	Corporate tax	Distance	Corruption I	Corruption II	Treaties I	Treaties II	EBRD index	NGOs/ pop	Reduction in emissions lead	Reduction in emissions CO_2	ESI overall	ESI env quality	ESI env stress
GDP	1.00														
Relative GDP per capita	-0.39	1.00													
Corporate tax	0.47	-0.17	1.00												
Distance	0.04	0.20	0.05	1.00											
Corruption I	0.39	0.28	0.08	0.31	1.00										
Corruption II	0.57	0.05	0.40	0.19	0.55	1.00									
Treaties I	0.22	-0.51	-0.47	-0.27	0.05	-0.15	1.00								
Treaties II	0.21	-0.22	-0.20	-0.18	0.37	0.13	0.79	1.00							
EBRD index	0.49	-0.65	0.40	-0.27	-0.42	0.08	0.12	-0.10	1.00						
NGOs/pop	-0.53	-0.28	-0.36	-0.20	-0.79	-0.36	0.05	-0.24	0.31	1.00					
Reduction in lead emissions	-0.24	0.07	0.19	-0.32	-0.25	-0.02	-0.08	0.11	0.39	0.24	1.00				
Reduction in CO_2 emissions	-0.07	-0.39	0.53	-0.16	-0.52	-0.25	-0.34	-0.50	0.65	0.34	0.44	1.00			
ESI overall	-0.37	-0.53	0.10	-0.33	-0.67	-0.59	0.23	0.05	0.35	0.53	0.34	0.66	1.00		
ESI env quality	-0.61	-0.04	0.14	-0.07	-0.61	-0.48	-0.40	-0.59	0.03	0.44	0.18	0.71	0.67	1.00	
ESI env stress	-0.55	0.21	-0.44	0.27	0.14	-0.26	-0.20	-0.37	-0.52	0.18	-0.37	-0.01	-0.02	0.41	1.00
ESI env institutions	-0.10	-0.65	0.15	-0.41	-0.79	-0.47	0.32	0.08	0.64	0.58	0.38	0.62	0.89	0.44	-0.37

[13]

The Unintended Disincentive in the Clean Air Act*

John A. List, Daniel L. Millimet, and Warren McHone

Abstract

The Clean Air Act and its subsequent amendments have been lauded as the primary stimulant to the impressive improvement in local air quality in the US since 1970. A key component of these regulations is the New Source Review (NSR) requirement, which includes the contentious stipulation that when an existing plant seeks to modify its operations, the entire plant must comply with current standards for new sources. This requirement was included to improve air quality in dirty areas, and prevent a deterioration of air quality in clean areas. Yet, whether NSR provides the proper plant-level incentives is unclear: there are strong disincentives to undertake major plant modifications to avoid NSR. In our examination of more than 2500 and 2200 plant-level modification decisions and closures, respectively, we find empirical evidence suggesting that NSR retards modification rates, while doing little to hasten the closure of existing dirty plants.

KEYWORDS: environmental regulations, Clean Air Act, New Source Review, propensity score matching

*John A. List is at AREC and the Department of Economics, University of Maryland, 2200 Symons Hall, College Park, MD 20742-5535; email: jlist@arec.umd.edu and NBER. Daniel L. Millimet (corresponding author) is at the Department of Economics, Southern Methodist University, Box 0496, Dallas, TX 75275-0496; email: millimet@mail.smu.edu. W. Warren McHone is at the Department of Economics, University of Central Florida, College of Business Administration, Orlando, FL 32816-1400; email: wmchone@bus.ucf.edu. The authors wish to thank Don Fullerton and two anonymous referees for careful comments. Thanks as well to Randy Becker, Wayne Gray, Shelby Gerking, Michael Greenstone, Vern Henderson, Randy Kroszner, Arik Levinson, Richard Newell, and Billy Pizer for useful comments throughout this research project. Seminar participants at various universities and conferences also lent useful insights throughout this line of research. All remaining errors are our own.

Recent reforms to the New Source Review (NSR) requirement in the Clean Air Act have emerged as a source of controversy.[1] These reforms, which relax some of the NSR requirements for the expansion of existing operations in certain industries, are, in part, a response to what has been a contentious issue surrounding NSR – the disparate treatment between existing facilities and new or expanding facilities. Existing facilities are "grandfathered" under much less stringent environmental standards, while new or expanding facilities are subjected to NSR and forced to install sufficient emissions control technologies. A provocative line of research has shown that the differential application of environmental regulations has led to unintended consequences among power plants: Nelson et al. (1993), Maloney and Brady (1988), Gollop and Roberts (1983), and Stewart (1981) present empirical evidence that suggests such uneven treatment has lengthened the lifetime of utility plants, admitting the possibility that air quality regulations have *increased* emissions of certain pollutants in the short run.

While these studies have provided invaluable insights, a related provision within the Clean Air Act (CAA) that has heretofore been ignored is the effect of NSR requirements on plant-level modification rates. Given that when an *existing* plant seeks to modify operations, the entire plant must comply with pollution control standards for new sources, it is quite possible that pollution-reducing modifications are deterred.[2] Accordingly, the CAA provides perspicuous incentives for existing plants: proceed with modification if the expected costs – potentially including installation of new capital that is less pollution-intensive if NSR is triggered – outweigh the expected benefits from more productive capital. Therefore, it is quite possible that this particular NSR requirement has contributed to *more*, rather than less, pollution in the short-run. This could be the case if NSR requirements deter modifications that would replace out-dated technology with new capital that, while potentially not as environmentally-friendly as technology mandated under NSR, is cleaner than the capital currently in use.

Our research goal in this study is not as ambitious as determining unequivocally whether NSR requirements governing modifications of existing plants have led to greater emissions. Rather, our main goal is to provide evidence on whether the NSR requirement for existing plants has retarded modification rates.[3] To our best

---

[1] Specifically, parts C and D of Title I of the Clean Air Act (CAA). This fact is highlighted when one considers the attention that the recent NSR reforms have received. For example, the New York Times reported on Nov. 22, 2002: "The Bush administration will ease clean air rules, allowing power plants and refineries to avoid new pollution controls when expanding operations… [T]he long-awaited rule changes will 'increase energy efficiency and encourage emissions reductions,' the Environmental Protection Agency said in a statement obtained by The Associated Press… Environmentalists and a group of Northeastern states said they plan to file suit immediately challenging the changes. The state officials contend that the easing of the clean air requirements 'will undermine efforts' to meet air quality standards."

[2] To be precise, only modifications of existing plants that are expected to lead to a net increase in emissions are subject to NSR. However, as explained more fully in Section I, the manner by which the expected change in net emissions is forecasted virtually guarantees that all modifications will fall within the bounds of NSR.

[3] More specifically, we examine whether attainment status influences modification and closure rates. We interchange NSR and attainment status throughout.

knowledge, current evidence is purely anecdotal. Yet in the absence of robust empirical evidence, these anecdotes have evidently taken on significant policy importance, as they have emerged in the most recent *Economic Report of the President* (2003; pp. 172; coauthored by List):

> The NSR permit process...might lead firms to delay or forgo plans to modernize their facilities in ways that would benefit the environment. To take just one example, a manufacturer that operates a process that includes a drying system determined that the system's energy efficiency could be improved if the existing drier nozzles were replaced with Teflon-coated nozzles. The firm found, however, that the replacement would be economical only if the expense of obtaining an NSR permit could be avoided. ... Since the firm could not readily discern whether the installation of new nozzles would be considered routine maintenance, repair or replacement, it decided not to proceed with the project. In this way, NSR deters firms from conducting needed repairs and often results in unnecessary emissions of pollutants. In this case NSR requirements actually made the environment worse off.

Our study fills part of this chasm by making use of the spatial variation in NSR compliance costs across counties in the US. In counties labeled "out-of-attainment" of federally mandated air quality levels modifying plants are compelled to install technology that achieves the Lowest Achievable Emissions Rate (LAER), *regardless of cost*. Alternatively, plants located in counties with air quality that is "in-attainment" of federal standards face a NSR requirement that is less onerous; modifying plants only need to install the Best Available Control Technology (BACT), which in most cases is much less costly than LAER equipment.

In our empirical analysis we use a series of empirical methodologies, including a difference-in-differences semi-nonparametric propensity score matching estimator, to analyze panel data on more than 2500 plant-level modification decisions in New York State from 1980 – 1990 to determine the effects of nonattainment status on plant-level modification decisions, given the way that NSR rules differ by attainment status.[4] To provide an explicit link to the capital turnover studies discussed above, we supplement this exercise with an empirical examination of more than 2200 plant-level closure decisions in the same state and time period to determine the impact of county-level attainment status on plant-level closure decisions in industries other than electric utilities. By combining results from these two data sets, we can examine if the differential application of NSR provisions based on a county's attainment status have unintentionally altered the turnover of antiquated manufacturing capital. If modification and closure rates of pollution-intensive plants have been affected by county-level attainment status, then one cannot rule out the possibility that NSR has had deleterious effects on air quality.

---

[4] See also Greenstone (2003) for another application of propensity score matching in the environmental literature.

Strikingly, the analysis is consonant with such a conclusion: our empirical estimates suggest that substantial modifications of existing pollution-intensive plants in nonattainment counties have been delayed, or foregone, without a concomitant increase in the turnover rate of dirty capital. In much the same manner that the Clean Air Act has altered new capital flows (e.g., Henderson, 1996; List et al., 2002), we find that the NSR requirement for existing plants has altered the equilibrium allocation of expenditures on capital improvement, conferring a competitive advantage to certain regions.

We view our work as fitting in broadly with the extant "pollution haven hypothesis" literature. While defined narrowly, the "pollution haven hypothesis" refers to the notion that multinational pollution-intensive firms will relocate to countries with weaker environmental standards, we take the hypothesis to represent a concept that is much more general: the tendency for pollution sources to locate, or relocate, in response to pollution policies. In this sense, our contribution is to examine one dimension of the pollution haven hypothesis that has received considerably less attention: the behavior of existing plants.

1. Data and Empirical Strategy

1.1. Data

The plant-level modification and closure data come from the Industrial Migration File (IMF), maintained until 1990 by the New York State (NYS) Department of Economic Development (DED). The goal of the IMF was to monitor all annual changes in manufacturing activity in NYS at the county-level. The IMF includes observations of individual plant openings, closings, and modifications and classifies plants by standard industrial classification (SIC) code.

Following previous studies, we focus on plants in pollution-intensive sectors most likely to be affected by county-level ozone attainment status.[5] Given that attainment status is determined by county-level air quality readings, we follow Greenstone (2002) and classify industrial sectors as "ozone pollution-intensive" if they emit at least 6% of the total industrial sector's emissions of nitrogen oxide or volatile organic compounds, the primary chemical precursors to ozone. Plants labeled pollution-intensive are in SIC

[5] The 1977 Clean Air Act Amendments set standards on five criteria air pollutants: sulfur dioxides (SO_2), carbon monoxides (CO), ozone (O_3), nitrogen oxides (NO_x), and total suspended particulates (TSP). Since ozone has attracted the most regulatory attention due to the limited progress that has been made to reduce concentration levels, we follow Henderson (1996), Becker and Henderson (2000), and List et al. (2002) and focus on county attainment status of ozone. Yet, we should note that if counties are in attainment for ozone and out-of-attainment for any (or all) other pollutants, then we are obtaining lower bound estimates of the treatment effect. Given that attainment status of the various individual pollutants affects abatement technology choice, however, we are comfortable with our modeling choice. We should note, in any event, that attainment status appears to be correlated across pollutants in our sample.

codes: 2611-31, 2711-89, 2812-19, 2861-69, 2911, 30, 32, 3312-3, 3321-5, 34, and 371. All remaining sectors are labelled non-pollution-intensive.

After classifying plant modifications by pollution intensity, the data are aggregated to the county-level. In total, we observe more than 2500 plant modifications: 742 (1788) pollution-intensive (non-pollution intensive) plant modifications across the 62 counties in NYS over the sample period. In addition, we observe 2269 plant closures: 561 (1708) pollution-intensive (non-pollution intensive) plant closures across the 62 counties in NYS over the sample period. We then combine these measures with each county's ozone attainment designation, and other county-level attributes (discussed below).[6] Over the sample period, slightly more than 25% of the observations are in nonattainment counties. Summary statistics are provided in Table 1.

Before proceeding, it is important to note in greater detail the costs associated with modifying operations in a nonattainment county. Before the most recent set of reforms mentioned in footnote 1, NSR required a source to obtain a permit prior to any major modification. The permit establishes various actions that the source must undertake to control its emissions of air pollution. For an existing source proposing a modification, NSR is only triggered if the modification will result in a net increase in emissions. However, the NSR program estimates post-modification emissions assuming the plant will operate 24 hours a day, year-round, even if the plant does not maintain non-stop production prior to the modification. As a result, even modest changes in a facility will most assuredly trigger NSR, even if those changes are unlikely to increase actual emissions. Once the NSR process is triggered, the time and monetary costs required prior to undertaking the modification may be extreme—permit haggling may take years, and the costs may reach millions of dollars.

We should note that we have used this data set for several other purposes. For example, List and McHone (2000), List et al. (2003a), and Millimet and List (2003) examine the influence of attainment status on plant births. Relatedly, List et al. (2003b) explore the effects of air quality regulation on the destination choice of relocating plants, while List et al. (2004) compare and contrast the effects of environmental regulations across foreign and domestic plant births. While there are certainly similarities across these studies—the methodologies, regressors, etc.—the outcomes of interest in these studies are new plant births or new plant relocations.[7] The current study analyzes a completely distinct issue: plant modifications (as well as closures). We link this issue to the current political debate concerning the NSR provision of the CAA.

---

[6] Although attainment status can range from in-attainment of the primary standard to out-of-attainment, with partial standards in between, ozone designation has typically been polar in nature; that is, a county is either in- or out-of-attainment. For a county to be labeled out-of-attainment, according to the CAA its second highest daily air quality reading must exceed 0.12 parts per million. This is currently being revised by the EPA, however.

[7] Millimet and List (2003) also utilize the plant closure data examined herein, but not the data on plant modifications.

1.2. The Empirical Models

1.2.1. Propensity Score Matching Method

A method of assessing the impact of the differential application of NSR provisions based on a county's attainment status, on plant modification and closure rates is the method of propensity score matching developed in Rosenbaum and Rubin (1983). While extensively used by statisticians, economic applications have proliferated only recently. A few notable examples include Heckman et al. (1997), Dehejia and Wahba (1999, 2002), Smith and Todd (2000), and List et al. (2002). Blundell and Costa-Dias (2002) provide an excellent introduction to the matching method, concluding, "….matching methods have been extensively refined in the recent evaluation literature and are now a valuable part of the evaluation toolbox" (p. 4).

The fundamental problem in identifying the effect of a binary variable, such as attainment status, is one of incomplete information. While the econometrician observes whether the "treatment" (in this case, nonattainment of the federal ozone standard) occurs and the outcome conditional on treatment assignment, the counterfactual is not observed. Let y_{i1} denote the outcome of observation *i* with the treatment ($T_i = 1$); y_{i0} denotes the outcome without the treatment ($T_i = 0$). If both states of the world were observable, the average treatment effect, τ, would equal $\bar{y}_1 - \bar{y}_0$, where the former (latter) average represents the mean outcome for the treatment (control) group. However, given that only y_1 or y_0 is observed for each observation, unless assignment into the treatment group is random, generally $\tau \neq \bar{y}_1 - \bar{y}_0$. In the application, y represents a count of either plant modifications or closures.

The solution Rosenbaum and Rubin (1983) advocate is to find a vector of covariates, *Z*, such that

$$(1) \qquad y_1, y_0 \perp T \mid Z, \qquad pr(T = 1 \mid Z) \in (0,1)$$

where $\perp$ denotes independence. If one is interested in estimating only the average treatment effect, the weaker condition

$$(1') \qquad E[y_0 \mid T = 1, Z] = E[y_0 \mid T = 0, Z] = E[y_0 \mid Z], \quad pr(T = 1 \mid Z) \in (0,1)$$

is required. For condition (1') to hold, *Z* should be multi-dimensional. Consequently, finding observations with identical values for all covariates in *Z* may be untenable. However, Rosenbaum and Rubin (1983) prove that conditioning on *p*(*Z*) is equivalent to conditioning on *Z*, where $p(Z) = pr(T = 1 \mid Z)$ is the propensity score. *p*(*Z*) is estimated via logit.

After estimating the propensity score, a matching algorithm is required to estimate the missing counterfactual, y_{0i}, for each treated observation *i*. The simplest algorithm is (single) nearest-neighbor matching, whereby each treated observation is paired with the control observation whose propensity score is closest in absolute value (Dehejia and Wahba, 2002).[8] As unmatched controls are discarded, the matching method identifies a sub-sample of controls that more closely resembles the treatment group in terms of pre-treatment attributes. The treatment effect on the treated (TT) is given by

$$\text{(2)} \qquad \tau_{TT} = E[y_1 \mid T = 1, p(Z)] - E[y_0 \mid T = 0, p(Z)] = E[y_1 - y_0 \mid p(Z)]$$

During the actual estimation here, the nearest-neighbor algorithm is amended along three dimensions. First, while a match exists for each out-of-attainment county, the propensity scores may still differ considerably. Because the unbiasedness of the matching estimator relies on the propensity scores being identical, pairs with scores significantly different are excluded. We present estimates for several cut-off values, which is a procedure known as "caliper matching" in the literature (Cochran and Rubin, 1973).

Second, to utilize our panel data fully, we amend the caliper matching method by restricting the pool of potential controls to which a given treated observation may be paired. Specifically, we match each treated observation twice: first, restricting matched pairs to be from the same year and region of the state (see Figure 1); second, restricting matched pairs to be the same county from different years. By matching within-year/within-region, or within-county, we explicitly remove any time-, region-, or county-specific unobservables not already controlled for by the propensity score. This is the matching method's analogy to fixed effects, and is similar in spirit to Smith and Todd's (2000) claim that matches used to identify the effect of employment programs should be from the same local labor market. Thus, the estimator in (2) becomes

$$\text{(3a)} \qquad \tau_{TT,tr} = E[y_1 \mid T = 1, p(Z), t, r] - E[y_0 \mid T = 0, p(Z), t, r] = E[y_1 - y_0 \mid p(Z), t, r]$$

$$\text{(3b)} \qquad \tau_{TT,i} = E[y_1 \mid T = 1, p(Z), i] - E[y_0 \mid T = 0, p(Z), i] = E[y_1 - y_0 \mid p(Z), i]$$

where t indexes year, r indexes region, and i indexes county.

Finally, as the usual matching estimators are applicable only to problems of "selection on observables," we amend the matching estimators in (3a), (3b) to remove explicitly certain unobservables by employing a difference-in-differences (DID) matching estimator. Since we have a count of both pollution-intensive *and* non-pollution-intensive plant modifications (closures) for each county-year observation, and the behavior of "clean" plants should not be affected by attainment status, any difference in the modification (closure) rates of "clean" plants across the matched

[8] Typically, nearest-neighbor matching is performed *with replacement*, implying that a given control observation may be matched with multiple treatment observations. Dehejia and Wahba (2002) verify that matching with replacement fares at least as well as matching without replacement, and possibly better.

treatment and control groups is assumed to reflect unobservable county-specific qualities that serve to influence plant modification (closure). Thus, in the spirit of similar estimators used in Eichler and Lechner (2002), Heckman et al. (1997), and List et al. (2003a), we define the DID counterpart to (3a), (3b) as

(4a) $\tau_{DID,tr} = \tau_{TT,tr} - \tau'_{TT,tr}$

(4b) $\tau_{DID,i} = \tau_{TT,i} - \tau'_{TT,i}$

where the prime refers to clean plants and so $\tau'_{TT,tr}$ ($\tau'_{TT,i}$) is the mean difference in the count of modifications (closures) of "clean" plants across the matched treatment and control groups. As the DID estimator only requires

(1") $E[y_0 - y'_0 \mid T = 1, Z] = E[y_0 - y'_0 \mid T = 0, Z] = E[y_0 - y'_0 \mid Z],\ pr(T = 1 \mid Z) \in (0,1)$

for identification, where y'_0 is the count of "clean" plant modifications (closures), Smith and Todd (2000) conclude that DID matching estimators are more robust.

Upon completing the matching estimation, balancing and specification tests are conducted. Balancing refers to the fact that after conditioning on the propensity score, the distribution of the conditioning variables, Z, should not differ across the treatment and control group in the matched sub-sample. Thus, after matching, we also test for differences in the mean of the Zs. The specification test proposed in Ham et al. (2001) requires testing for mean differences in the lagged outcome across the matched treatment and control groups. In the present context, this test serves two purposes. First, if the lagged outcome differs across the treatment and control groups, then the presence of uncontrolled unobservables may bias the estimated treatment effect. Second, since lagged plant modifications (closures) may affect current plant births due, for example to agglomeration externalities, differences in lagged modifications may have a direct effect on the outcome as well, further biasing the estimator (Henderson, 1997; 2003).

1.2.2. Parametric Approach

For comparison, we also estimate several parametric models. The count of (pollution-intensive) plant modifications (closures) in county i at time t, y_{it}, is assumed to be given by

(5) $y_{it} = \exp(X_{it}\beta)\eta_{it} + \varepsilon_{it}$

where X_{it} is a vector of county attributes, including attainment status, η_{it} captures all unobservable, time-varying attributes of county i and is potentially correlated with some of the variables in X_{it}, and ε_{it} is an error term satisfying $E[\varepsilon \mid X, \eta] = 0$. Given the inclusion of a constant in X_{it}, E[η] = 1 can be assumed without loss in generality (Mullahy, 1997).

Equation (5) is estimated via the fixed effects (FE) Poisson model of Hausman, Hall and Griliches (1984; hereafter HHG). Resulting estimates of β will be consistent if $E[\eta \mid X] = E[\eta]$; thus, regressors in X may not be correlated with time-specific, county-specific unobservables.[9] If $E[\eta \mid X] \neq E[\eta]$, then another estimator is needed. Mullahy (1997) and Windmeijer and Silva (1997) discuss various instrumental variables (IV) and two-step solutions, given a sufficient number of instruments, denoted by vector W_{it}.

If the source of endogeneity is a dichotomous treatment variable, as is the case with county-level attainment status, a straightforward two-step solution is available. If one envisions plant modification (closure) decisions as dependent upon the latent variable underlying attainment status, consistent estimates of the effect of that latent variable is obtained by replacing the treatment variable in (5) with its predicted linear index.[10] The estimating equation then becomes

$$(5') \quad y_{it} = \exp(\delta T^{*}_{it} + X'_{it}\beta')\eta_{it} + \varepsilon_{it}$$

where T^{*}_{it} is a latent variable, such that if $T^{*}_{it} > 0$ then a county is out-of-attainment ($T_{it} = 1$); if $T^{*}_{it} \leq 0$, then a county is in attainment ($T_{it} = 0$); δ is the parameter of interest; X'_{it} contains the remaining variables in X_{it} excluding attainment status; and β' is the corresponding parameter vector. Consistent estimates can be obtained by replacing T^{*} with $W\gamma$, where $T^{*}_{it} = W_{it}\gamma + \upsilon_{it}$. Assuming γ is unknown, it can be estimated via logit. The second-stage standard errors are obtained via bootstrap.

Prior to continuing, it is important to note the major differences between the matching and parametric Poisson estimators. First, matching estimators entail fewer distributional assumptions. Second, matching estimators allow one to use additional endogenous variables that are difficult to incorporate into standard parametric count models: for example, lagged values of the dependent variable. In a parametric framework, one requires valid exclusion restrictions that are uncorrelated with these omitted endogenous variables. Third, matching allows for nonparametric interactions between the covariates in Z in the determination of the outcome of interest (Bratberg et al. 2002). Fourth, matching estimators utilize only a sub-sample of the control group that is most "similar" to the treatment group, whereas parametric models utilize all available observations. This may present problems in matching applications when the treatment group is small. Finally, the matching estimators employed herein yield an estimate of the *treatment effect on the treated* (TT), whereas the parametric approach estimates the expected treatment effect for a random observation.

---

[9] HHG estimates are also consistent in the presence of correlation between variables in X and time-specific unobservables that are constant across counties if X includes period-specific dummies.

[10] The usual two-step solution of replacing the dichotomous variable with its predicted probability will not produce consistent estimates of β in a Poisson framework (Windmeijer and Silva, 1997).

1.3. The Use of Count Data

One potential shortcoming to models as specified in Section 1.2, however, is that the data contain only the *count* of pollution-intensive plant modifications and closures; not the *percentage* of plants modifying or closing. More formally, we examine $y = f(X)$, where y is the modification or closure count and X is a vector of county attributes including attainment status, rather than $y/N = g(X)$, where N is the total stock of pollution-intensive plants. Nonetheless, assume that absent regulation: (i) all plants pollute the same amount per unit of output on average, and (ii) the size distribution of plants is roughly constant across counties. Then, because attainment status is related to pollution levels, it follows that being out-of-attainment implies a higher stock of existing pollution-intensive plants, N. Consequently, if we find that nonattainment status reduces the modification (closure) count, y, then nonattainment must also be associated with a lower percentage of modifications (closures), y/N. Alternatively, if we find no impact of attainment status on modifications (closures), then it is possible for nonattainment to have either no effect or a negative effect on the percentage of modifications (closures). Finally, if we find a positive impact of attainment status on modifications (closures), then we obtain no information concerning the effect of nonattainment on the percentage of modifications (closures).

2. Empirical Results

2.1. Modification Rates

2.1.1. Propensity Score Matching Estimates

Table 2 presents the first-stage logit estimates used to form the propensity score. The specification is similar to the first-stage equation used in the two-step HHG model (discussed below) with the inclusion of higher order terms and interactions to facilitate the balancing of the covariates across the matched treatment and control groups (Dehejia and Wahba, 2002).

The first matching algorithm pairs each out-of-attainment county with the in-attainment county from the same year and region with the nearest propensity score. Although there are 176 treatment (nonattainment) observations, only 76 have a possible match.[11] Of the 76 matched pairs, we retain those with propensity scores that differ by less than 1%, 5%, 10%, and 20%. Under these cut-offs, 8, 16, 20, and 27 pairs are obtained. Using the same cut-offs but matching each treatment county to itself at a different point in time yields 9, 11, 18, and 25 matches.[12]

---

[11] All counties in five regions are in nonattainment in various years. Thus, there is no control observation for these counties to serve as a match.

[12] There are at most 143 possible matches using the within-county matching algorithm as three counties are in nonattainment all eleven years of the sample.

Table 3 presents estimated mean differences between the matched treatment and control groups, along with p-values associated with the null hypothesis that the means are equal, for each of the eight matched sub-samples. Examination of the table yields three important insights. First, within-year, within-region matching algorithm balances the mean of all the covariates (at the $p < 0.05$ level) across the treatment and control groups using the 1% and 5% calipers; the within-county algorithm balances the means using the 1%, 5%, and 10% calipers. This includes the variables specifically controlled for in the first-stage logit, as well as property taxes, the proportion of the population with a high school diploma, and highway expenditures. Thus, these algorithms satisfy the balancing test.

Second, the five algorithms passing the balancing test also pass the specification test proposed in Ham et al. (2001) as lagged pollution-intensive plant modifications is balanced. Finally, of these five algorithms, all yield negative point estimates for the average treatment effect on the treated. Moreover, the three within-county estimates are statistically significant (at the $p < 0.05$ level) despite the sample size. These point estimates imply that in a nonattainment county roughly three pollution-intensive plants per year forego modification due to the county's attainment status and the potential added cost that this status may trigger under NSR (1%: $\tau_{DID,i}$=-3.56, p=0.04; 5%: $\tau_{DID,i}$=-3.09, p=0.04; 10%: $\tau_{DID,i}$=-2.33, p=0.03). Given that the average county experiences 1.09 major (pollution-intensive) modifications per year, the foregone activity suggests that the NSR regulation may have a sizeable deterrent effect.

2.1.2. Parametric Results

For comparison, we also utilize several parametric estimators. Empirical results are displayed in Table 4. Columns (1) and (2) display the results using equation (5), thereby treating attainment status as exogenous. Borrowing from the literature on plant location, the specification in column (1) includes county and time fixed effects, county-level attainment status, and a measure of scale (manufacturing employment). The specification in column (2) adds additional controls for real manufacturing wages, population, and real property taxes. Both specifications indicate a statistically insignificant effect of attainment status on plant modifications, and the point estimates are extremely close to zero.[13]

Because the most likely explanation for the divergence between the HHG and matching estimates is that the parametric estimates are biased due to the presence of unobservables correlated with attainment status and plant modifications, we turn to the two-step model in (5'). Consistency of the estimates relies on finding a valid instrument for attainment status. Finding a valid exclusion restriction is not trivial, since attainment status depends on the lagged level of air quality, which in turn depends on

[13] As the standard errors from the Poisson model are sensitive to over- or under-dispersion in the data, we also estimate negative binomial models, which relax the restriction of equality of the conditional mean and variance functions. The results, available upon request, differed very little.

lagged manufacturing activity, which in turn may affect current plant behavior due to positive Marshall-Arrow-Romer externalities or negative Jacobs externalities.

In an attempt to circumvent this issue, we follow List et al. (2002) and exploit a natural phenomenon: wind direction. Since county-level attainment status is not based on own emissions, but rather observed local air quality readings, emissions from neighboring counties may influence attainment status due to transboundary effects. Since the jet stream flows from west to east, we use the proportion of contiguous Western neighbors that are out-of-attainment to identify the model.[14] Before examining the two-step results, we note that (Western) neighboring attainment status is a highly significant determinant of own attainment status ($p < 0.01$; see Table 2); thus, we are not concerned about problems associated with weak instruments (Stock et al., 2002). Moreover, the positive coefficient is consonant with our logic that transboundary pollution affects the attainment status of Eastern neighboring counties.

In terms of the actual results – displayed in Columns (3) and (4) of Table 4 – the point estimates are –0.11 and –0.13, respectively, and 90% confidence intervals (obtained via 1000 bootstrap repetitions) both encompass zero.[15] When interpreting the magnitude of the two-step coefficients, one must note that the two-step model does not provide an estimate of the treatment effect *per se* since the coefficient refers to a one-unit increase in latent attainment status. Since the predicted mean of latent attainment status (i.e., the mean of $Z_{it}\hat{\gamma}$) is –2.3, a 2.3 unit increase in latent attainment is required to move the average county from in- to out-of-attainment. Multiplying the two-step coefficient in column (4) yields an estimated treatment effect of roughly –0.30, with a 90% confidence interval of [-0.69, 0.05]. This implies a deterrent effect of roughly 0.3 pollution-intensive plant modifications per annum from being out-of-attainment.

Given that the discrepancy between the parametric and matching estimates cannot be explained by the treatment of attainment status as exogenous, at least three possible explanations remain: (i) the matching algorithm, by utilizing only a sub-sample of the control group, produces an estimator that is not subject to "outliers;" (ii) the parametric assumptions of the Poisson model are invalid; or, (iii) the matching algorithm yields an estimate of the average treatment effect on the treated, whereas the one-step parametric models estimate the treatment effect on a random observation from the population and the two-step HHG model estimates the Local Average Treatment Effect (LATE), the effect of nonattainment on those counties whose nonattainment status is driven by the instrument (i.e., "compliers"). If the effect of nonattainment varies across the "compliers" and those counties in nonattainment regardless of the value of the instrument (i.e., "always-takers"), then the two-step model will not identify the average treatment effect on the treated (Angrist et al., 1996).

Unfortunately, we cannot easily discern among these explanations. There is no formal specification test for the Poisson model (other than testing for over- or under-

[14] For counties located on the NYS border, we obtained data on the attainment status of neighboring counties in other states to form the appropriate instrument.

[15] The estimate in column (3) is statistically different from zero using a *one-sided* alternative at the $p < 0.12$ confidence level; the estimate in column (4) at the $p < 0.07$ confidence level.

dispersion), the matched samples are too small to estimate the parametric models in a meaningful fashion, and there is no formal test of the assumption needed for the two-step approach to identify the average effect as opposed to the LATE. One additional parameter we can estimate using the method of propensity score matching for comparison to the above results is the population average treatment effect. If we label "in-attainment" as the treatment and re-do the matching by pairing each in-attainment county with an out-of attainment county, we can obtain estimates of the average treatment effect *on the untreated.* The weighted average of the treatment effect on the treated and untreated constitutes the overall average treatment effect. Performing this exercise using the within-county algorithm with 1%, 5%, and 10% calipers yields estimates of the average treatment effect of –1.64, -1.25, and –0.87, respectively.[16] Although closer to the two-step HHG estimates reported in Table 4, a sizeable discrepancy persists.

2.2. Closure Rates

In this section we summarize empirical estimates using the spatial distribution of the 2269 plant closures across New York State from 1980-1990. Results in this section have stand-alone importance as they provide an explicit link to the empirical work that has shown uneven regulation leads to a lengthening of the time before power plants are retired (see, for example, Nelson et al., 1993; Maloney and Brady, 1988; Gollop and Roberts, 1983; Stewart, 1981). The intuition behind this finding is that uneven regulation—commonly termed "new source bias"—confers a competitive advantage to existing plants, as they face lower compliance expenditures, *ceteris paribus*. While this finding is robust in the electric utility industry, an important extension is to examine whether these results apply to other stationary sources. This is exactly what this section offers.

2.2.1. Propensity Score Matching Estimates

Table 5 presents estimated mean differences between the matched treatment and control groups for four measures of plant closures, along with p-values associated with the null hypothesis that the means are equal, for each of the eight matched sub-samples presented in Table 3. The four types of closures are: (i) complete closures, (ii) closures due to bankruptcy, (iii) closures accompanied by the plant relocating to another county in New York State, and (iv) closures accompanied by the plant relocating (partially or fully) out of state. As the matching algorithms are identical to those used in Table 3, we focus on the within-year, within-region matching algorithm using the 1% and 5% calipers, and the within-county matching algorithm using the 1%, 5%, and 10% calipers.

---

[16] The fact that the average treatment effect is smaller in absolute value than the average treatment effect on the treated implies that the cost of nonattainment (in terms of foregone plant modifications) is greater for those counties actually out-of-attainment.

Of the five algorithms satisfying the balancing test, all yield positive point estimates for the average treatment effect on the treated for each of the four types of closures. However, the estimated effect of NSR on the each of the first three types of closures, which together constitute nearly 80% of all pollution-intensive plant closures, is insignificant. The only statistically significant results (at the $p < 0.05$ level) are for the three within-county estimates for closures accompanied by a relocation out of state. The point estimates for these closures imply that approximately one pollution-intensive plant per county per year closes and relocates out of state due to the fact that the plant is located in a nonattainment county in New York State (1%: $\tau_{DID,i}$=1.22, p=0.00; 5%: $\tau_{DID,i}$=1.00, p=0.01; 10%: $\tau_{DID,i}$=0.72, p=0.01). As the average county experiences 0.17 (pollution-intensive) closures with an accompanied (partial or complete) relocation out of state per year, this does represent a non-negligible effect of the NSR regulation. While the results are mixed, overall NSR does not appear to have an economically meaningful impact on the closure of dirty plants.[17]

2.2.2 Parametric Results

For completeness, we examine closures using the parametric estimators. Empirical results are displayed in Table 6. Columns (1) and (2) display the results treating attainment status as exogenous; Columns (3) and (4) present the two-step estimates using the attainment status of Western neighbors as an instrument.

In terms of the models treating attainment status as exogenous, all of the point estimates are positive, suggesting that NSR may, in fact, increase the closure of dirty plants. However, all of the estimates are statistically insignificant at the $p < 0.10$ level except for the models analyzing closures with a bankruptcy filing. While this is consonant with the viewpoint of NSR advocates, bankruptcy filings constitute less than 10% of all pollution-intensive plant closures in the sample. Turning to the two-step HHG model results, the point estimates are negative for three of the four closure measures—the closures accompanied with a partial or complete move out of state is the exception. Further, the 90% confidence intervals that were obtained via 1000 bootstrap repetitions always encompass zero. The exogenous and two-step HHG results do not support the position that NSR regulation facilitates the turnover of dirty capital.[18]

Overall, our empirical evidence on the distribution of closure rates does not match the results reported in the literature on capital turnover among electric utilities. There are several potential reasons for this difference: (i) aggregate data for pollution-intensive manufacturing may mask important inter-industry and inter-plant

[17] Pooling all closures together and examining the combined effect of nonattainment on aggregate closures yields a statistically significant estimate only using the within-county matching algorithm and a 1% caliper ($\tau_{DID,i}$=1.91, p=0.04).

[18] Pooling all closures together and examining the combined effect of nonattainment on aggregate closures does not alter the findings. For the model in column (3), the estimated coefficient on attainment status is –0.11 (90% confidence interval: [-0.28, 0.08]); for the model in column (4), the estimated coefficient on attainment status is –0.09 (90% confidence interval: [-0.29, 0.08]).

heterogeneity, (ii) plant closure in the manufacturing sector may be a distinctively different decision process compared to capital turnover among power plants, and (iii) the extant empirical evidence for power plants may be due to stagnation in the average heat rate and technical change for electric utilities (see Joskow, 1987).

3. Concluding Remarks

Grandfathering of existing firms has played an important role in regulatory programs in the US in the past several decades. Despite the widespread use of differentiated regulations, little attention has been paid to their effects on plant-level modification decisions in the regulated sector. This particular area of study is acutely important for health and safety regulation, as modest plant-level modifications may yield large declines in the regulated externality.

As a case study into this issue, we examine whether the NSR provision in the Clean Air Act had important effects on plant-level modification decisions in the manufacturing sector. To date, data limitations have precluded any systematic effort to model the effects of NSR on the behavior of existing plants, rendering anecdotes as the sole pieces of evidence concerning the potentially deleterious effects of NSR. Our study offers such an empirical analysis. Examining more than 2500 plant-level modification decisions, we find that a county's attainment status and the potential added cost that this status may trigger under NSR had deleterious effects on plant-level modification decisions. Further, our examination of 2200 closure decisions did not find any consistent evidence that NSR affected the rate of closure of existing dirty plants.

In sum, then, the finding that NSR deterred plant modifications that could have led to reduced emissions, in combination with the nonexistent effects of NSR on pollution-intensive plant closures, suggests the possibility that in the short run the NSR stipulation for existing manufacturing plants has led to *more*, rather than less, pollution.

References

Angrist, J.D., G. Imbens, and D.B Rubin, "Identification of Causal Effects Using Instrumental Variables," *Journal of the American Statistical Association* **91**, 444-472 (1996).

Becker, R. and J.V. Henderson, "Effects of Air Quality Regulations on Polluting Industries," *Journal of Political Economy* **108**, 379-421 (2000).

Blundell, R. and M. Costa-Dias, "Alternative Approaches to Evaluation in Empirical Microeconomics," *Portuguese Economic Journal* **1**, 91-115 (2002).

Bratberg, E., A. Grasdal, and A.E. Risa, "Evaluating Social Policy by Experimental and Nonexperimental Methods," *Scandanavian Journal of Economics* **104**, 147-171 (2002).

Cochran, W. and D. Rubin, "Controlling Bias in Observational Studies," *Sankyha* **35**, 417-446 (1973).

Dehejia, R.H. and S. Wahba, "Casual Effects in Nonexperimental Studies: Reevaluating the Evaluation of Training Programs," *Journal of the American Statistical Association* **94**, 1053-1062 (1999).

Dehejia, R.H. and S. Wahba, "Propensity Score Matching for Nonexperimental Causal Studies," *Review of Economics and Statistics* **84**, 151-161 (2002).

Economic Report to the President, Washington, DC: United States Government Printing Office (2003).

Eichler, M. and M. Lechner, "An Evaluation of Public Employment Programmes in the East German State of Sachsen-Anhalt," *Labour Economics* **9**, 143-186 (2002).

Gollop, F.M. and M.J. Roberts, "Environmental Regulations and Productivity Growth: The Case of Fossil-Fueled Electric Power Generation," *Journal of Political Economy* **91**, 654-74 (1983).

Greenstone, M., "The Impacts of Environmental Regulations on Industrial Activity: Evidence from the 1970 and 1977 Clean Air Act Amendments and the Census of Manufactures," *Journal of Political Economy* **110**, 1175-1219 (2002).

Greenstone, M., "Did the Clean Air Act Cause the Remarkable Decline in Sulfur Dioxide Concentrations?," *Journal of Environmental Economics and Management*, forthcoming (2003).

Ham, J.C., X. Li, and P.B. Reagan, "Matching and Selection Estimates of the Effect of Migration on Wages for Young Men," mimeo, Ohio State University (2001).

Hausman, J., B. Hall, and Z. Griliches, "Econometric Models for Count Data with an Application to the Patents – R&D Relationship," *Econometrica* **52**, 909-938 (1984).

Heckman, J.J., H. Ichimura, and P.E. Todd, "Matching as an Econometric Evaluation Estimator: Evidence from Evaluating a Job Training Program," *Review of Economic Studies* **64**, 605-654 (1997).

Henderson, J.V., "Effects of Air Quality Regulation," *American Economic Review* **86**, 789-813 (1996).

Henderson, J.V., “Externalities and Industrial Development,” *Journal of Urban Economics* **42**, 449-470 (1997).

Henderson, J.V., “Marshall's Scale Economies,” *Journal of Urban Economics* **53**, 1-28 (2003).

Joskow, P.J., “Productivity Growth and Technical Change in the Generation of Electricity,” *Energy Journal* **8**, 17-38 (1987).

List, J.A. and W.W. McHone, “Measuring the Effects of Air Quality Regulations on “Dirty” Firm Births: Evidence from the Neo- and Mature- Regulatory Periods,” *Papers in Regional Science* **79**, 177-190 (2000).

List, J.A., D.L. Millimet, P.G. Fredriksson, and W.W. McHone, “Effects of Environmental Regulations on Manufacturing Plant Births: Evidence from a Propensity Score Matching Estimator,” *Review of Economics and Statistics* **85**, 944-952 (2003a).

List, J.A., Millimet, D.L. and McHone, W.W. "Effects of Air Quality Regulation on the Destination Choice of Relocating Plants," *Oxford Economic Papers*, 55 (4): 657-678 (2003b).

List, J.A., D.L. Millimet, and W.W. McHone, "Effects of Environmental Regulation on Foreign and Domestic Plant Births: Is There a Home Field Advantage?." mimeo, University of Maryland (2004).

Maloney, M.T. and G.L. Brady, “Capital Turnover and Marketable Pollution Rights,” *Journal of Law and Economics* **31**, 203-226 (1988).

Millimet, D.L. and J.A. List, "The Case of the Missing Pollution Haven Hypothesis," mimeo, University of Maryland (2003).

Mullahy, J., “Instrumental-Variable Estimation of Count Data Models: Applications to Models of Cigarette Smoking Behavior,” *Review of Economics and Statistics* **79**, 586-593 (1997).

Nelson, R., T. Tietenberg, and M. Donihue, “Differential Environmental Regulation: Effects on Electric Utility Capital Turnover and Emissions,” *Review of Economics and Statistics* **75**, 368-373 (1993).

Rosenbaum, P. and D. Rubin, “The Central Role of the Propensity Score in Observational Studies for Causal Effects,” *Biometrika* **70**, 41-55 (1983).

Smith, J. and P. Todd, "Does Matching Address Lalonde's Critique of Nonexperimental Estimators," *Journal of Econometrics*, forthcoming (2000).

Stewart, R.B., "Regulation, Innovation, and Administrative Law: A Conceptual Framework," *California Law Review* **69**, 1256-1270 (1981).

Stock, J., J. Wright, and M. Yogo, "A Survey of Weak Instruments and Weak Identification in Generalized Method of Moments," *Journal of Business and Economic Statistics* **20**, 518-529 (2002).

Windmeijer, F.A.G. and J.M.C. Silva, "Endogeneity in Count Data Models: An Application to Demand for Health Care," *Journal of Applied Econometrics* **12**, 281-94 (1997).

Table 1 Description of Variables

Variable	Full Sample	In-Attnmt.	Non-Attnmt.	Definition and Source
Pollution-Intensive Plant Modifications Per County Per Year	1.09 (2.31)	0.81† (1.73)	1.89† (3.36)	Actual count of plant modifications from 1980-1990 labelled as having production activities that is pollution-intensive. Industrial Migration File. New York State Dept. of Economic Development (DED).
Non-Pollution-Intensive Plant Modifications Per County Per Year	2.62 (4.93)	1.91† (3.20)	4.66† (7.71)	Actual count of plant modifications from 1980-1990 labelled as having production activities that is non-pollution-intensive. Industrial Migration File. New York State Dept. of Economic Development (DED).
Attainment Status	0.26 (0.44)	--	--	Intensity of county-level pollution regulations. Dichotomous variable = 1 if county is out-of-attainment of federal standards for ozone, 0 otherwise. Federal register Title 40 CFR Part 81.305.
Employment (100,000s)	1.50 (3.57)	0.08† (1.45)	3.40† (6.22)	Total employment in manufacturing. *County Business Patterns.*
Wage (1000s)	17.01 (4.1.2)	17.34† (3.84)	16.08† (4.74)	Total annual manufacturing payroll divided by the number of employees by county, adjusted for inflation. *County Business Patterns.*
Firms	11.71 (5.39)	12.23† (5.49)	10.22† (4.84)	Total number of manufacturing firms. *County Business Patterns.*
Population (100,000s)	2.87 (4.78)	1.82† (3.40)	5.87† (6.60)	County population. *Current Population Reports.* U.S. Bureau of Census.
Property tax (1000s)	0.56 (0.24)	0.57 (0.27)	0.54 (0.17)	Real property tax collected per capita. *Census of Governments.*
Per Capita Income (1000s)	13.50 (4.76)	13.58 (4.30)	13.24 (5.88)	Real per capita income. *County Business Patterns.*
Highway Expenditures	0.17 (0.09)	0.18† (0.10)	0.15† (0.08)	Total highway expenditures. *County Business Patterns.*
High School Graduates	70.05 (6.81)	70.91† (6.75)	67.58† (6.35)	Percent of population with high school diploma. *Current Population Reports.*

Notes:

1. Data are for the 62 New York counties from 1980-1990. N = 682 (176 out-of-attainment).
2. Numbers represent sample means. Standard deviations are in parentheses.
3. † indicates difference in means across in- and out-of-attainment samples is statistically significant at $p < 0.05$ level using a two-sided t-test.

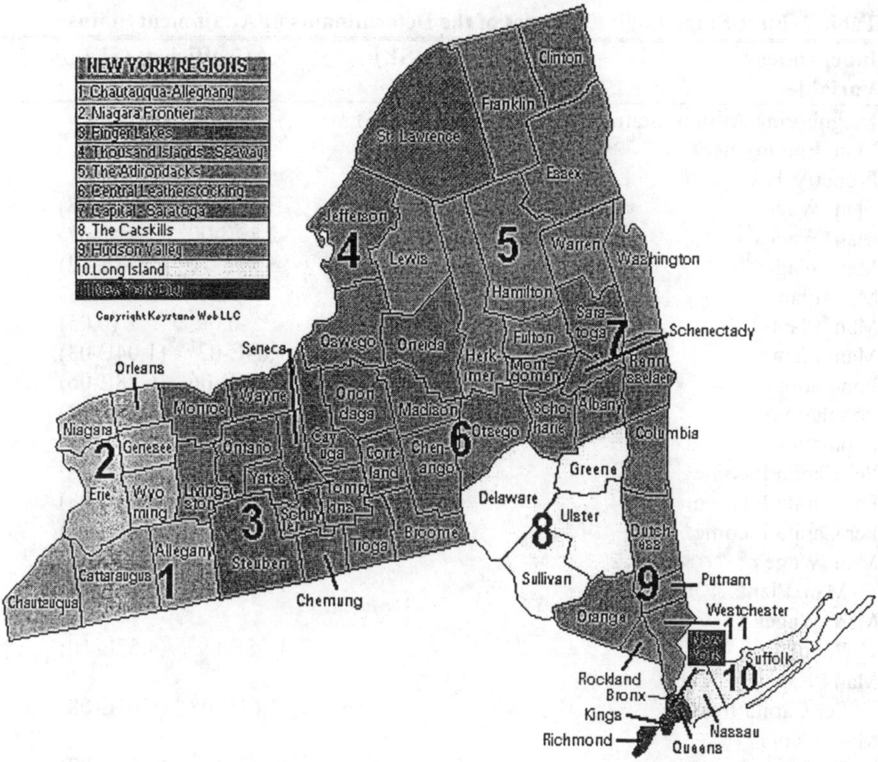

Figure 1. Regional Breakdown, New York State.

Table 2 First-Stage Logit Estimates of the Determinants of Attainment Status

Independent Variable	Coefficient (1)	(SE)	Coefficient (2)	(SE)
Neighboring Attnmt. Status	2.85*	(0.33)	—	
Man. Employment	1.99E-06	(1.29E-06)	—	
Property Taxes	-1.85E-03*	(8.75E-04)	—	
Man. Wages	-3.95E-06	(7.08E-05)	3.63E-03	(2.55E-03)
Man. Wages^2			-2.23E-07	(1.41E-07)
Man. Wages^3			4.27E-12	(2.74E-12)
Man. Plants			1.40*	(0.58)
Man. Plants^2			-0.09*	(0.05)
Man. Plants^3			1.84E-03*	(1.04E-03)
Population	1.62E-06*	(5.09E-07)	-1.85E-06	(6.28E-06)
Population^2			7.37E-12	(6.12E-12)
Population^3			-3.14E-18*	(1.82E-18)
Per Capita Income			4.73E-03*	(1.25E-03)
Per Capita Income^2			-1.86E-07*	(9.64E-08)
Per Capita Income^3			2.63E-12*	(1.40E-12)
Man. Wages * Man. Plants			-9.57E-06	(3.20E-05)
Man. Wages * Population			1.08E-09*	(4.53E-10)
Man. Wages * Per Capita Income			-1.61E-08	(6.61E-08)
Man. Plants * Population			-8.61E-07*	(3.54E-07)
Man. Plants * Per Capita Income			1.67E-05	(3.04E-05)
Population * Per Capita Income			-8.88E-10*	(4.10E-10)
Time Effects	Yes		Yes	
Log-likelihood	-180.7		-145.8	
Pseudo R^2	0.54		0.63	
N	682		682	

Notes:

1. Dependent variable is equal to one if county is out-of-attainment of federal ozone standards during the year, zero otherwise. Neighboring attainment status is the percentage of Western contiguous neighbors that are out of attainment.
2. Standard errors are in parentheses beside the coefficient estimates and are adjusted for clustering within counties. * indicates significant at the 10% level using a two-sided alternative.
3. Model (1) is used in the two-step FE Poisson estimation. Model (2) is used to generate the propensity score estimates.
4. Time effects jointly significant at the 1% level.

Table 3a Propensity Score Estimates of Attainment Status Effect on Plant Expansions: Within Region and Year Matching

Independent Variable	Maximum Difference (0.01)	(0.05)	(0.10)	(0.20)
Propensity Score	0.00 (p=0.98)	0.00 (p=0.98)	0.01 (p=0.90)	0.04 (p=0.62)
Dirty Plant Expansions ($\tau_{TT,\bullet}$)	**0.25** **(p=0.72)**	**-0.44** **(p=0.70)**	**-0.55** **(p=0.55)**	**-0.52** **(p=0.45)**
Clean Plant Expansions	1.25 (p=0.34)	0.13 (p=0.92)	-0.10 (0.92)	-0.04 (p=0.96)
Net Plant Expansions ($\tau_{DID,\bullet}$)	**-1.00** **(p=0.21)**	**-0.56** **(p=0.22)**	**-0.45** **(p=0.26)**	**-0.48** **(p=0.15)**
Lagged Dirty Plant Expansions (1 year)	0.71 (p=0.53)	-0.29 (p=0.79)	-0.24 (p=0.79)	-0.39 (p=0.56)
Man. Wages (1000s)	-0.06 (p=0.98)	-0.91 (p=0.44)	-0.88 (p=0.36)	-0.72 (p=0.33)
Man. Employment (1000s)	29.94 (p=0.63)	4.05 (p=0.93)	-0.37 (p=0.99)	1.16 (p=0.97)
Man. Plants	-0.79 (p=0.82)	-2.37 (p=0.26)	-2.21 (p=0.23)	-1.97 (p=0.24)
Population (1000s)	59.49 (p=0.55)	4.61 (p=0.96)	0.31 (p=1.00)	2.43 (p=0.97)
Per Capita Income (1000s)	-0.66 (p=0.79)	-0.61 (p=0.66)	-0.55 (p=0.62)	-0.40 (p=0.63)
Property Tax	-389.13 (p=0.06)	-186.81 (p=0.10)	-148.00 (p=0.10)	-102.04 (p=0.13)
High School Graduates (%)	-3.61 (p=0.32)	-3.39 (p=0.12)	-4.96 (p=0.01)	-5.80 (p=0.00)
Highway Expenditure	-0.16 (p=0.09)	-0.07 (p=0.16)	-0.05 (p=0.24)	-0.02 (p=0.44)
Number of Matched Pairs	8	16	20	27
Number of Unique Controls	8	15	17	18

Notes:

1. Figures represent mean difference between treatment counties (out-of-attainment) and control counties (in-attainment). P-values in parentheses are for the tests that the mean difference across the treatment and controls groups are equal.
2. "Dirty" plants are those defined as pollution intensive (see text); "clean" are all remaining manufacturing plants.
3. "Unique controls" reports the number of control counties that are matched with at least one treatment county.

Table 3b Propensity Score Estimates of Attainment Status Effect on Plant Expansions: Within County Matching

Independent Variable	Maximum Difference (0.01)	(0.05)	(0.10)	(0.20)
Propensity Score	0.00 (p=1.00)	0.01 (p=0.97)	0.04 (p=0.75)	0.06 (p=0.60)
Dirty Plant Expansions ($\tau_{TT,\bullet}$)	**-1.89 (p=0.53)**	**-1.73 (p=0.50)**	**-1.83 (p=0.31)**	**0.32 (p=0.84)**
Clean Plant Expansions	1.67 (p=0.69)	1.36 (p=0.70)	0.50 (p=0.84)	2.84 (p=0.42)
Net Plant Expansions ($\tau_{DID,\bullet}$)	**-3.56 (p=0.04)**	**-3.09 (p=0.04)**	**-2.33 (p=0.03)**	**-2.52 (p=0.30)**
Lagged Dirty Plant Expansions (1 year)	0.25 (p=0.94)	0.67 (p=0.81)	0.86 (p=0.62)	1.71 (p=0.31)
Man. Wages (1000s)	0.54 (p=0.60)	-0.01 (p=0.99)	-0.76 (p=0.37)	-1.56 (p=0.04)
Man. Employment (1000s)	3.11 (p=0.98)	2.53 (p=0.98)	-4.33 (p=0.95)	-18.47 (p=0.79)
Man. Plants	0.59 (p=0.70)	0.48 (p=0.74)	0.29 (p=0.83)	0.22 (p=0.89)
Population (1000s)	-0.65 (p=1.00)	-0.31 (p=1.00)	0.19 (p=1.00)	-3.89 (p=0.98)
Per Capita Income (1000s)	0.33 (p=0.84)	-0.20 (p=0.90)	-0.82 (p=0.44)	-1.60 (p=0.09)
Property Tax	1.22 (p=0.98)	1.00 (p=0.98)	-9.39 (p=0.79)	-13.64 (p=0.78)
High School Graduates (%)	-1.09 (p=0.70)	-0.89 (p=0.71)	-1.16 (p=0.50)	-1.18 (p=0.45)
Highway Expenditure	-0.00 (p=0.97)	-0.00 (p=0.92)	-0.01 (p=0.67)	-0.01 (p=0.37)
Number of Matched Pairs	9	11	18	25
Number of Unique Controls	6	7	10	12

Notes: See Table 3a.

Table 4 Parametric Estimates of the Determinants of County-Level Plant Expansions

Independent Variable	Fixed Effects Poisson			
	HHG		Two-Step HHG	
	(1)	(2)	(3)	(4)
Non-Attainment	1.21E-03	-0.02	-0.11	-0.13
	(0.13)	(0.13)	[-0.28,0.04]	[-0.30,0.02]
ln(Employment)	-1.33	0.44	-1.70	0.34
	(1.03)	(1.39)	[-3.61,-0.61]	[-2.79,3.44]
ln(Wage)	---	-4.35*	---	-4.31*
		(1.36)		[-7.66,1.03]
ln(Population)	---	-0.86	---	-2.09
		(3.12)		[-9.55,5.42]
ln(Prop. Tax)	---	-0.97	---	-1.57
		(1.34)		[-4.16,0.75]
Period Effects	Yes	Yes	Yes	Yes
County Effects	Yes	Yes	Yes	Yes
Log-likelihood	-582.2	-576.2	-580.7	-574.2
N	682	682	682	682

Notes:

1. Dependent variable is the count of annual pollution-intensive manufacturing plant expansions from 1980-1990.
2. Non-Attainment equals 1 if county is out-of-attainment of federal ozone standard, 0 otherwise.
3. Models (1) and (2) treat attainment status as exogenous (estimated using HHG (1984) quasi-maximum likelihood (QML)); Models (3) and (4) treat attainment status as endogenous (estimated using a two-step QML estimation procedure). The proportion of Western neighboring counties that are out-of-attainment is used as the instrument (see Table 2).
4. Standard errors are in parentheses beneath coefficient estimates; 90% bootstrap confidence intervals – based on 1000 repetitions – in brackets. *indicates significant at the 10% level using a two-sided alternative.
5. ln indicates the natural logarithm of the variable.
6. Time effects are jointly significant in models (1) and (2) at the $p < 0.01$ level.

Table 5a Propensity Score Estimates of Attainment Status Effect on Plant Closures: Within Year and Region Matching

Independent Variable	Maximum Difference (0.01)	(0.05)	(0.10)	(0.20)
Complete Closures				
Dirty Plant Closures ($\tau_{TT,\bullet}$)	**0.50 (p=0.02)**	**0.06 (p=0.82)**	**0.05 (p=0.83)**	**0.00 (p=1.00)**
Clean Plant Closures	-0.13 (p=0.76)	0.56 (p=0.27)	0.45 (p=0.27)	0.30 (p=0.34)
Net Plant Closures ($\tau_{DID,\bullet}$)	**0.63 (p=0.19)**	**-0.50 (p=0.29)**	**-0.40 (p=0.30)**	**-0.30 (p=0.30)**
Lagged Dirty Plant Closures (1 year)	-0.14 (p=0.69)	0.00 (p=1.00)	0.06 (p=0.86)	0.09 (p=0.74)
Closures with Bankruptcy				
Dirty Plant Closures ($\tau_{TT,\bullet}$)	**-0.13 (p=0.33)**	**0.00 (p=1.00)**	**0.00 (p=1.00)**	**0.00 (p=1.00)**
Clean Plant Closures	0.00 (p=1.00)	0.00 (p=1.00)	-0.05 (p=0.32)	-0.07 (p=0.16)
Net Plant Closures ($\tau_{DID,\bullet}$)	**-0.13 (p=0.33)**	**0.00 (p=1.00)**	**0.05 (p=0.57)**	**0.07 (p=0.33)**
Lagged Dirty Plant Closures (1 year)	0.00 (p=1.00)	0.00 (p=1.00)	0.00 (p=1.00)	0.09 (p=0.15)
Closures Accompanied with Move In State				
Dirty Plant Closures ($\tau_{TT,\bullet}$)	**0.13 (p=0.33)**	**0.06 (p=0.33)**	**0.05 (p=0.32)**	**0.04 (p=0.32)**
Clean Plant Closures	0.00 (p=1.00)	-0.06 (p=0.33)	-0.10 (p=0.15)	-0.07 (p=0.31)
Net Plant Closures ($\tau_{DID,\bullet}$)	**0.13 (p=0.33)**	**0.13 (p=0.17)**	**0.15 (p=0.09)**	**0.11 (p=0.18)**
Lagged Dirty Plant Closures (1 year)	0.00 (p=1.00)	0.00 (p=1.00)	0.00 (p=1.00)	0.04 (p=0.32)
Closures Accompanied with Partial or Complete Move Out of State				
Dirty Plant Closures ($\tau_{TT,\bullet}$)	**0.13 (p=0.33)**	**0.06 (p=0.33)**	**0.05 (p=0.32)**	**0.04 (p=0.32)**
Clean Plant Closures	-0.13 (p=0.66)	0.00 (p=1.00)	0.05 (p=0.76)	0.07 (p=0.55)
Net Plant Closures ($\tau_{DID,\bullet}$)	**0.25 (p=0.44)**	**0.06 (p=0.76)**	**0.00 (p=1.00)**	**-0.04 (p-0.78)**
Lagged Dirty Plant Closures (1 year)	-0.14 (p=0.34)	0.00 (p=1.00)	0.00 (p=1.00)	0.00 (p=1.00)

Notes: See Table 3a. Sample sizes are identical to Table 3a.

Table 5b Propensity Score Estimates of Attainment Status Effect on Plant Closures: Within County Matching

Independent Variable	Maximum Difference (0.01)	(0.05)	(0.10)	(0.20)
Complete Closures				
Dirty Plant Closures ($\tau_{TT,\bullet}$)	**1.44 (p=0.30)**	**1.09 (p=0.34)**	**0.72 (p=0.32)**	**0.80 (p=0.17)**
Clean Plant Closures	0.67 (p=0.59)	0.55 (p=0.60)	0.72 (p=0.32)	2.12 (p=0.09)
Net Plant Closures ($\tau_{DID,\bullet}$)	**0.78 (p=0.19)**	**0.55 (p=0.30)**	**0.00 (p=1.00)**	**-1.32 (p=0.19)**
Lagged Dirty Plant Closures (1 year)	0.88 (p=0.49)	0.98 (p=0.36)	0.87 (p=0.20)	0.97 (p=0.08)
Closures with Bankruptcy				
Dirty Plant Closures ($\tau_{TT,\bullet}$)	**0.11 (p=0.33)**	**0.09 (p=0.33)**	**0.17 (p=0.07)**	**0.16 (p=0.04)**
Clean Plant Closures	0.00 (p=1.00)	0.00 (p=1.00)	0.06 (p=0.56)	0.12 (p=0.30)
Net Plant Closures ($\tau_{DID,\bullet}$)	**0.11 (p=0.33)**	**0.09 (p=0.33)**	**0.22 (p=0.05)**	**0.04 (p=0.71)**
Lagged Dirty Plant Closures (1 year)	0.00 (p=1.00)	0.00 (p=1.00)	0.11 (p=0.41)	-0.38 (p=0.00)
Closures Accompanied with Move In State				
Dirty Plant Closures ($\tau_{TT,\bullet}$)	**0.11 (p=0.33)**	**0.09 (p=0.33)**	**0.06 (p=0.32)**	**0.12 (p=0.25)**
Clean Plant Closures	-0.22 (p=0.15)	-0.18 (p=0.15)	-0.11 (p=0.15)	0.08 (p=0.39)
Net Plant Closures ($\tau_{DID,\bullet}$)	**0.33 (p=0.09)**	**0.27 (p=0.09)**	**0.17 (p=0.09)**	**0.04 (p=0.74)**
Lagged Dirty Plant Closures (1 year)	-0.13 (p=0.23)	-0.10 (p=0.36)	-0.06 (p=0.36)	0.06 (p=0.61)
Closures Accompanied with Partial or Complete Move Out of State				
Dirty Plant Closures ($\tau_{TT,\bullet}$)	**0.33 (p=0.18)**	**0.27 (p=0.18)**	**0.11 (p=0.41)**	**-0.44 (p=0.23)**
Clean Plant Closures	-0.89 (p=0.04)	-0.73 (p=0.06)	-0.61 (p=0.03)	-0.40 (p=0.43)
Net Plant Closures ($\tau_{DID,\bullet}$)	**1.22 (p=0.00)**	**1.00 (p=0.01)**	**0.72 (p=0.01)**	**-0.04 (p=0.92)**
Lagged Dirty Plant Closures (1 year)	0.00 (p=1.00)	0.00 (p=1.00)	-0.05 (p=0.72)	0.18 (p=0.41)

Notes: See Table 3a. Sample sizes are identical to Table 3b.

Table 6 Parametric Estimates of the Determinants of County-Level Plant Closures

Independent Variable	Fixed Effects Poisson			
	HHG		Two-Step HHG	
	(1)	(2)	(3)	(4)
Complete Closures				
Non-Attainment	0.12	0.14	-0.16	-0.13
	(0.21)	(0.21)	[-0.29,0.12]	[-0.31,0.13]
Period Effects	Yes	Yes	Yes	Yes
County Effects	Yes	Yes	Yes	Yes
Log-likelihood	-298.14	-296.75	-297.47	-296.38
Closures with Bankruptcy				
Non-Attainment	1.38*	1.31*	-0.27	-0.29
	(0.67)	(0.66)	[-1.97,0.03]	[-2.50,0.15]
Period Effects	Yes	Yes	Yes	Yes
County Effects	Yes	Yes	Yes	Yes
Log-likelihood	-66.08	-62.88	-67.94	-64.52
Closures Accompanied with Move In State				
Non-Attainment	0.20	0.51	-0.19	-0.10
	(0.35)	(0.41)	[-0.63,0.34]	[-0.63,0.54]
Period Effects	Yes	Yes	Yes	Yes
County Effects	Yes	Yes	Yes	Yes
Log-likelihood	-133.81	-130.60	-123.33	-121.62
Closures Accompanied with Partial or Complete Move Out of State				
Non-Attainment	0.13	0.29	0.11	0.09
	(0.35)	(0.36)	[-0.45,0.22]	[-0.52,0.24]
Period Effects	Yes	Yes	Yes	Yes
County Effects	Yes	Yes	Yes	Yes
Log-likelihood	-152.15	-149.65	-152.09	-149.89

Notes:

1. Specifications (1) and (3) include log manufacturing employment as an additional control. Specifications (2) and (4) include log manufacturing employment, log manufacturing wage, log population, and log property taxes as additional controls. $N = 682$ in each model.
2. See Table 4.

[14]

Trade Pessimists vs Technology Optimists: Induced Technical Change and Pollution Havens*

Corrado Di Maria and Sjak A. Smulders

Abstract

Our paper focuses on the role of endogenous technology and technology spillovers in explaining cross country differences in pollution and the pollution haven effect of international trade. In our North-South trade model, technology is endogenously developed by the North and imitated by the South. Environmental regulators choose national environmental policies by trading off the income gains and the disutility from a rise in pollution. Differences in environmental stringency are entirely driven by differences in investment opportunities and distortions that follow from the difference in intellectual property rights protection. We show that without goods trade and in the absence of technology subsidies, the North imposes more stringent environmental regulation than the South. When opening up to trade, the South experiences a rise in prices for pollution-intensive goods and tends to raise pollution as in a standard trade model. Induced technical change, however, may reverse this pollution haven effect.

KEYWORDS: Pollution Havens, Endogenous Technical Change, International Trade

*CentER and Department of Economics, Tilburg University, Warandelaan 2, P.O. Box 90153, 5000 LE Tilburg, the Netherlands. The authors wish to thank Larry Karp, Don Fullerton, Harry Huizinga, Emiliya Lazarova, Edwin van der Werf, seminar participants in Ascona, Budapest and Venice, and two anonymous referees for valuable comments.

In the debate on pollution havens, trade pessimists disagree with technology optimists.[1] From the perspective of standard trade theory, the international flows of goods and investment across countries with different environmental standards are likely to concentrate pollution in countries with lax environmental regulations; these countries acquire a (real or apparent) comparative advantage in pollution-intensive activities. The natural concern is that international trade would allow rich countries in the North to clean up their environment at the expense of environmental quality in poorer countries in the South. Opposite to this view is the argument that pollution in different countries is much more technology-driven than trade-driven: while trade liberalization has been important over the last decades, the role of technological change in determining trade patterns and income levels has been at least as important. Technology transfer by multinational firms and the diffusion of clean technologies are often claimed to be a powerful counterbalance to the pollution haven tendencies. When trade induces the North to specialize in clean production, it might also shift its innovation efforts towards cleaner technology. If this technology diffuses to the South, the environment might benefit.

This paper formalizes the interconnection between trade, technological change, and environmental regulation. We model two regions which we call the North and the South. Each region produces two tradable goods, with pollution stemming from only one of them. In both regions local environmental regulators choose environmental policy by trading off the income gains from a rise in pollution against the disutility due to the lower environmental quality. Regulators, households and firms all take prices as given, with the exception of producers of intermediate goods in the North. They set their own price and invest in new technology to maximize profits. A fraction of these new technologies diffuse to the South since firms there can copy technologies at no cost.

We show that indeed trade may induce the North to develop pollution-saving technologies and the South to reduce pollution, thus reversing traditional reallocation and specialization effects from trade. However, such support for technology optimism is not the only possible outcome. If it is hard to find substitutes for pollution-intensive goods, an increase in innovation efforts by the North in clean sectors results in an increased demand for pollution-intensive goods from the South; in this case, trade induces pollution-using technical change and the induced technology response to trade reinforces the incentive for South to increase pollution. Low substitution thus results in technology pessimism.

Our analysis involves two main steps. First, we study how costs and benefits of environmental policy in the innovating region differ from those in the imitating region. The resulting differences in environmental stringency provide one of the regions with a comparative advantage in pollution-intensive production as a basis for trade. Second, we study whether international trade leads to more or less pollution in the South when the technologies developed in the North diffuse to the South. We identify two effects by which trade affects environmental policy: The conventional *terms of trade effect*, by which trade

[1] For a comprehensive review of this literature see, for example Zarsky (1999). Antweiler et al. (2001) and Copeland and Taylor (2004) provide a thorough discussion of the role on international trade on the environment, both from the theoretical and the empirical point of view.

affects domestic producers' prices and induces them to reallocate production according to comparative advantage; and the *induced technology effect*, which arises since trade affects profits in the two sectors in the North so that innovation effort shifts from one sector to another. This affects the mix of technologies that are available not only in the North, but also - through imitation - in the South.

The existing literature mainly focuses on trade aspects, identifying different sources of comparative advantages. First, if environmental quality is a normal good, increases in the level of income will induce a higher demand for environmental quality. Richer countries will thus tend to have more stringent environmental regulation relative to poorer countries, and to develop a comparative advantage in the production of less polluting goods (see Cole (2004)). Second, the comparative advantage for the South in the polluting (resource-intensive) sector may arise as a consequence of ill-defined property rights on the common pool resource (as in Chichilnisky (1994)). As South does not regulate access to resources, it over-exploits them, leading to the emergence of an apparent comparative advantage vis-à-vis an otherwise identical country. Finally, comparative advantage in pollution-intensive production may originate from differences in the relative endowments of productive factors (Copeland and Taylor (2004)). If capital-intensive goods are also relatively more pollution-intensive, and rich countries are relatively more endowed with capital, they might enjoy a comparative advantage in the pollution-intensive good. This might explain the fact that most production of pollution-intensive goods takes place in developed countries, in spite of their stringent regulation.

We complement this literature with our finding that differences in investment-innovation opportunities and distortions between the innovating North and the imitating South provide a source of comparative advantage in pollution-intensive goods. We isolate this effect by abstracting from the sources of comparative advantage discussed in the previous paragraph. In particular, our assumptions on preferences imply that being richer does not lead *per se* to more stringent environmental policy; we assume that the property rights on the resource base are perfectly defined and enforced; and that the relative factor endowments are identical across regions. Instead, in our setting comparative advantage stems from the difference in enforcement of intellectual property rights in the two regions: the North protects innovators and generates innovation, the South cannot protect innovators and imitates technology from the North.[2]

By almost exclusively relying on trade theory, the theoretical literature on the pollution haven hypothesis seems to have placed insufficient weight on the technology aspects of the debate. Indeed, there is also a small literature that deals with endogenous innovation and pollution havens, e.g. Golombek and Hoel (2004) and Ben Youssef (2003). We differ from these papers in that we do not assume *a priori* that technological change always results in cleaner production. Instead we derive the nature of technological change endogenously from profit incentives (following Acemoglu's (2002) theory of directed technological change). We also differ from these papers by allowing firms, rather than a planner, to

[2] To avoid confusion, notice that here we refer to intellectual property rights protection, while the above discussion refers to resource property rights.

decide on innovation, so that innovation externalities and second-best policies play a role.

The organization of the paper is as follows: section 1 introduces the model, while section 2 discusses the equilibrium. In particular, we address environmental regulation in section 2.5. Section 3 discusses the pure terms of trade effect by modelling innovation opportunities such that trade induces neither pollution-saving nor pollution-using technical change. Section 4 is instead devoted to the analysis of the induced technology effect. Finally, in section 5 we collect and compare our main results and conclude the paper.

1. The Model

Our economy is made up of two regions, each comprising a set of small countries, which we call the North and the South. These two regions only differ in the institutions regulating intellectual property rights (IPR's) protection. In particular, we assume that IPR's are perfectly enforced in the North, while they are not enforced in the South.

The economy admits a representative consumer who derives utility (U) at each moment in time from produced consumption goods, C, and from environmental quality, $E = \bar{E} - R$, where $\bar{E}$ is the maximum level of environmental quality and R is resource use. Here we view environmental quality as the amount of natural resources ($\bar{E}$) not devoted to productive use (R) at each point in time. We model it as a flow variable, since at every moment in time it returns to its maximum level, $\bar{E}$. Hence, our variable R represents pollution, or more precisely, extractive use of natural resources - clean water and clean air, say - for use in production. We specify the following (intertemporal) CRRA utility function:

$$\int_0^\infty \frac{(C(t)(\bar{E}-R(t))^\phi)^{1-\varsigma}-1}{1-\varsigma} e^{-\rho t} dt, \tag{1}$$

where ρ is the rate of time preference and ς is the inverse of the intertemporal elasticity of substitution.[3]

Each consumer maximizes the utility in (1) subject to the budget constraint[4]:

$$C+M+D \leq Y \equiv \left(Y_L^{\frac{\varepsilon-1}{\varepsilon}} + Y_R^{\frac{\varepsilon-1}{\varepsilon}} \right)^{\frac{\varepsilon}{\varepsilon-1}}, \tag{2}$$

where M is physical investment (machines), and D is the total amount of research and development (R&D) expenditure. The production function in (2) shows that final output (Y) is obtained as a CES aggregate of two intermediate goods, Y_R and Y_L, with an elasticity of substitution equal to ε. Moreover expression (2) states that consumption, investment and R&D expenditure are all the possible uses of the final good.

[3] As is usual in growth theory, the Cobb-Douglas/CRRA structure allows for a balanced growth path with constant environmental quality and a constant rate of growth of consumption. We touch again on this issue on page 11 below. For a complete discussion see Bovenberg and Smulders (1995).

[4] To simplify notation, as long as no confusion arises, we suppress time arguments from now on.

The pollution-intensive good (Y_R) is produced using resources and a set of differentiated man-made inputs which we refer to, for simplicity, as "machines", $m_R(j)$. The range of machines that can be used to produce pollution-intensive goods is indicated by N_R. The labour-intensive good (Y_L) is produced using labour and a different set of machines, whose range is N_L. The production functions for the two intermediate goods are:

$$Y_R = \frac{1}{1-\beta}\left(\int_0^{N_R} m_R(j)^{(1-\beta)}dj\right)R^{\beta}, \tag{3}$$

and

$$Y_L = \frac{1}{1-\beta}\left(\int_0^{N_L} m_L(j)^{(1-\beta)}dj\right)L^{\beta}. \tag{4}$$

For simplicity we have modeled pollution, R, as an input. Stokey (1998) has shown, however, that the Cobb-Douglas specification in (3) can be seen as the reduced form of a technology with pollution as an output and abatement possibilities. The key property is that a reduction of pollution can be achieved either by a reduction of market output or by an increase in machine inputs.

Technological change arises from costly innovation: private firms invest in R&D labs in which blueprints for new machine varieties are developed (as in Rivera-Batiz and Romer (1991)). Since the R&D stage has to be finished before production of new machines can take place, R&D costs are sunk. As a consequence, only innovators who expect to wield some monopoly power in the future will actually engage in R&D activities. This means that innovation will only take place in the North, where intellectual property rights are protected, while southern producers are able to copy these innovations.

Since R&D generates new blueprints, the total range of machines, N, increases with R&D investments, D. We consider two alternative specifications. In the first, R&D decisions affect the rate at which new blueprints are developed, but innovators cannot know in advance in which of the two sectors their specific innovation will prove most useful. In this case, which we label as "undirected technical change", blueprints in sector i expand at the following rate:[5]

$$\dot{N}_i = \frac{\gamma_i D}{(N_R+N_L)^{\psi}/\eta}, \tag{5}$$

where D is total R&D investment in the economy and the γ_i is the exogenous probability ($\gamma_R + \gamma_L = 1$) that an individual researcher's effort results in a successful new machine design for sector i. The denominator at the right-hand side of (5) represents the "difficulty of R&D" (cf. Segerstrom (1998)), which depends on two opposing effects. On the one hand, we assume that the more innovations in any sector already have been generated, the harder it is to come up with the next innovation. We capture this by the factor $(N_R+N_L)^{\psi}$, where $\psi > 0$. On the other hand, new innovation opportunities may arise exogenously

[5]Throughout the paper dotted variables indicate time derivatives, that is, $\dot{x} = dx/dt$.

over time. This can be viewed as reflecting any kind of innovation not driven by sector-specific profit incentives. We capture the level of innovation opportunities by parameter η, which we assume to grow at an exogenous rate ψg.[6]

In our second specification of innovation possibilities, technical change is directed, following Acemoglu (2002), so that innovators can target their efforts at a specific sector, and blueprints in sector i expand at the following rate:

$$\dot{N}_i = \frac{D_i}{(N_R + N_L)^{\psi}/\eta}. \tag{6}$$

While the denominator is the same and motivated by the same arguments as in (5), the rate of innovation in each sector depends on the total R&D effort taking place in it, D_i, with $D_L + D_R = D$.

Although we think the directed technical change variant is more realistic, contrasting undirected to directed technical change allows us to disentangle the terms of trade effect and the induced technology effect.

Finally, we assume that labour supply is exogenously fixed at L, while the environmental regulator in each country puts an endogenously determined cap on pollution R.

2. Equilibrium

Firms maximize profits and households maximize utility. They all take as given prices and factor rewards, with one notable exception: producers of machines in the North can set their own price as they operate under monopolistic conditions. Finally the environmental regulator in each country maximizes the utility of the representative agent by choosing the nation-wide level of pollution (resource supply), taking as given international prices. All markets clear and the resource constraint is satisfied. The final good is chosen as the numeraire.

In what follows we only focus on the long-run balanced growth path of this economy, where prices and the amount of natural resources used in production are constant and where output, consumption, investment and R&D outlays, as well as N_R and N_L, all grow at the (exogenous) rate g.[7]

---

[6] Assuming $\psi = 0$, we would have an endogenous growth model, in which the long-run growth rate is endogenous. Although generating very similar results, the fully endogenous growth version of this model involves a more complicated analysis so that we ignore it here.

[7] With constant factor supply R and L, the constant returns to scale production functions (2)-(4) imply that output grows at the same rate as N_R and N_L. The goods market equilibrium condition (2), moreover, requires Y, C, M and D to grow at a common rate. The innovation functions (5) and (6) imply that the growth rate of N_i can be constant only if $(N_R + N_L)^{\psi}/\eta$ is constant (cf. Jones (1995)). Hence, the growth rate of this economy equals $(\dot{N}_i/N_i) = (\dot{\eta}/\eta)/\psi = g$. Given the choice of the final good as numeraire, prices are constant along the balanced growth path.

2.1 Intermediate Goods and Machines

We start from the production of intermediate goods. To simplify the exposition, we let $S_R \equiv R$ and $S_L \equiv L$. Firms that employ factor i ($i = L, R$) maximize their profits choosing the amount of factor S_i to employ and the amount of machines $m_i(j)$ of each type to use. Their maximization problem is then:

$$\max_{S_i,\{m_i(j)\}} p_i Y_i - w_i S_i - \int_0^{N_i} p_{m_i}(j) m_i(j) dj, \tag{7}$$

subject to the production functions (3) and (4), and taking as given goods' prices p_i, machine prices $p_{mi}(j)$, and factors' prices w_i.[8] The demands for machines resulting from the above maximization are:

$$m_i(j) = \left(\frac{p_i}{p_{m_i}(j)}\right)^{\frac{1}{\beta}} S_i. \tag{8}$$

Consider first the supply of machines in the North. Since intellectual property rights are perfectly enforced, ownership of blueprints allow northern producers to act as monopolists. Assuming that υ units of the final good are required to produce each machine and that machine production is subsidized at rate τ_m, the expression for the profits of a monopolist supplying the i-complementary machine j is given by $\pi_i(j) = (p_{m_i}(j) - \upsilon(1-\tau_m))\, m_i(j)$, $j \in (0, N_i]$. Given the demand function in (8), the profit-maximizing price will be set as a mark-up over marginal cost, and will equal $p_{m_i}(j) = \upsilon(1-\tau_m)/(1-\beta)$. To simplify the algebra, we assume υ to be equal to $1-\beta$, so that $p_{m_i}(j) = (1-\tau_m)$. Substituting for machine prices and demand functions in the expressions for the profits of the technology monopolist, one gets:

$$\pi_i(j) = (1-\tau_m)^{-(1-\beta)/\beta} \beta p_i^{1/\beta} S_i. \tag{9}$$

Using machine demands (8) and prices in the sectorial production functions (3) and (4), we obtain the supply functions for northern firms, that we identify with an n superscript:

$$Y_i^n = \left(\frac{1}{1-\beta}\right)^{1/\beta} \left(\frac{1-\beta}{1-\tau_m}\right)^{(1-\beta)/\beta} (p_i^n)^{(1-\beta)/\beta} N_i^n S_i^n. \tag{10}$$

Next, turn to the South. IPR's are not enforced in the South, patent protection is not effective and the sunk costs associated with the development of new blueprints cannot be recouped. This effectively rules out the possibility that a local R&D sector may arise in the South. We assume, though, that southern producers can copy at no cost some of the

[8] For simplicity, we assume that machines depreciate fully after use. As discussed by Acemoglu (2002), assuming slow depreciation of machinery would not change the balanced growth equilibrium path.

blueprints developed in the North. In particular, a fraction $\delta \in (0,1)$ of the blueprints from North becomes available in the South, so that

$$N_i^s = \delta N_i^n.$$

Diffusion is incomplete due to a time lag between innovation and imitation, or because some blueprints are inherently too complex (and thus too costly) to copy.

As no institutional arrangement protects the monopoly power of machine producers, and no sunk costs prevent copying by more than one producer, perfectly competitive markets ensue. The price of machines in the South will then equal marginal cost: $p_{m_i}(j) = (1-\beta)$. Substituting machine demands (8) and prices into the sectorial production functions (3) and (4), we obtain:

$$\text{(11)} \qquad Y_i^s = \left(\frac{1}{1-\beta}\right)^{1/\beta} (p_i^s)^{(1-\beta)/\beta}\, \delta N_i^n S_i^s.$$

Comparing (11) and (10), we see that for $\delta < ((1-\beta)/(1-\tau_m))^{(1-\beta)/\beta}$, productivity (per unit of factor S_i) is lower in the South than in the North, so that the diffusion parameter δ realistically allows the South to be poorer than the North.

2.2 Final Goods and the Production Elasticity of Pollution

Final goods' producers demand intermediate goods up to the point where marginal productivity equals price. This implies $\partial Y/\partial Y_i = p_i$, which from (2) implies the following relative demand for intermediate goods:

$$\text{(12)} \qquad \frac{p_R}{p_L} = \left(\frac{\Upsilon_R}{\Upsilon_L}\right)^{-\frac{1}{\varepsilon}},$$

where Υ_i represents the demand for intermediate good i.

In the rest of the analysis, the production elasticity of polluting inputs will play a central role. We can write this elasticity as

$$\text{(13)} \qquad \frac{\partial Y}{\partial R}\frac{R}{Y} = \beta\theta_R = \beta\frac{p_R Y_R}{Y},$$

where we made use of (3) to write $(\partial Y_R/\partial R)(R/Y_R) = \beta$ and we introduced the production elasticity (and cost share) of pollution-intensive goods in final goods production, $\theta_R \equiv (\partial Y/\partial Y_R)(Y_R/Y) = p_R Y_R/Y$. Since the production elasticity of pollution and θ_R move together for any given β, we will find it easier to refer to the latter in the discussion that follows. Moreover, note that since (2) features constant returns to scale, $1-\theta_R = p_L Y_L/Y$. Using (10) or (11), we can express the relative costs share $p_R Y_R/p_L Y_L$ in each region as:

$$\text{(14)} \qquad \frac{\theta_R}{1-\theta_R} = \left(\frac{p_R}{p_L}\right)^{1/\beta} \frac{N_R R}{N_L L}.$$

This reveals that the share of pollution-intensive goods in production, θ_R, depends on relative prices, relative factor supply, and technology.

2.3 R&D and Innovation

Firms in the North have an incentive to innovate whenever innovation earns a rate of return that is at least equal to the market interest rate. This implies the following arbitrage equation for innovator j:

$$\frac{\tilde{\pi}_j}{v_j} + \frac{\dot{v}_j}{v_j} = r \tag{15}$$

where $\tilde{\pi}_j$ is the per period flow of expected revenues (dividends) for the holder of a blueprint j, v_j is the market value of blueprint j, and r is the interest rate. In the model of undirected technical change, expected dividends equal $\tilde{\pi}_j = \gamma_R\pi_R + \gamma_L\pi_L$; in the model of directed technological change they equal $\tilde{\pi}_j = \pi_i$ if j's effort is directed to sector $i = R, L$.

We assume free entry in research activities, which implies that the value of a blueprint cannot exceed its cost. From (5) and (6), this implies $v_j \leq (N_R + N_L)^{\Psi}/\eta$. In particular, in the undirected technical change model, this condition must hold with equality whenever innovation occurs, that is, whenever $D > 0$; in the directed technical change model, there are two type of innovators, $j = R, L$, and the expression holds with equality whenever innovation occurs in sector j (i.e. whenever $\dot{N}_j > 0$).

In the undirected technical change model, the long-run ratio N_R/N_L equals γ_R/γ_L, as follows from (5). Hence, the relative supply of machine variety only depends on research technology and cannot be affected by e.g. trade. Once in a balanced growth path, only shocks to N_i or γ_i can bring the economy from its balanced growth path. This allows us to treat N_R/N_L as a constant in the undirected technical change model.

In the directed technical change model, along the balanced growth path innovation takes place in both sectors, the cost and value of a blueprint is the same in both sectors ($v_R = v_L$) and to satisfy the condition (15), we must have $\pi_R = \pi_L$. After substitution of (9), this implies the following "no-arbitrage" condition:

$$\left(\frac{p_R}{p_L}\right)^{1/\beta} \frac{R}{L} = 1. \tag{16}$$

The relative profitability of innovations in each sector thus increases with the price of the intermediate good they are used to produce (the price effect), and with the relative supply of the factor they complement (the market-size effect). By these two channels trade or changes in environmental policy affect innovation.

2.4 Households

Households maximize their lifetime utility (1), subject to the usual intertemporal budget constraint. This results in the Keynes-Ramsey rule stating that consumption grows at a rate proportional to the difference between the interest rate and consumer's rate of time preference.

Along the balanced growth path, consumption and output grow at the same rate, g, so that we may express this equation as:

$$r = \rho + \varsigma g. \tag{17}$$

2.5 Environmental Regulation

We assume that the environmental regulator determines the level of pollution in the economy.[9] We model the regulator assuming that she only aims at correcting the environmental externality from production mentioned above. She chooses the supply of pollution to maximize the utility of the representative agent at each moment in time, taking as given the choices made by the other economic agents in the economy, concerning the level of consumption and of investment in both machines and R&D.

To gain some intuition, we first derive the environmental policy rule in general terms. As in a static context, the maximization of $U(C, \bar{E} - R)$ subject to a budget constraint of the form $C = F(R, \cdot) - D - M$, where D and M are the (given) amounts of investment in R&D and in machines, yields the following first-order condition

$$\left(\frac{\partial U}{\partial C}\right)\frac{\partial Y}{\partial R} - \left(\frac{\partial U}{\partial E}\right) = 0,$$

that we can rewrite, in terms of elasticities, as

$$\left(\frac{\partial Y}{\partial R}\frac{R}{Y}\right)\frac{Y}{C} = \left(\frac{(\partial U/\partial E)E}{(\partial U/\partial C)C}\right)\frac{R}{E - R}. \tag{18}$$

This equation balances the marginal benefits from pollution (in the form of additional consumption goods) and its marginal costs (in the form of lower environmental amenities), both measured in terms of consumption.

Equation (18) also shows how the supply of pollution depends, all other things equal, on the production elasticity of polluting inputs, on the consumption to output ratio, and, finally, on the share of environmental amenities in utility. First, a higher production elasticity of resources, $\left(\frac{\partial Y}{\partial R}\frac{R}{Y}\right)$, makes resources more valuable in production and increases the marginal benefits of pollution. In other words, the costs of reducing pollution become larger. Second, a lower consumption to output ratio C/Y increases the equilibrium supply of pollution by increasing its marginal benefits. Intuitively, in an economy with lower consumption per unit of output consumption is perceived as relatively scarce, raising the marginal value of production. Finally, a higher share of amenities in utility, $\frac{(\partial U/\partial E)E}{(\partial U/\partial C)C}$,

[9] We also assume that environmental policy is perfectly enforced in both regions.

increases the marginal costs of pollution, shifts the *MC* curve up, and hence decreases equilibrium pollution.

We can rewrite (18) noting that the share of environmental amenities equals ϕ, see (1), and that the production elasticity of pollution is given in (13). This yields the following expression for the marginal benefits and the marginal costs of pollution:

$$\underbrace{\frac{\theta_R \beta}{C/Y}}_{MB} = \underbrace{\phi \frac{R/L}{\overline{E}/L - R/L}}_{MC}. \tag{19}$$

Rearranging terms and subtracting R/L from both sides of the previous equation we obtain,

$$\left(\frac{\beta}{\phi C/Y}\right)\frac{\overline{E}}{L} - \left(\frac{\beta}{\phi C/Y}+1\right)\frac{R}{L} = \left(\frac{1-\theta_R}{\theta_R}\right)\frac{R}{L},$$

which we can rewrite, using (14), and letting $\frac{\beta}{\phi C/Y} = \Omega$,[10] as:

$$\Omega\frac{\overline{E}}{L} - (\Omega+1)\frac{R}{L} = \left(\frac{p_R}{p_L}\right)^{-1/\beta}\left(\frac{N_R}{N_L}\right)^{-1}. \tag{20}$$

This expression will prove more tractable in the analysis that follows than (19), but bear in mind that this condition still states that pollution will be set at a level that equates marginal benefits to marginal costs. We will use this expression to determine the level of pollution chosen by each region's regulator under autarchy and under free trade, and to discuss how these decisions will be influenced by the directedness of technical change.

In the introduction, we pointed out that our model rules out international differences in environmental policy due to differences in income, factor endowments, or resource property right regimes. We do so in order to isolate the effect of endogenous technology from other mechanisms explaining pollution haven effects in the existing literature. Equation (20) reveals how we do this. First, if output and consumption grow at the same rate, $\Omega \equiv \beta/(\phi C/Y)$ is constant over time. With growth, then, the equilibrium supply of pollution must be the same for each level of income. This is due to the Cobb-Douglas specification of the utility function (1), which implies that the larger demand for environmental quality due to higher income is exactly offset by the higher benefit of pollution generated by the increasing productivity of polluting inputs that drives growth. Second, we assume that relative factor endowments, $\overline{E}/L$, and preferences, reflected by ϕ, are the same in both regions. This leaves us with three determinants for the differences in R/L, as shown in equation (20): differences in the consumption-to-output ratio across countries, prices, and technology.[11]

[10] It is important for the analysis that follows to bear in mind that Ω is a constant along the balanced growth path in each country. Yet, its value depends on the country-specific level of C/Y. Thus, a country with more consumption per unit of final output, will have a lower Ω.

[11] From (20), it is straightforward to see that our analysis using the C/Y ratios can be alternatively carried out through differences in ϕ and $\overline{E}/L$. As long as North has higher ϕ and/or lower $\overline{E}/L$, the analysis is qualitatively similar to the analysis that follows where the North has an higher C/Y ratio than the South.

In the next section we discuss the differences in the C/Y ratios between North and South. We will show that they do not depend either on the trade regime, or on the assumptions concerning the innovation process. With this in mind, we can explain the different outcomes of the model under the different trade regimes in terms of price and technology effects only.

In the remainder of the paper we will be able to separate price and technology effects since we will first focus on undirected technological change - in which case the bias N_R/N_L is exogenous - and later on directed technological change - with the bias N_R/N_L endogenously determined. For each technology regime, moreover, we will discuss and compare autarchy, where prices differ across countries, and free trade, where price equalization obtains.

2.6 The Consumption to Output Ratio

As mentioned above, the consumption to output ratio constitutes one of the central elements of our analysis. Here we provide an expression for this ratio in each country that we can use to substitute into expression (20). The Appendix shows that the the C/Y ratio in the South is

$$\left(\frac{C}{Y}\right)^s = 1-(1-\beta) = \beta, \tag{21}$$

while its northern homologue is

$$\left(\frac{C}{Y}\right)^n = 1-(1-\beta)\left(\frac{1-\beta}{1-\tau_m}\right) - (1-\beta)\beta\frac{g}{r}. \tag{22}$$

Comparing the two expressions above, we note two differences. First, the North will invest less in physical capital, since northern producers of intermediate goods face higher machine prices than their southern counterparts. Indeed, in the North monopolistic competition, fostered by IPR's enforcement, drives prices above marginal cost, while this cannot happen in the South where patent protection is not enforced. This first effect is reflected by the second term on the right-hand side of (22). As long as the monopolistic distortion is not offset by an appropriate subsidy ($\tau_m = \beta$), the North will spend less for machines than the South. Second, the North consumes less than the South out of its gross output, since part of it is invested in developing new technology (research expenditures, the second term at the right-hand side of the equation above). Northern innovators find it profitable to forego some consumption in order to invest in R&D, whereas this is impossible for southern (potential) innovators. It is clear that these two effects work in opposite directions, and that, depending on the level of the subsidy, either of the two can dominate. In particular, it is easy to show that $(C/Y)^n > (C/Y)^s$ whenever $\tau_m < \frac{\beta(r-g)}{r-g\beta}$.

3. Undirected Technical Change

In this section we analyze the pollution supply by the two regions assuming that innovators cannot change the long-run composition of technology N_R/N_L. We analyze both the autarchy case, where the only interdependence between the two regions stems from the international diffusion of technology, and the free trade situation. In this context the only effect of trade is through changes in prices. This section will provide us with a benchmark for the analysis of the directed technical change case, in the following section.

3.1 Autarchy

Under autarchy, domestic relative demand for final goods equals domestic supply. Substituting supply (10) or (11) into the expression for demand (12), we find that the domestic price ratio is given by the following function of relative factor supply and technology:

$$\frac{p_R}{p_L} = \left(\frac{N_R R}{N_L L}\right)^{-\beta/\sigma}, \tag{23}$$

where $\sigma = 1 + (\varepsilon - 1)\beta$ is the elasticity of substitution between R and L in aggregate production. Substituting this into (20), we obtain the following expression that determines pollution supply under autarchy:

$$\Omega\frac{\overline{E}}{L} - (\Omega+1)\frac{R}{L} = \left(\frac{N_R}{N_L}\right)^{\frac{1-\sigma}{\sigma}}\left(\frac{R}{L}\right)^{\frac{1}{\sigma}}. \tag{24}$$

We have thus obtained a (relatively) simple equation in the relative factor supply only. We note that $\overline{E}/L$, N_R/N_L, and σ are the same across countries, while Ω differs. Recalling that $\Omega = \beta/(\phi C/Y)$, it is straightforward to see that the region with the highest C/Y has the lowest Ω, and thus chooses the lowest R/L, according to (24).

Since a closed form solution cannot be obtained analytically, we derive further results graphically. The left-hand side of the equation above is linear in pollution supply and negatively sloped, while the right-hand side is an increasing function of R/L with a curvature that depends on σ. We plot the left- and the right-hand sides of (24) in Figure below, as *LHS* and *RHS*, *aut*, respectively.[12] The solid lines describe the situation under autarchy.

It is easy to see from (24) that the line depicting the left-hand side will be steeper and higher for larger Ω (that is, for smaller consumption to output ratio). In the picture we depicted the situation in which the North has the largest C/Y and thus the lowest line.

[12] The figure is drawn for the case where $\sigma < 1$. The curves depicting the right-hand side would be concave for $\sigma > 1$, but qualitatively nothing would change in our analysis.

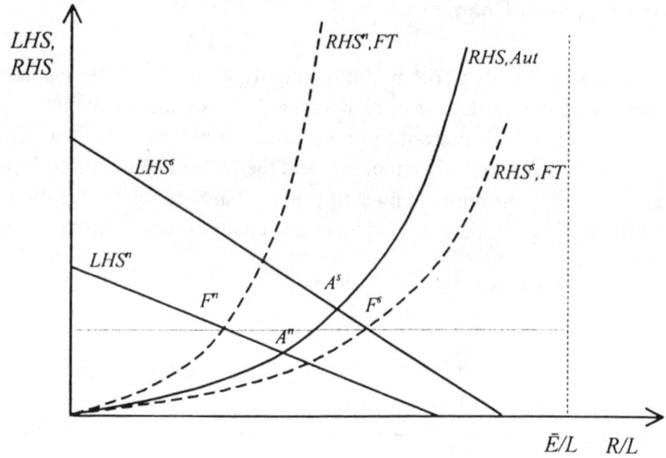

Figure 1: Equilibrium with Undirected Technical Change under Autarchy and Free Trade

The right-hand side is shared by the two countries, so that the equilibrium will obtain at a point characterized by less pollution in the North than in the South.

Intuitively, the reason for this result lies in the difference in the marginal benefits from pollution. Remembering the first-order condition in (19), it is immediate to see that, while the marginal cost term (on the right-hand side) is the same for every country, the marginal benefit term (on the left-hand side) depends negatively, *ceteris paribus*, on the C/Y ratio. Loosely speaking, the more consumption each country obtains from any unit of output, the lower the marginal benefit (in terms of consumption) it derives from any marginal increase in pollution.

3.1 International Trade

Once goods are traded internationally without frictions, a single world price p_i^w for each good $i = R, L$ will prevail. This implies that in equilibrium the right-hand side of (20) has the same value across countries. Equating the left-hand side of (20) for both regions, and rearranging, we find:

$$\frac{R^s}{L^s} = \frac{R^n}{L^n} + \left(\frac{\Omega^s - \Omega^n}{1 + \Omega^s}\right)\left(\frac{\overline{E}}{L} - \frac{R^n}{L^n}\right), \tag{25}$$

This shows that, as long as $\Omega^s > \Omega^n$ (that is, as long as $(C/Y)^s < (C/Y)^n$), the South will pollute more than the North also under free trade. More generally, we can conclude that countries that consume relatively less tend to impose laxer environmental regulation and hence to pollute more. From now on we state that $\Omega^s > \Omega^n$ corresponds to a situation in which South has a comparative advantage in polluting goods (if $\Omega^n > \Omega^s$, the North has it).

When international trade is allowed, the equilibrium world relative price is determined by the market clearing condition on the world markets, that is, from (12) and relative supply: $(p_R/p_L)^{-\varepsilon} = (Y_R^n + Y_R^s)/(Y_L^n + Y_L^s)$.

Substituting (10) and (11), we find the following solution for the world relative price:

$$\frac{p_R}{p_L} = \left(\frac{N_R}{N_L}\left[\lambda^n \frac{R^n}{L^n} + \lambda^s \frac{R^s}{L^s}\right]\right)^{-\frac{\beta}{\sigma}}, \tag{26}$$

where $\lambda^n = 1 - \lambda^s = [1 + (\frac{1-\tau_m}{1-\beta})^{(1-\beta)/\beta}\delta L^s/L^n]^{-1}$. The term in the brackets on the right-hand side in (26) is the world supply of pollution relative to labour, in efficiency terms, written as a weighted average of the national relative factor supplies R^n/L^n and R^s/L^s.

By substituting (26) and (25) into (20), we find:

$$\Omega^n \frac{\overline{E}}{L} - (\Omega^n + 1)\frac{R^n}{L^n} = \left(\frac{N_R}{N_L}\right)^{\frac{1-\sigma}{\sigma}} \left[\frac{R^n}{L^n} + \lambda^s \frac{\Omega^s - \Omega^n}{1+\Omega^s}\left(\frac{\overline{E}}{L} - \frac{R^n}{L^n}\right)\right]^{\frac{1}{\sigma}}. \tag{27}$$

By interchanging the s and n superscripts we get the condition that determines equilibrium pollution in the South. To understand the effect of free trade when technical change is undirected, we need to compare this expression with the corresponding one for autarchy, (24). We do this using once more a graphical treatment for clarity. First of all, we notice that the left-hand side does not change, so that the relevant curves are still the solid downward sloping lines in Figure . As for the right hand side, all else equal it will be below the corresponding autarchy curve whenever the own country's Ω is larger than the foreign one. Assuming once more that the North consumes relatively more than the South, this means that the northern curve will shift up, while the southern one will shift down, relative to autarchy. This is represented by the two dotted lines in of Figure .[13] Under these conditions, including $\Omega^s > \Omega^n$, free trade in Figure increases pollution in the South and decreases it in the North. Thus the case of undirected technical change corresponds to the position of the trade pessimist.

North and South are interdependent through goods trade and diffusion of technology from North to South. The literature on international technology diffusion argues that the two are connected: international communication and contacts stimulate technology

[13] To construct the diagram in Figure , it would be sufficient to draw only one of the dotted lines. As noted above, with international price equalization, the right-hand side of (20) must be the same for every country, in equilibrium. Thus both crossing points F^n and F^s will lie on the same horizontal line, as drawn in the figure above. We will make use of this in the next section.

spillovers, e.g. Keller (2004). We therefore now study how our results depend on our parameter for technology diffusion, δ. An increase in δ, the fraction of northern technologies copied in the South, raises production levels in the South, see (11). In autarchy, this raises aggregate production without changing relative variables. In particular, the equilibrium pollution/labour ratio is independent of δ, see (20). In free trade, the increased supply from the South affects the world price: if the South has a comparative advantage in pollution-intensive goods ($\Omega^s > \Omega^n$), the world relative supply of these goods increases with δ (for given relative factor supplies in the regions) and their relative price falls. This lower price for polluting goods makes it less attractive for both countries to pollute. Indeed, in (27) an increase in diffusion parameter δ lowers λ^n and increases λ^s so that both dashed curves in Figure shift up and both regions pollute less. A change in δ, however, can never reverse our finding that the country with comparative advantage in pollution-intensive goods pollutes more in free trade than in autarchy; this result holds for any $0 < \lambda^n = 1 - \lambda^s < 1$ and hence for any δ, see (27).

4. Directed Technical Change

When innovators can direct innovation at a particular sector, the nature of technological change becomes endogenous and responds to shocks, for example to changes in international prices. International trade thus naturally affects environmental policy in both regions, not only directly through prices, but also indirectly, through the induced changes in technology. Furthermore, even when there is no international trade in goods, innovation in the North affects technology and environmental policy in the South, through the international diffusion of northern technologies.

4.1 Pollution-using vs Pollution-saving Technical Change

Under directed technical change, international trade endogenously induces either pollution-using or pollution-saving technical change. In this subsection we provide a definition these concepts.

If innovation occurs at a faster pace in one of the sectors, the N_R/N_L ratio changes. The effect of this change on the composition of production in the long run depends on the elasticity of substitution between factors of production, σ. In general, a change in the composition of technology affects the production elasticity of pollution. This either increases or reduces the benefits of pollution in terms of output. When this elasticity increases, the marginal benefits from pollution increase and thus the supply of pollution will increase in equilibrium, recall (18). We call this kind of technical change *pollution-using*. The opposite occurs when the elasticity decreases, making pollution less beneficial. In this case we speak about *pollution-saving* technical change. The two concepts correspond to higher and lower abatement costs, respectively. An inspection of equations (24) and (27) shows that the right-hand side of both expressions increases with $(N_R/N_L)^{(1-\sigma)/\sigma}$. We can interpret this as showing that an increase in this term, which we label the technology bias, represents pollution-saving technical change.

We can make this point more formal by recalling, from our discussion in section 2.2, that we can proxy the production elasticity of pollution by the cost share of polluting goods θ_R. Using equation (14) and substituting the autarchy prices from (23), for example, we get

$$\frac{\theta_R}{1-\theta_R} = \left(\frac{N_R}{N_L}\right)^{-\frac{1-\sigma}{\sigma}} \left(\frac{R}{L}\right)^{-\frac{1-\sigma}{\sigma}}, \tag{28}$$

which clearly shows that, when the technology bias $(N_R/N_L)^{(1-\sigma)/\sigma}$ increases, θ_R falls for given factor supply, which implies pollution-saving technical change.[14] This also shows that the nature of technical change not only depends on the direction (the change in N_R/N_L) but also on the elasticity σ. For example if firms in the pollution-intensive sector innovate more than in the other sector, the ratio N_R/N_L increases. Whether this raises or reduces the production elasticity of pollution, depends of the elasticity of substitution σ . If the two goods are gross substitutes ($\sigma > 1$), the relatively fast productivity improvements in the pollution-intensive sector induce a shift towards this sector and technological change is pollution-using. If however gross complementarity applies ($\sigma < 1$), the same productivity improvements trigger a relative increase in the demand for the complementary input and technological change is labour-using, and hence pollution-saving.

4.2 Autarchy

The long-run technology bias is determined in the North by the condition that the profits from innovation are equal across sectors, that is, from the no-arbitrage equation derived in section 2.3. We first focus on autarchy, finding the following expression for the technological differences across sectors using (16) and (23):

$$\left(\frac{N_R}{N_L}\right)^{\frac{1-\sigma}{\sigma}} = \left(\frac{R^n}{L^n}\right)^{-\frac{(1-\sigma)^2}{\sigma}}. \tag{29}$$

This equation shows that under autarchy the bias of technology in both regions depends only on pollution policy in the North, R^n/L^n. Moreover, we can immediately see that a decrease in the relative supply of pollution results in pollution-saving technological change. Hence, the more stringent the environmental policy in the North, the more pollution-saving technological change.

To determine equilibrium pollution supply under autarchy, we can substitute the long-run bias derived above into the environmental policy rule (24). The expression differs

[14] The corresponding expression for θ_R under free trade is obtained by substitution of the international prices from (26). In the resulting expression θ_R still decreases with $(N_R/N_L)^{(1-\sigma)/\sigma}$ for given factor supply. Our classification is thus independent of the trade regime.

across countries. For the North the equilibrium condition reads:

$$\Omega^n \frac{\overline{E}}{L} - (\Omega^n + 1)\frac{R^n}{L^n} = \left(\frac{R^n}{L^n}\right)^{2-\sigma}, \tag{30}$$

while for the South we get,

$$\Omega^s \frac{\overline{E}}{L} - (\Omega^s + 1)\frac{R^s}{L^s} = \left(\frac{R^n}{L^n}\right)^{2-\sigma} \left(\frac{R^s/L^s}{R^n/L^n}\right)^{1/\sigma}. \tag{31}$$

In Figure we construct a diagram analogous to the previous one, to illustrate the situation in this case. Once again, we indicate the curves that are relevant for the discussion of the autarchy case with solid lines.

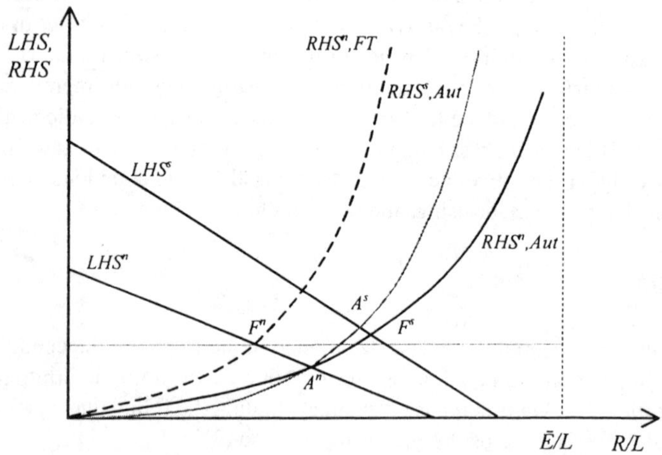

Figure 2: Equilibrium under Directed Technical Change under Autarchy and Free Trade

We start from the curves for the North. The curve that depicts the left-hand side is identical to the straight line we drew in Figure . The right-hand side is increasing for $\sigma < 2$ and decreasing for $\sigma > 2$. In the picture we depict the case where $\sigma < 1$ and $\Omega^n < \Omega^s$. The intersection point A^n represents the equilibrium for the North.[15] We can

[15]As long as $\sigma \leq 2$ this equilibrium always exists and is unique. When $\sigma > 2$, the line describing the right-hand side is decreasing and convex, it can intersect the straight line either once, twice, or never. When multiple crossings occur, we must rule out the intersection at which *RHS* cuts *LHS* from above, since starting

use the corresponding value of R^n/L^n to construct the curve for the South. This curve intersects the northern one at A^n, and is drawn as the lighter curve in the figure. Point A^s represents the equilibrium for the South.

This picture enables us to draw some conclusions for the autarchy case.

In the first place, we notice that, as in the undirected technical change case, also here the country with the highest Ω will pollute more. In this sense the regime of technical progress does not influence the outcome. As long as the North enjoys a higher C/Y ratio, it will choose a lower equilibrium level of pollution than the South.

Second, comparing equations (24) and (30), we immediately conclude that the line depicting the right-hand side for the North is steeper under undirected technical change than under directed technical change. This has the important consequence that any shock (such as changes in Ω through preferences or subsidies, or in relative endowments, $\overline{E}/L$) has a larger impact on pollution supply under directed technical change than under undirected technical change. Intuitively, this happens because under directed technical change the adjustment occurs not only through policy but also through changes in the composition of technology. For example, assume that preferences become greener (ϕ increases), this leads to the adoption of more stringent environmental policy and to a reduction in pollution supply. Under directed technical change this leads firms to develop pollution-saving technologies, which would not have occurred had technical change been undirected, and thus to curb pollution further.

Third, shocks in the North affect the South even without international trade in goods, through the international technology spillovers. Indeed, once a shock hits the North, both pollution and technology adjust. Technology diffusion will then induce the South to modify its pollution decisions according to the new available technology. This means, for example, that the South will benefit from pollution-saving technical change that occurs in the North following a tightening in the stance of environmental policy.

4.3 International Trade

When technical change is directed, the effects of a liberalization in international trade are twofold. First, price changes affect pollution supply through the traditional terms-of-trade mechanism as with undirected technological change. Second, price changes now also affect the direction of technological change. In this section we will show that, while the terms-of-trade effect tends to increase differences in pollution supply, the effects of changes in technology can either exacerbate or dampen this tendency. In particular, we show that the effect of the induced technical change may prove strong enough for the non-innovating country to reverse the terms-of-trade effect. Thus, provided that factors

from such a point, slightly increasing pollution supply raises marginal benefits relative to marginal costs of pollution. In other words, second order conditions are violated. Moreover, the case of no intersection is not of interest to our discussion, as it implies a corner solution in which both regions produce with zero pollution in the long run. In the rest of the paper we thus focus on interior solutions, which can always be constructed by choosing low enough a value for ϕ.

of production are sufficiently good substitutes, both regions might end up changing their environmental policy in the same direction.

As in the previous section, we obtain the long-run technology bias from the no-arbitrage condition (16), but we now use (26) to substitute for prices. This yields:

$$\left(\frac{N_R}{N_L}\right)^{\frac{1-\sigma}{\sigma}} = \left(\frac{R^n}{L^n}\right)^{1-\sigma} \left[\lambda^n \frac{R^n}{L^n} + \lambda^s \frac{R^s}{L^s}\right]^{-\frac{1-\sigma}{\sigma}}. \tag{32}$$

This equation shows that changes in the relative factor supply affect the bias of technological change through two terms: $(R^n/L^n)^{1-\sigma}$, which represents the market-size effect; and $[\lambda^n R^n/L^n + \lambda^s R^s/L^s]^{-(1-\sigma)/\sigma}$, which captures the price effect.

The market-size effect relates the direction of technical change to the potential market for factor-specific innovations. For example, for given prices, a reduction in pollution (relative to a constant labour supply) reduces the potential profits from innovations in the pollution-intensive sector, and directs R&D expenditure to the labour-intensive sector. The price effect works in the opposite direction. Imagine, for example, a reduction in pollution at the world level for given domestic supply (because pollution is reduced elsewhere). This leads to higher prices for dirty goods. As a consequence, innovation in the pollution-intensive sector becomes more attractive.

As we discussed above, under autarchy a pollution reduction in the North always results in pollution-saving technological change. In the presence of international trade, this is no longer necessary. In this case, indeed, it is the world supply of factors rather than the northern one to determine the price level. Accordingly, changes in the northern pollution supply have a relatively weaker impact on prices than before. As northern pollution is reduced, innovation in the pollution-intensive sector becomes less attractive through the market-size effect and this tends to reduce N_R/N_L. Compared to autarchy, the mitigating effect of an increase in the relative price of the polluting goods is now less salient. Hence, more stringent environmental policy makes N_R/N_L decrease more under free trade than under autarchy. Whether this implies a pollution-saving or pollution-using technical change relative to autarchy depends once more on whether polluting inputs and labour are gross substitutes or gross complements, see section 4.1.

Substituting (25) into (32) and using the new expression to substitute for the technology bias in (27), we find the condition that determines equilibrium pollution supply in the North:

$$\Omega^n \frac{\overline{E}}{L} - (\Omega^n + 1)\frac{R^n}{L^n} = \left(\frac{R^n}{L^n}\right)^{1-\sigma} \left[\frac{R^n}{L^n} + \lambda^s \frac{\Omega^s - \Omega^n}{\Omega^s + 1}\left(\frac{\overline{E}}{L} - \frac{R^n}{L^n}\right)\right] \tag{33}$$

This equation solves for pollution in the North under free trade and can be compared to the corresponding equation for autarchy, (30). Notice that, for given R^n/L^n, the right-hand side of (33) is larger than the one in (30), provided that $\Omega^s > \Omega^n$ (and smaller otherwise). Hence, when opening up to trade, the North reduces pollution supply if it was relatively clean in autarchy. Thus, the effect of trade on pollution in the North is similar to the undirected technology case. The reason is that the technology response to trade in the North is governed by the same incentives as the environmental policy response: to exploit the comparative advantage and benefit from the trade-induced increase in the relative price for labor-intensive goods, the North shifts production to the the labour-intensive sector not only by polluting less but also by innovating more in that sector.

In contrast, the effects of trade in the South might change due to fact that trade induces technical change and that this change is determined by conditions in the North.

In terms of Figure , when opening up to trade, the right-hand side curve of the North (the dashed line) shifts up so that the intersection necessarily moves to the left for North. The equilibrium under trade in the North is indicated by the point F^n. Just as in section 3, we know that prices and technology are common to the two regions. To determine the equilibrium for the South we just need to find the intersection of the LHS^s curve with the horizontal line through F^n. It is clear that this intersection might be above or below A^s, implying opposite effects of international trade on pollution in the South. The rest of this section is devoted to discussing this point.

Let's continue assuming $\Omega^s > \Omega^n$ for concreteness. This implies that the North pollutes relatively less than the South in the autarchy equilibrium. When the South opens up to trade with the relatively cleaner North, the price of the pollution-intensive goods increases for any given level of pollution in the South, see (26). This is the conventional terms-of-trade effect that makes it attractive for the South to specialize in the production of the dirty good and hence to increase pollution.

The increase in p_R/p_L might be mitigated and even offset by a fall in the technology bias N_R/N_L, see (20). Indeed, when the North opens up to trade, N_R/N_L falls.[16] The North changes the direction of technical change towards the sector in which it has a comparative advantage. When this happens, the South faces pollution-using technical change if $\sigma < 1$, but pollution-saving technical change if $\sigma > 1$. From (27) we see that we need an increase in the technology bias $(N_R/N_L)^{(1-\sigma)/\sigma}$ (that is, pollution-saving technical change) to offset the terms of trade effect (i.e. the fact that the term in brackets falls when opening up to trade).

This means that, as long as the goods are gross complements, the induced technical change effect has the same direction as the terms-of-trade effect, while it has an opposite effect when $\sigma > 1$. In other words, if $\sigma < 1$, not only for the North but also for the South, trade has the same effect as with undirected technological change. When goods are gross substitutes, instead, it is possible that the effects of international trade on pollution decisions in the South are reversed.

In Figure we present numerical examples to show that it is indeed possible that the South reduces pollution upon opening up to trade. The parameters values we used were such that $\Omega^s > \Omega^n$, so that the South pollutes more than North.[17] Above the line,

---

[16]We know that when the North opens up to trade its pollution supply (R^n/L^n) falls. From (16) this implies that the northern relative price of pollution-intensive goods p_R/p_L will then rise. Finally, from (20), this implies that N_R/N_L falls.

[17]For the simulations, the following values of the parameters were used: $L^n = 1, L^s = 3, \bar{E}/L = 0.5, g =$

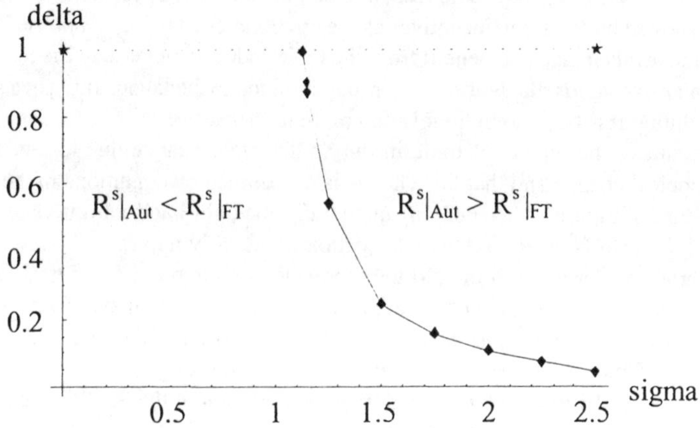

Figure 3: Pollution Policy in the South under Free Trade

combinations of the elasticity of substitution σ and the diffusion parameter δ are such that the South pollutes less in free trade than in autarchy; below the line, the opposite holds. As derived above, if $\sigma < 1$, trade always increases pollution in the South. The figure further reveals that better substitution or more diffusion of technologies make it more likely that trade improves the environment in the South. The reason is, first, that with a larger elasticity of substitution trade results in smaller price changes so that the conventional terms-of-trade effect that drives pollution havens is small. Second, with more diffusion (larger δ), the effective size of the South, as measured by λ^s, is larger and any change in Northern pollution supply has a smaller effect on world prices and a relatively larger effect on the market size for innovations in the labour-intensive sector. Thus, more diffusion implies a large change in the technology bias.[18]

We end this section by discussing the case with $\Omega^n > \Omega^s$, in which the North pollutes more than the South, reversing our earlier assumption on comparative advantage. Indeed, now the North has a comparative advantage in the production of polluting goods and starts innovating more in the pollution-intensive sector, once trade is opened. As a result, N_R/N_L now rises. If $\sigma < 1$, this implies pollution-saving technical change, and the South reduces pollution after trade is opened. This result is qualitatively similar to the one

$0.015, \beta = 0.2, \rho = 0.025, \varsigma = 1, \phi = 0.06, \tau_m = 0.$

[18]In Figure , a larger σ makes the *RHS* curves flatter, while a larger δ, by increasing λ^s, shifts the RHS^n, FT curve up. As a result, the vertical distance between the autarchy points A^n and A^s is small and the vertical distance between A^n on the one hand, and F^n and F^s on the other hand, is large, such that it becomes more likely that F^s is above A^s.

obtained with terms-of-trade effects only in section 4. If $\sigma > 1$, however, trade induces pollution-using technical change, so that the technology effect might offset the terms-of-trade effect. For sufficiently large σ and δ, the South will increase pollution. Hence, if the innovating region has a comparative advantage in pollution-intensive goods, if these goods are good substitutes for other goods, and if a large fraction of the new technologies diffuse to the imitating region, then trade is bad for the environment in both regions.

5. Summary and Conclusions

In our model the North innovates and technologies diffuse to the South. Since only a fraction of the technologies diffuse, the South is less productive and has lower income than the North. By construction, this income difference does not lead to differences in environmental stringency in our model. Instead, a country that consumes a relatively large part of income has a relatively large demand for environmental quality, as produced consumption goods and environmental amenities are traded off in utility. Whether the South or the North has higher demand for environmental quality thus depends on investment opportunities and distortions. On the one hand, the North has to incur the cost of innovation (while the South gets innovations for free), which makes consumption scarce and reduces demand for environmental quality in the North. On the other hand, intermediates are priced above cost in the North. This introduces a distortion not present in the South and reduces investment. Which of the two effects will prevail will depend on the level of subsidies to technological change.

In the paper we have discussed the two possibilities separately, one in which the North pollutes less than the South, and the other in which it pollutes more. These situations correspond to a comparative advantage for pollution-intensive goods in the South or in the North, respectively. When trade is opened, prices change and the regions exploit their comparative advantage. By this terms-of-trade effect, environmental quality always improves in the country that has most stringent environmental policy in autarchy and always deteriorates in the other. Induced technological change never reverses this pattern in the North, but may do so in the South provided that pollution-intensive and labour-intensive goods are gross substitutes ($\sigma > 1$).

While the terms-of-trade effect has opposite effects on environmental quality in the two regions, the induced technology effect affects environmental quality in both regions in the same direction. The reason is that technology is determined by profit conditions in the North and then diffuses to the South. Northern innovation activities will shift in the direction of the sector that uses intensively the relatively abundant factor (pollution or labour) and thus increases the relative supply from this sector. If these goods are gross substitutes for goods from the other sector, world demand shifts away from goods in which the South has a comparative advantage. Hence, if the South has a comparative advantage in pollution-intensive goods, the induced technology effect nevertheless makes production of these goods less profitable and tightening environmental policy become less costly in the South. Thus, effectively, abatement costs have fallen. The net effect may be that trade induces more stringent environmental policies in both regions. However, if the North started as the dirty region, the net effect may be that trade induces more pollution in both regions by raising abatement costs.

Technology diffusion from North to South is therefore not unambiguously good for the environment in the South, since whether technologies are pollution-using or pollution-saving depends on the profitability of different innovation projects, which in turn depends on comparative advantages and substitutability between goods with different pollution-intensity. Our results highlight that endogenous technological change is potentially but not necessarily a blessing. The main reason is that the lack of intellectual property rights protection in the South creates distortions. Innovating firms cannot recoup their costs in the South and only direct their efforts to Northern markets, so that the resulting technologies cannot be in the interests of the South in all respects. Moreover, innovation costs are asymmetrically born by Northern consumers, which creates asymmetries in the costs associated with environmental policies, the driving force behind pollution havens.

Appendix: The Consumption to Output Ratio in the Two Regions

Consider the final good market equilibrium condition (2). In both the North and the South, the total cost of producing machines in sector i amounts to $\upsilon m_i N_i$ $(i = R,L)$, where υ represents the unit cost of production and $m_i N_i$ is the total amount of machines produced. Since the machines share in intermediate producers' output is given by $1-\beta$, we rewrite total cost as $(\upsilon/p_{m_i})(1-\beta)p_i Y_i$. Substituting the appropriate expressions for prices and marginal costs in the two regions, $p^n_m = (1-\tau_m)$ and $p^s_m = \upsilon = 1-\beta$, and summing over the two sectors, we find total expenditure in machines in the North and in the South, respectively, as:

$$M^n = (1-\beta)\left(\frac{1-\beta}{1-\tau_m}\right)Y^n, \text{ and } M^s = (1-\beta)Y^s.$$

In the South, whatever is not needed for physical investment will be consumed. The C/Y ratio in the South will then be given by:

$$\left(\frac{C}{Y}\right)^s = 1-(1-\beta) = \beta.$$

To characterize the same ratio in the North, we need to derive the equilibrium R&D investment outlays. Starting with the directed technical change case, from (6) it is possible to write the R&D investment in each sector as $D_i = gN_i(N_R+N_R)^{\Psi}/\eta$, where g is the growth rate of the economy. From (9), (10), and the definition of the interest rate along the balanced growth path $(r = \eta(N_R+N_R)^{-\Psi}\pi_i)$, we find $N_i/(\eta(N_R+N_R)^{-\Psi}) = p_i Y_i \beta(1-\beta)/r$. Summing over both sectors, we can write the total expenditure in research and development, $D = D_L + D_R$, as

$$D = (1-\beta)\beta\frac{g}{r}Y.$$

Following the same steps it is straightforward to obtain the same expression for the case of undirected technical change.

Accordingly, we can write the consumption to output ratio in the North as follows:

$$\left(\frac{C}{Y}\right)^n = 1-(1-\beta)\left(\frac{1-\beta}{1-\tau_m}\right)-(1-\beta)\beta\frac{g}{r}.$$

References

Acemoglu, D. (2002) "Directed Technical Change", *The Review of Economic Studies*, 69: 781-809.

Antweiler, W., Copeland, B. R. and M. S. Taylor (2001) "Is Free Trade Good for the Environment?", *American Economic Review*, 91: 877-908.

Ben Youssef, S. (2003) "Transboundary Pollution, R&D Spillovers and International Trade", *FEEM working paper* 39.2003.

Bovenberg, L. A. and S. Smulders (1995) "Environmental Quality and Pollution-augmenting Technological Change in a Two-sector Endogenous Growth Model", *Journal of Public Economics*, 57: 369-391.

Chichilnisky G. (1994) "North-South Trade and the Global Environment", *American Economic Review*, 84: 851-874.

Cole, M. A. (2004) "Trade, the Pollution Haven Hypothesis and the Environmental Kuznets Curve: Examining the Linkages", *Ecological Economics*, 48: 71-81.

Copeland, B. R. and M. S. Taylor (2004) "Trade, Growth, and the Environment", *Journal of Economic Literature*, 42: 7-71.

Golombek, R. and M. Hoel (2004) "Unilateral Emission Reductions and Cross-Country Technology Spillovers", *Advances in Economic Analysis & Policy*, 4(2). http://www.bepress.com/bejeap/advances/vol4/iss2/art3

Jones, C. I. (1995) "R&D Based Models of Economic Growth", *Journal of Political Economy*, 103: 759-784.

Keller, W. (2004) "International Technology Diffusion", *Journal of Economic Literature*, forthcoming.

Rivera-Batiz, L. A. and P. M. Romer (1991) "Economic Integration and Endogenous Growth", *Quarterly Journal of Economics*, 106: 531-555.

Segerstrom, P. S. (1998) "Endogenous Growth Without Scale Effects", *American Economic Review*, 88: 1290-1310.

Stokey, N. L. (1998) "Are There Limits to Growth?", *International Economic Review*, 39: 1-32.

Zarsky, L. (1999) "Havens, Halos and Spaghetti: Untangling the Evidence about Foreign Direct Investment and the Environment", in *Foreign Direct Investment and the Environment*, OECD, pp. 47-74.

[15]

Does Trade Promote Environmental Coordination?: Pollution in International Rivers*

Hilary Sigman

[Corrected Version]

Abstract

This paper examines whether trade relationships facilitate resolution of international environmental spillovers. Trade might promote cooperation by providing opportunities for implicit side payments, allowing linkage between environmental and trade concessions, providing direct leverage over other countries' production, or instilling a perception of shared goals. Using data from the UN's Global Environmental Monitoring System (GEMS) on water quality in international rivers, the paper examines the influence of bilateral trade on pollution in rivers that cross international borders. It reports evidence of lower water pollution in rivers shared between countries with more extensive trade. Improved coordination from expanded trade may thus represent a benefit to weigh against the environmental costs of the pollution havens effect.

KEYWORDS: Environment, Transboundary Pollution, Water Quality

*A previously published version of this paper contained simple programming errors. This version has the corrected results; the previous version is stored for archival purposes and is available at the same web address. I appreciate the careful work of Conan Crum, who helped identify the programming problem. I am grateful to Alexandra Miltner for research assistance and to Scott Barrett, Don Fullerton, Arik Levinson, Gilbert Metcalf, Kerry Smith, and seminar participants at the NBER and University of Michigan for comments. This material is based upon work supported by the National Science Foundation under grant number SES–9876498.

International trade may help resolve transboundary environmental disputes for several reasons. First, trade provides opportunities for implicit side payments if explicit side payments are politically difficult. Second, trade may allow agreements with "linkage" between environmental and trade concessions, providing economic threats to support bargaining over environmental objectives (Limão, 2002). Third, integrated economies allow countries direct leverage over each other's production, for example through the pollution content tariffs suggested by Copeland (1996). Finally, economic interactions may instill a perception of shared goals that helps resolve disputes in other arenas (Neumayer, 2002).

This paper is the first to study the empirical relationship between bilateral trade and transboundary pollution specifically between the trading partners. The paper examines whether the intensity of bilateral trade affects pollution in rivers shared by the two countries.

Rivers offer several advantages as a context in which to examine the effect of bilateral trade relationships. First, different rivers each represent a distinct resource problem and thus a separate observation; regional or global pollution problems, such as acid rain or greenhouse gases, have only a few outcomes at any one time. Second, rivers are shared by a small and well-defined group of countries, making it possible to choose a limited number of bilateral trade values relevant to the resource. Third, a country may share different rivers with different neighbors, giving rise to cross-sectional variation even within countries. Finally, trade will only play a role for pollution problems with some hope of coordination. The small number of parties make Coasean bargaining more feasible for rivers than for regional and global pollution problems involving large numbers of countries.

A number of treaties address transboundary freshwater agreements.[1] However, I focus on the effects of trade on revealed pollution levels, rather than participation in explicit agreements. Explicit agreements may be public relations exercises rather than binding environmental policy. Several earlier studies suggest that environmental treaties do not constrain behavior (Conte Grande, 1999; Murdoch, Sandler, and Sargent, 1997; Congleton, 1995). In addition, countries may reach accommodations that they do not formalize with a treaty.

The United Nations' Global Environmental Monitoring System (GEMS) provides panel data on pollution levels at 247 river monitoring stations around the world, 72 of which are shared between two countries. I matched data from GEMS to data from Statistics Canada (1998) on bilateral trade between the countries sharing the river. Other explanatory variables include country characteristics (such as income, political rights, and general openness to trade), population upstream from the station (calculated using a Geographic Information System (GIS)), and other river characteristics that may affect pollution. The data are analyzed using country and monitoring-station fixed effects to address unobserved heterogeneity. Robustness analyses vary the measures of the intensity of bilateral trade and address the potential endogeneity of trade, using a gravity model of trade. I study biochem-

[1] Hamner and Wolf (1998) describe 6 (4%) of 145 treaties as principally addressing pollution. However, a current web version of their database now lists 100 agreements with some bearing on water quality out a total of over 400 (Wolf, 2004).

ical oxygen demand (BOD), a common form of pollution for which earlier research shows evidence of free riding by upstream countries (Sigman, 2002).

The results support the hypothesis that trade facilitates environmental cooperation. When country effects are included in the equation, trade intensity reduces pollution levels at downstream stations. The results are robust to varying the definition of trade intensity and accounting for the endogeneity of trade. Station fixed effects equations do not yield effects of trade, but this may reflect the limitations of using time-series variation to identify the effect of trade on environmental relationships.

The outline of the paper is as follows. Section 1 discusses the possible relationship between shared environmental quality and the level of trade between the environmental partners. Section 2 describes the water pollution data, trade data and other variables assembled for the project. Section 3 presents estimated equations, using fixed effects and a variety of measures of trade intensity. Section 4 briefly concludes.

1. Agreements, trade, and transboundary pollution

International trade between neighbors may affect transboundary pollution because of its effects on policy coordination or the effects of trade on overall pollution levels. This section provides more discussion of these two types of effects.

1.1 Coordination

Trade may facilitate coordination of policies for four related reasons. First, environmental agreements for rivers are likely to require side payments from the downstream countries to the upstream polluting countries (Mäler, 1990).[2] Governments may find it politically difficult to make substantial direct payments to other governments or to pollution sources in other countries; these payments seem to validate a right of upstream countries to pollute that downstream countries may perceive as unfair. Greater potential gains from trade and more extensive economic contacts may facilitate implicit side payments. These side payments may take the form not only of trade preferences, but also other rewards, such as government purchasing contracts, that will be available if the countries' economies are well integrated.

Whether direct side payments are politically difficult is an empirical question. Direct side payments for pollution reduction do sometimes occur. For example, a convention concerning chlorine pollution of the Rhine involved direct payments from downstream countries to facilities in upstream countries (Bernauer, 1996). Treaties that allocate water supplies frequently involve direct payments: Hamner and Wolf (1998) find 44 examples of such payments among the 54 water allocation treaties. Thus, we might fail to find a

[2]Most of the growing theoretical literature on international environmental agreements focuses on global pollutants, such as greenhouse gases, which differ from the regional and directional pollution studied here. On the role of side payments and sanctions in that context, see Hoel, 1992; Carraro and Siniscalco, 1993; Chandler and Tulkens, 1994; and Barrett, 2001.

relationship between bilateral trade and shared resources if this history indicates that the political hurdles to direct payments are not very high.

Second, trade may facilitate cooperation as a threat to support demand for greater provision of shared environmental goods. This use of trade as a threat, sometimes called "linkage," promotes environmental cooperation in some models (Limão, 2002). Again, countries with extensive economic contacts may have more opportunities for linkage than countries with less contact.

Third, trade contacts allow countries direct leverage over each others' economy. In the absence of agreements, importing countries may impose environmental charges on imported goods based on their contribution to the pollution, such as the pollution content tariffs studied by Copeland (1996). Extensive trade raises the likelihood that the downstream country imports the polluting good.

Finally, trade may assist in environmental coordination because trade relationships have cultural implications. Upstream countries are more likely to make sacrifices on behalf of downstream countries who are viewed as friends. Trade may follow from such goodwill and also promote it, because economic activity requires personal contacts between people in the two countries. Neumayer (2002) looks at the association between general trade openness and multilateral environmental agreements in an attempt to test this goodwill link; however, this test lacks the specificity of the bilateral trade and environmental relationships studied here.

Any effects of trade on coordination would result from the extent of economic interaction between the countries, which depends not only on trade frictions but also on potential gains from trade given by the preferences and endowments of the two countries. In this way, these trade effects differ from the composition, scale, and technique effects from reducing trade frictions. One implication is that the trade intensity measure for the current study should reflect the amount of trade, not trade barriers as in the literature on trade and the environment.[3] As a second implication, the current study does not need to condition on the complementarity between the two economies: a large natural trade relationship is consistent with the hypotheses suggesting trade may be associated with better environmental coordination.

1.2 Scale, composition, and technique effects

A drawback of examining pollution levels rather than explicit agreements is that bilateral trade may affect the uncoordinated equilibrium as well as the results of policy coordination. Transboundary pollution levels may change with trade just as local pollution levels change from the composition, scale, and technique effects of overall trade that have been the focus of a growing empirical literature.[4] The estimated equations include a variable

[3]For example, Reppelin-Hill (1999) and Damania et al. (2003) use measures based on import and export duties, in addition to openness.

[4]For recent empirical work, see Antweiler, Copeland and Taylor (2001) and Frankel and Rose (2002). For surveys, see Copeland and Taylor (2003) and Dean (1999).

for overall openness to distinguish the effects of bilateral trade from the general effects of trade on the environment.

However, special scale and composition effects may also result from bilateral trade. One scale effect of bilateral trade that may appear in these data results from the location of activity. If production of traded goods occurs near the border with the trading partner, then rivers that cross the border may have higher pollution with more trade. This scale effect will yield a positive link between exports and pollution, tending to counteract the policy coordination effects. As a result, any negative association between trade and pollution in the empirical analysis may understate the true effects of trade on coordination.

Nonuniformly mixed transboundary pollutants may also have a special composition effect. For these pollutants, damages from pollution are higher the farther upstream (or upwind) the activity is located because of greater exposed population. From the perspective of global welfare, downstream countries have a relative advantage in pollution-intensive activities because the farther downstream a pollution source is located downstream, the lower the external costs, all else equal. Thus, reducing trade frictions would move pollution downstream if environmental policies are coordinated and vary with exposure. This effect would appear in the empirical analysis as an effect of trade that raises pollution downstream and lower pollution upstream.

2. Data

2.1 Water pollution data

The United Nations' Global Environmental Monitoring System (GEMS) provides annual or triennial average pollution levels at river monitoring stations around the world from 1979 through 1996. Using a GIS, I coded the countries that share the river for each GEMS station. Of the 291 stations in GEMS reporting the pollutant studied here, 96 are "international" in the sense that they are upstream or downstream of international borders or located on a river when it forms a border between two countries. Lack of one or more of the explanatory variables reduced the sample to 247 stations, including 72 that are international (see Table 1). The largest number of stations excluded (23) are international stations that lack trade data.[5]

Figure 1 shows the international distribution of stations. As the map shows, the stations are located around the world, with especially heavy concentration in Europe. Overall 33% of stations and 22% of observations used in the analysis are in Europe. The European share of international stations is even larger: 47% of stations and 44% of observations.

Sample selection may be a problem in the GEMS data, although they have been used often in the environmental economics literature for lack of alternatives. Participating coun-

[5]Trade data are missing for several reasons. Two stations report pollution only in 1979 before the trade series began. In addition, no data are available for trade between Belgium and Luxembourg and among the former USSR countries, which are grouped together in the data set. A few other countries, especially in Africa, lack data, presumably because their governments do not collect this information.

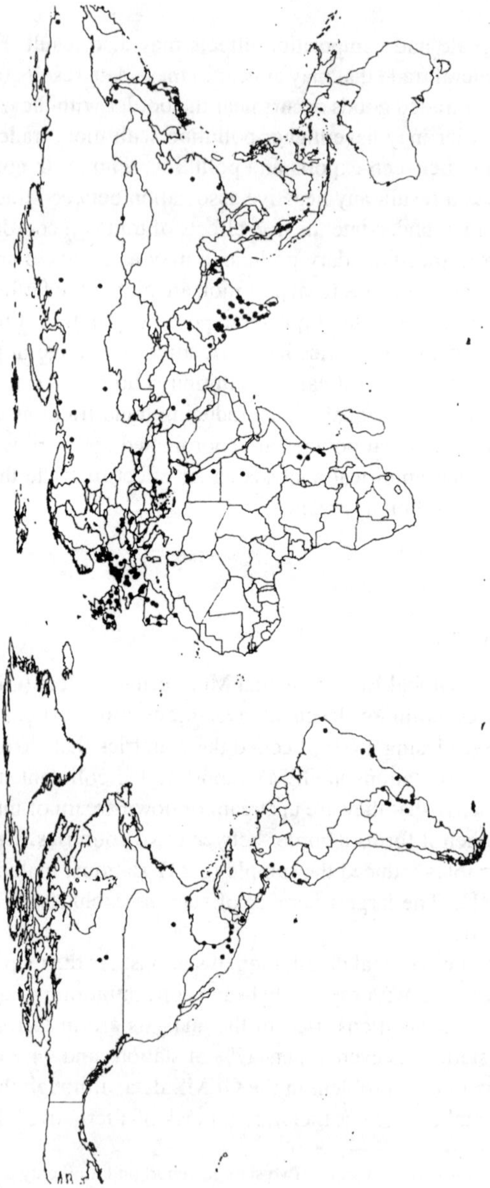

Figure 1: GEMS river monitoring stations used in the analysis

tries select monitoring stations without restrictions, but with the request from the GEMS program that they choose locations with and without extensive human influence. For the current research, the greatest concern is that countries may strategically chose monitoring locations to reflect their bargaining interests. Upstream countries have an incentive to report cleaner water selectively, but downstream countries have the reverse incentive. Thus, the empirical analyses may provide evidence of such a bias by distinguishing upstream and downstream locations on international rivers.[6]

I use biochemical oxygen demand (BOD) as the pollution measure for several reasons. First, BOD is among the most common water quality measures in GEMS, providing many observations for analysis. Second, BOD levels are easily measured by standard procedures, which helps assure consistency in data quality across countries. Other pollutants, such as metals, are more difficult to measure, giving rise to greater data quality concerns. Third, elevated BOD is attributable to a range of human activities, especially sewage, so heterogeneity in local industrial activity is not very important to BOD levels. Fourth, BOD may travel reasonably far downstream, allowing significant spillovers at many stations on international rivers. It would be interesting to look at other pollutants, such as pathogens, that have a direct effect on human health. However, pathogens do not travel more than a few miles downstream, so cross-border spillovers are less important. Fifth, BOD (and oxygen depletion more generally) is a major focus of regulation. In the United States, for example, many more permits for point sources of water pollution address BOD than any other single pollution measure.[7] Thus, governments influence the level of this pollutant and can coordinate their choices.

Table 1 reports summary statistics for the observations used in the analysis, dividing the sample into international and domestic stations. The first column in Table 1 first reports average BOD concentrations. The water quality appears to be somewhat better on domestic rivers (BOD concentration of 5.6 milligrams per liter (mg/l)) than international ones (8.7 mg/l), as free riding might suggest. To help interpret these concentrations, consider that rivers with BOD higher than 4 mg/l are not acceptable for any recreational use (including boating) in the United States (Vaughan, 1986).

The reported pollution levels are annual mean concentrations at the station from 1979 and 1990 and triennial means for the next six years (for some stations, triennial means start in 1988). The second row of Table 1 reports that the means are based on an average of 14.8 measurements at international stations and 16 measurements at domestic stations, with substantial variability across stations.

---

[6]A more thorough solution would model the selection of sites explicitly. However, such a model is difficult to estimate because the set of potential monitoring locations is infinite (anywhere along a river anywhere in the world). In addition, few characteristics are available for the unmonitored places.

[7]BOD accounts for 17% of National Pollution Discharge Elimination System (NDPES) permits in the 2003 Permit Compliance System. Dissolved oxygen accounts for the next largest percentage, 1%.

Table 1: Means and standard deviations of the data

	International stations		Domestic stations	
Variable	Mean	S.D.	Mean	S.D.
Mean BOD concentration (mg/l)	8.7	31.6	5.6	15.2
Number of measurements per observation	14.8	12.7	16.0	18.2
Downstream station (share)	.765		–	
Trade intensity if downstream (% of GDP)	1.62	1.88	–	
Upstream station (share)	.243		–	
Trade intensity if upstream (% of GDP)	7.15	6.46	–	
Border station (share)	.045		–	
Trade intensity if border (% of GDP)	18.7	39.1	–	
Upstream population	1.40	3.74	.82	2.91
GDP per capita (thousand 1996 dollars)	11.43	6.47	7.66	6.90
Political rights (1 (low) – 7 (high))	2.66	1.90	2.56	1.49
Other openness (% of GDP)	36.93	20.71	34.8	27.0
Flow (m^3/sec)	4603	8818	1210	3772
Flow data missing	.233		.087	
Deoxygenation rate ($days^{-1}$)	.373	.104	.415	.133
Total observations	395		992	
Number of stations	72		175	

Note: Standard deviations for continuous variables only.

2.2 Explanatory variables

Statistics Canada (1998) provides data on bilateral trade flows by country in current US dollars from 1980 through 1995. These data were deflated by the US Producer Price Index and merged with the GEMS data, using my coding of the upstream, downstream, and border countries for each station.

The analysis includes several measures of the importance of the trade with the downstream country to the upstream country's economy. Most equations use the bilateral version of the standard openness measure: the ratio of exports plus imports to total GDP. Trade intensity between an upstream country, u, and downstream country, d, is:

$$\text{Trade\_intensity}_{ud} = \frac{\text{Exports}_{ud} + \text{Imports}_{ud}}{\text{GDP}_u}, \tag{1}$$

where Exports_{ud} and Imports_{ud} are exports and imports from the upstream country to the downstream country and GDP_u is the upstream country's GDP.[8] Trade is scaled by the upstream country's GDP because the goal is to measure the downstream country's influence on the upstream economy. The station may be located in either the upstream or downstream country, so GDP_u may be own or upstream country's GDP. Robustness checks later in Section 3 use alternative trade intensity measures, such as export intensity only and total trade rather than its share of upstream GDP.

Table 1 reports the mean of this trade intensity variable separately for stations that are upstream and downstream of borders and for stations on a border. The estimated equations distinguish trade intensity by upstream and downstream location of the station for a few reasons. First, as mentioned above, distinguishing upstream and downstream stations helps address strategic reporting. Second, several stations are both upstream and downstream of borders. It is not clear which intensity measure to select for these stations. Distinguishing the two effects solves this problem by allowing a variable for each relationship to be included in the equations. Third, distinguishing these effects allows the possibility of different effects at upstream and downstream stations. At upstream stations, any elevated pollution results from free riding by that country. At downstream stations, the pollution reflects the higher endowment of the pollution from upstream less any additional pollution control that the downstream country exercises in response to the dirtier water it receives. As a result, cooperation may have smaller effects on downstream pollution levels than upstream levels. In addition, as argued above, the composition effect of trade may shift pollution to the downstream country.

Several of the stations are located on a river when it forms a border between two countries. These stations present a different sort of negotiation regime than the remaining stations. Each country is both a polluter and a victim, perhaps giving rise to a richer set of negotiated solutions and reducing the need for nonenvironmental linkages. These stations also present a technical problem for the analysis: the trade intensity measure requires

[8] The denominator is the only thing that keeps this definition of trade intensity from being symmetric between the upstream and downstream country because $\text{Exports}_{ud} = \text{Imports}_{du}$ and $\text{Imports}_{ud} = \text{Exports}_{du}$.

selecting a country for GDP in the denominator. To avoid arbitrary assignments, I defined trade intensity for border stations as the average intensity between the two countries.[9]

Table 1 shows that the majority of international stations (76.5%) are downstream of a border. The mean trade intensity for these stations is 1.6% of the upstream country's GDP (with a mean of 3.3% when the borders are between EU countries and .5% when the borders they are not). This value ranges in the data from slightly more than nearly zero to 8%. A higher mean, 7.2%, characterizes the upstream stations, where the maximum trade intensity is 18%.

Table 1 also reports a number of other explanatory variables. These variables were chosen to reflect the costs and benefits of achieving given pollution levels on a river. They include upstream population, country characteristics, and river characteristics.

The 1994 population upstream of the station is included as a measure of uncontrolled pollution at the station. I used a GIS to construct this variable from the 1994 Gridded Population of the World (Tobler et al., 1995) and data on flow direction from the US Geologic Survey's Global Hydro1K database. Hydro1K contains the direction that water flows in a grid of 1 km by 1 km cells (based on the cell's altitude relative to its neighbors). The "flow accumulation" function of ArcView then makes it possible to estimate the upstream drainage area for all cells in the grid. Weighting the upstream cells by their estimated population causes "flow accumulation" to return a grid of total upstream population, which was then used to attribute upstream population to the location of the station. The upstream population values are noisy because the calculated location of the river (based on flow accumulation) sometimes does not correspond to its actual location.[10] However, these miscalculations do not seem to introduce any systematic bias from the perspective of the current analysis. Table 1 reports higher upstream population along international rivers than domestic rivers, as might be expected.

Several characteristics of countries are included. National per capita income may affect the costs of pollution control and the benefits of water quality. The Penn World Table (Heston et al., 2001) provides annual income levels standardized for cross-country comparisons.[11] Participating countries are relatively high income: as Table 1 reports, income averages over $11,000 per capita in 1996 dollars at international stations and $7,000 at domestic stations. International stations have higher income because of the dominance of European observations within this sample. Some previous studies have found that pollution rises and then falls with income, a pattern sometimes called the "environmental

[9]The numerator of this measure is the same for both stations because of the symmetry of exports plus imports, so the averaging just affects the denominator.

[10]I believe these differences result from insufficiently fine resolution of the Hydro1K flow direction grid. An alternative approach to calculating upstream population would be to count upstream cities manually. However, this approach is difficult for complex river systems, with many tributaries to take into account. It also might introduce systematic error by overrepresenting population in places with high urban concentration relative to more rural areas.

[11]When the Penn World Table 6 does not provide data, I have interpolated it by rescaling data from earlier Penn World Table versions or from the World Bank.

Kuznets curve" (Selden and Song, 1994; Grossman and Krueger, 1995). The estimated equations include a quadratic in income to control for these effects.

In addition, the equations include a measure of general openness to trade. An open country is likely to trade extensively with its neighbors. If a country's openness also affects pollution levels (as previous research on trade and the environment suggests), then failure to include openness could bias the coefficients on the bilateral trade measures. Therefore, the estimated equations include an adjusted version of the standard openness measure, total exports plus total imports divided by GDP, for the country in which the station is located. To reduce collinearity, the adjusted version of openness excludes bilateral trade with the country with which the river is shared.

A final country attribute is a measure of the political structure of the country. Earlier research has suggested that more responsive governments are likely to choose lower pollution than autocratic regimes (Congleton, 1992; Barrett and Graddy, 2000). Freedom House (2004) annually evaluates countries' "political rights" on a scale from 1 (lowest) to 7 (highest). Despite the high income of countries in the data, the average political rights score is only 2.6.

Geography may affect the country attributes relevant at a given station. At stations downstream of borders, conditions in the upstream country as well as the country of the station may be influential. For this reason, most of the estimated equations also include GDP, openness, and political rights of the upstream country as well as the country of the stations. At stations on borders, two countries' data are equally relevant. To avoid arbitrarily selecting one country, all the country-level variables for border stations (GDP, openness, and political rights) are the mean of the values for the two neighbors.

Finally, several river characteristics are included. First, the river flow determines dilution rates and thus the effect of waste input on in-stream pollution concentrations. This dilution would lead one to expect an inverse relationship between river flow and pollutant levels. Second, the time rate of exponential decay of BOD (known as the "deoxygenation rate," k) is used as a measure of the speed of natural attenuation. I calculated this value from the GEMS data on river temperature, using a nonlinear function from the scientific literature (Bowie et al., 1985).

3. Results

The estimated equations have the form

$$\log b_{it} = g(TI_{it}, Pop_i, CountryChar_{it}, RiverChar_{it}, t) + \varepsilon_{it}, \tag{2}$$

where b_{it} is the mean pollution concentration at station i in year t. TI_{it} (which is TI_{ud} for the upstream and downstream countries at station i in year t) is a matrix of trade intensity variables. Except with station fixed effects, equations also include dummy variables for the location of the station — whether it is upstream, downstream or on a border river — so that zeros can be entered for irrelevant trade intensity variables (e.g., for all trade intensity variables at domestic stations). Pop_i is upstream population for the station in

1994. Country characteristics, $CountryChar_{it}$, include income, openness, and political rights. River characteristics, $RiverChar_{it}$, include flow and the deoxygenation rates. A trend, t, is included to capture changes over time in pollution control technology and environmental preferences. Some of the equations also include country or station fixed effects.

A log-log functional form was chosen for the estimated equation because factors that affect the uncontrolled pollution levels, such as upstream population and river flow, seem likely to have multiplicative effects. The log-log specification has two exceptions. GDP variables enter the equation with a quadratic in levels to follow approximately the specifications used earlier literature such as Grossman and Krueger (1995) and Antweiler et al. (2001). The deoxygenation rate, k_{it}, enters in levels because of the exponential decay function for BOD.

The error, ε_{it}, has a few characteristics that are taken into account in the estimation. First, BOD levels for any observation are the mean of multiple measurements. Because differences in the number of measurements cause heteroskedastic errors, the equations are estimated by weighted least squares with the number of measurements as weights. Second, errors at a single station across multiple years are likely to be correlated. To address this problem, the standard errors are adjusted for clustering at the station level.

3.1 Basic equations

Table 2 reports the results of estimates of equation (2). I will discuss the estimates of the coefficients on the trade measures first, followed by the other covariates.

3.1.1 Trade measures

The first equation in Table 2 does not include country effects. In this equation, the estimated coefficients on bilateral trade are statistically insignificant for all locations. The point estimates on trade at downstream stations and upstream stations have the expected negative sign, but a positive effect is estimated at borders.

The second equation in Table 2 adds country effects to account for heterogeneity across countries not adequately captured by the explanatory variables. Although addressing this heterogeneity is desirable, country effects may absorb some relevant variation in pollution levels. If countries must set environmental policies at a national level, they may not be able to reduce pollution only on rivers shared with countries with whom they have an extensive trade relationship. However, they might still free ride by choosing lower national pollution control than socially optimal. The effect of trade on these deviations will not be detected with country effects.

With the inclusion of country effects in column 2, a negative coefficient is estimated on trade at both downstream stations (−.140) and upstream stations (−.265). The coefficient is statistically significant at the 5% level at downstream stations only. The effects at upstream stations have large standard errors, perhaps because the sample has substantially fewer upstream stations than downstream stations.

Table 2: Weighted least squares estimates with and without country fixed effects

	Dependent variable: Log (Mean BOD)					
	(1)		(2)		(3)	
Country fixed effects?	No		Yes		Yes	
Downstream station	.485	(.313)	.939	(.322)	-.290	(.666)
Log(Trade intensity) if downstream station	-.047	(.043)	-.140	(.040)	-.293	(.078)
Upstream station	.531	(.462)	.208	(.774)	-.291	(.634)
Log(Trade intensity) if upstream station	-.076	(.131)	-.265	(.256)	-.189	(.236)
Border station	-.630	(.480)	-.851	(.318)	-1.20	(.441)
Log(Trade intensity) if border station	.308	(.138)	.190	(.103)	.290	(.123)
Log(Upstream population)	.052	(.025)	.076	(.026)	.076	(.026)
Own country charactistics:						
GDP per capita	.054	(.060)	.109	(.075)	.136	(.078)
GDP per capita squared	-.003	(.002)	-.0027	(.0021)	-.0031	(.0022)
Log(Other openness)	.214	(.139)	.003	(.111)	.011	(.112)
Log(Political rights)	-.241	(.134)	-.185	(.122)	-.118	(.134)
Upstream country char.:						
GDP per capita	–		–		.197	(.074)
GDP per capita squared	–		–		-.0055	(.0019)
Log(Other openness)	–		–		.250	(.173)
Log(Political rights)	–		–		.440	(.192)
Log(Flow)	-.104	(.039)	-.076	(.037)	-.071	(.036)
Flow missing	-.124	(.448)	-.056	(.516)	-.009	(.517)
Deoxygenation rate, k	1.263	(.559)	-2.01	(1.14)	-1.98	(1.13)
Year	.022	(.009)	.0018	(.0087)	-.0019	(.0091)
Intercept	-1.57	(.725)	–		–	
R^2	.19		.42		.42	
Number of observations	1387		1387		1381	
Number of stations	247		247		247	

Notes: Weighted by number of measurements.
Standard errors (in parentheses) robust to clustering at the station level.

The equations are consistent with the hypothesis that free riding occurs but is mitigated by trade. The upstream and downstream station dummies both have positive coefficients, suggesting a detrimental effect of the border with no trade, whereas trade seems to lower pollution. Combining the two effects for downstream stations, pollution is still slightly higher at downstream stations at mean trade intensity. At the 95th percentile value of trade intensity at downstream stations, the net effect of the two variables is about a 40% reduction in pollution. Although this combined effect is inconsistent with the coordination hypothesis, it is not statistically different than zero. Thus, an inference is that free riding occurs, but sufficient trade overcomes it.

At the point estimates, the estimated elasticity is higher at upstream stations than downstream stations. Although this difference is not statistically significant, it does contradict the special composition effect described in section 1 in which freer trade allows industry to move to the lower reaches of a river where exposure is lower. This effect would have given rise to a negative coefficient upstream and a positive coefficient downstream because polluting industries should end up downstream.

The coefficient on intensity at border stations has a positive, but not statistically significant coefficient (.190 in the equation in column 2). This positive coefficient could be the scale effect mentioned above: production of traded goods occurs near the relevant border, resulting in higher pollution. This effect may be strongest for border rivers because the area near the river accounts for a larger share of the border region than for upstream or downstream stations. For upstream-downstream relationships, places where highways — rather than rivers — cross the border may be a greater focus of trade-directed production. However, the data contain only a small number of border stations (11 stations), so results for borders should be treated as somewhat anecdotal. They are not statistically significant at 5% in column 2 and in several equations reported later.

Column 3 adds characteristics of the upstream country for downstream stations. The variables may affect the condition of the river bestowed upon a country from its upstream neighbors. Their inclusion strengthens the point estimate of the effect of trade at downstream stations (from −.140 in column 2 to −.293 in column 3).

3.1.2 Other determinants

Upstream population has a statistically significant and positive coefficient in all the equations in Table 2. The estimated elasticity of pollution to population is fairly low, however, only .076 in column 2. This coefficient may reflect downward bias from measurement error, given the problems calculating this value discussed above.

The country characteristics do not have strong effects on pollution levels. The relationship between the log of pollution and log of GDP follows an inverted U-shaped pattern, but the coefficients are not individually or jointly statistically significant at 5% in any of the equations. Equations estimated with higher order polynomials in income (not shown in Table 2) also did not have statistically significant coefficients on income and had similar coefficients on trade intensity. The coefficient on other openness is not statistically significant at the 5% level in any of the estimated equations.[12] Political rights also do not have a statistically significant effect on pollution, although the point estimate is negative, as expected.

When the characteristics of any upstream country are added in column 3, upstream GDP and political rights have statistically significant coefficients with the expected signs.[13]

The river characteristics enter with the expected signs. The coefficient on river flow rate is negative across all equations as expected. It is statistically significant at 5% in columns 1 and 2. A negative coefficient is expected on the deoxygenation rate, k, because the variable is a decay rate for BOD. Except in column 1, the estimated coefficient is negative and statistically significant at 10%.

Finally, the time trend is statistically significant only when no fixed effects are included in column 1. In this column, the time trend is positive. These results suggest that GEMS expanded over time to include countries with more polluted rivers, but do not indicate dramatic improvement within countries in pollution concentrations.

3.2 Station fixed effects

Table 3 contains estimated coefficients with effects for station-level unobserved heterogeneity, such as local economic activity and river conditions. Station fixed effects should absorb most of the heterogeneity that might arise from differences in land use upstream from the station, which does not change very rapidly over time.[14] The upstream population estimate used in earlier equations cannot be included with fixed effects because it does not vary over time. To measure some population trends, this equation instead includes the population of the country of the station and, for downstream stations, the upstream country. Excluding upstream population permits a somewhat larger sample; 19 stations excluded earlier for lack of these data are now included, bringing the number of stations up to 266 for the station fixed-effects equation.

With these station fixed effects, all trade intensity variables have statistically insignificant coefficients. The failure to find effects may support the hypothesis that the observed effect of trade on pollution is a coordination effect rather than a scale effect. Short run fluctuations in trade are unlikely to influence the ability of countries to coordinate and thus the purely time-series identification in Table 3 would fail to yield results. By contrast, if the

[12]The equations were also estimated with an interaction between this other openness variable and GDP, in the spirit of the equations of Antweiler et al. (2001). The interaction is included because countries at different income levels may have different techniques of production and thus experience different effects from trade openness. The interaction was not statistically significant in any equation and is not shown in the tables.

[13]Although it may appear that upstream country characteristics are more likely to matter than own country characteristics, this inference fails to recognize the difference in identification of the two sets of coefficients. The equation includes own country fixed effects, so the identification for own country effects comes only from time series variation, whereas the upstream country effects also rely on cross sectional variation.

[14]With station-level fixed effects, Sigman (2003) finds no effect of changes the share of land in agriculture or urban uses on water quality at river monitoring stations in the U.S.

Table 3: Weighted least squares estimates with station fixed effects

	Dependent variable: Log (Mean BOD)	
Log(Trade intensity) if downstream station	-.062	(.061)
Log(Trade intensity) if upstream station	.001	(.204)
Log(Trade intensity) if border station	.199	(.136)
Own country charactistics:		
GDP per capita	.033	(.050)
GDP per capita squared	-.0006	(.0014)
Log(Other openness)	.053	(.091)
Log(Political rights)	-.124	(.060)
Log(Population)	.803	(.462)
Upstream country characteristics:		
GDP per capita	.196	(.089)
GDP per capita squared	-.0066	(.0032)
Log(Other openness)	-.062	(.186)
Log(Political rights)	.283	(.213)
Log(Population)	.607	(.512)
River characteristics:		
Log(Flow)	-.013	(.020)
Deoxygenation rate, *k*	-.636	(.437)
Year	-.013	(.009)
R^2	.86	
Number of observations	1500	
Number of stations	266	

Note: Weighted by number of measurements.

negative association results from more direct production-related factors, we would expect an immediate pollution response and an effect in Table 3. Only the positive coefficient on trade at border stations retains its magnitude from Table 2 to Table 3. The robustness of this coefficient is consistent with the hypothesis advanced above that the coefficient is driven by the scale effect.

Few of the other determinants have statistically significant coefficients in Table 3. Country population enters with a positive coefficient (as expected) that is statistically significant at 10% and large in magnitude. Of the country characteristics, only the political rights variable is statistically significant; it is negative, which is consistent with political improvements reducing pollution. The upstream country GDP variables are individually statistically significant, but collectively upstream country characteristics are not. Turning to the river characteristics, flow does not have a statistically significant coefficient perhaps because of data limitations. Time-varying flow data are available only for the last six years and, even then, for a subset of the stations. The deoxygenation rate, which depends on temperature, continues to have the expected sign but is also not statistically significant.

3.3 Alternative specifications

Table 4 reports the results for alternative specifications, including different measures of trade intensity. The covariates in the equations are the same as in column 3 of Table 2, but have been suppressed in Table 4 to allow a focus on the coefficients of interest.

In the first row of Table 4, trade intensity depends only upon exports from the upstream country to the downstream country. Upstream politicians are more likely to respond favorably to offers that increase their exports than their imports. The estimated coefficient on trade at downstream stations continues to be statistically significant at the 5% level; the coefficient at upstream stations continues to be statistically insignificant but similar in magnitude to earlier estimates with the more inclusive trade intensity measure.

The second row of Table 4 redefines trade intensity to depend on the absolute value of exports plus imports between the two countries, rather than their values relative to upstream GDP as in Table 2. Pollution reduction requires an expenditure for which the upstream country would like to be compensated, so the ability to provide a sufficient absolute sum could be more relevant than the share of GDP. This redefinition brings the coefficients on trade intensity closer to zero; the estimated effect is downstream stations remains negative, but becomes statistically insignificant. The latter result raises some questions about the robustness of the estimated relationship; however, the trade intensity measure used in Table 2 is more compelling than this measure.

The next two rows in Table 4 address the geography of the relationships in more detail. The third row excludes observations on rivers when they form country borders. Border stations do not fit into the basic conceptual framework in which the downstream country demands — and the upstream country supplies — environmental quality. At border stations, trade may play less of a role because both countries are in both positions with perhaps more room for agreements without linkages. In addition, it is technically difficult

Table 4: Weighted Least Squares estimates for alternative specifications

Specification	Trade intensity at downstream station	Trade intensity at upstream station	Trade intensity at border station
Trade intensity= $\frac{\text{Exports}_{ud}}{\text{GDP}_u}$	-.179	-.256	.208
	(.077)	(.259)	(.117)
Trade intensity= $\text{Exports}_{ud} + \text{Imports}_{ud}$	-.074	-.171	.114
	(.060)	(.214)	(.076)
Exclude border stations	-.302	-.181	–
	(.077)	(.242)	
Trade intensity only for stations within 100 miles of border	-.330	-.214	.081
	(.103)	(.244)	(.094)
Distinguish borders within EU from all others			
EU	-.166	-.243	.151
	(.067)	(.249)	(.128)
Non-EU	-.250	.253	.189
	(.071)	(.172)	(.227)
p-value of test of equality	.04	.03	.78
Distinguish by upstream political conditions			
Low political rights	-.351	-.173	.309
	(.096)	(.235)	(.166)
High political rights	-.341	-.384	.330
	(.092)	(.278)	(.111)
p-value of test of equality	.69	.04	.28

Notes: Full equations as in column 3 of Table 2.
Standard errors (in parentheses) robust to clustering at the station level.
Excluding border stations reduces the sample to 1307 observations and 236 stations.

to chose intensity measures for border stations. For example, when border stations are also downstream, the intensity measure includes trade between the upstream country and the reporting country, but not its neighbor's trade with the same country. Eliminating border stations gives results similar to those in column 3 of Table 2.

The fourth row of Table 4 addresses the effect of distance to the border. If trade influences pollution through coordination, this effect should only be seen where transboundary spillovers occur. Far upstream of a border, the vast majority of the pollution costs are borne by the upstream country both because its exposure is greater and because BOD attenuates as water flows downstream. Similarly, far downstream of a border, most pollution from the upstream country should have attenuated, so we would not expect an effect of coordination. To address this geography, the fourth row shows the equations when the upstream and downstream trade intensity variables enter only if the station is within 100 miles of the border. Although one might expect more negative coefficients on trade with this modification, the estimates in Table 4 remain similar in magnitude to those in Table 2. At downstream stations, the point estimate goes from −.293 (column 3 of Table 2) to −.330 (4th row of Table 4) and also strengthens only slightly at upstream stations.

The last two sets of rows in the Table 4 report interaction effects of particular interest. One set of interactions distinguishes the effects of trade inside and outside the European Union. Trade may have less of an effect inside the EU than outside for several reasons. First, centralized environmental institutions within the EU may make bilateral side payments or threats unnecessary. Supporting this view, Sigman (2002) finds that EU stations exhibit less evidence of free riding than other stations on international rivers. Second, explicit side payments may be more readily available for countries already cooperating on other policy matters, as the example of chlorine pollution in the Rhine suggests.

At downstream stations, we see the anticipated pattern. When the border is within the EU, the coefficient on trade intensity is negative (−.166) and statistically significant, but lower in absolute value than the coefficient at borders that are not between EU members (−.250). The difference between the two coefficients is statistically significant. However, at upstream stations, the results are less supportive. Neither coefficient is now statistically significant at 5%. The point estimate within the EU is more consistent with an effect of trade on policy coordination, whereas the point estimate is positive outside the EU. The data set includes very few upstream stations outside the EU, so one explanation is that the data are not rich enough for a reliable test of the hypothesis at upstream stations. Statistically insignificant results are found at border stations, again perhaps because of a paucity of data.

The final set of interactions in Table 4 consider the effects of differences in political regimes. Countries with more responsive governments may be more willing to reach negotiated solutions to environmental problems with neighbors. Dividing the trade effects according to whether the country has a closed regime or open regime (where closed regimes have Freedom House political rights scores of two or less), the coefficients are statistically significantly different from each other for upstream stations only. However, neither upstream coefficient is statistically significant, so it is difficult to interpret this difference as support for any hypothesis.

3.4 Endogeneity of bilateral trade

The equations estimated above do not determine the direction of causality of any link between trade and pollution in shared rivers. A negative coefficient on trade might indicate that trade relationships facilitate reaching common environmental goals. However, it may also indicate that countries that have achieved environmental coordination expand their trade. If the causality goes this direction, it still suggests a role for trade in supporting environmental negotiations. Nonetheless, one could not then draw the policy implication that expanded trade will help resolve international environmental disputes.

To try to distinguish the first effect from the second, this subsection presents results for equations that instrument for trade intensity. The instruments are based on "gravity" models that use geographic characteristics of trading partners to predict their trade. Frankel and Romer (1999) use two measures of the size of the two countries, area and population, and their landlocked status.[15] Thus,

$$TI_{it} = g(A_{ui}, A_{di}, P_{uit}, P_{dit}, LL_{ui}, LL_{di}, CountryChar_{it}) \quad (3)$$

where A_u and A_d represent the land area of the upstream and downstream country at station i, and P their populations, and LL their landlocked status. Country population is the only time-varying instrument. The exclusion restrictions to identify the model require that these country-level measures do not affect pollution in a specific river. This assumption is easiest to maintain for country areas. For population, the restriction requires that the station-specific upstream population measure accurately captures the influence of population and that country-level measures do not add relevant information.

Instrumental variables estimates require instruments for three variables: trade intensity at upstream stations, downstream stations and border stations. Thus, the full set of instruments includes area and country population and landlocked status of the country of the monitoring station, any upstream country, any downstream country, or any border country. The variables have values of zero when not relevant for any station. The instrumental variables equation is the same equation as in column 3 of Table 2, including country fixed effects and upstream country characteristics.

Table 5 presents the instrumental variables estimates, which remain similar to their earlier values. Trade intensity at downstream stations continues to have a statistically significant coefficient, which has become more negative (−.524) than before (−.293). The coefficient on trade intensity at upstream stations also becomes more negative but remains statistically insignificant; the coefficient at border stations remains positive and is

[15] In addition, they use a series of interactions of these variables for whether countries are neighbors. These variables are not available because all relationships are between neighbors in this study. Frankel and Romer sum predictions of bilateral trade across many partners to form an aggregate predicted openness, which is their instrument. However, the bilateral relationships are included in the main equation here, eliminating the need for this intermediate step.

Table 5: Instrumental variables estimates

	Dependent variable: Log (Mean BOD)	
Downstream station	-.378	(.828)
Log(Trade intensity) if downstream station	-.524	(.160)
Upstream station	2.786	(4.70)
Log(Trade intensity) if upstream station	-.994	(1.10)
Border station	-1.67	(.655)
Log(Trade intensity) if border station	.382	(.209)
Log(Upstream population)	.075	(.027)
Own country charactistics:		
GDP per capita	.193	(.085)
GDP per capita squared	-.0047	(.0024)
Log(Other openness)	-.065	(.127)
Log(Political rights)	-.112	(.133)
Upstream country characteristics:		
GDP per capita	.321	(.148)
GDP per capita squared	-.0079	(.0039)
Log(Other openness)	.387	(.244)
Log(Political rights)	.715	(.309)
River characteristics:		
Log(Flow)	-.067	(.038)
Flow missing	.043	(.498)
Deoxygenation rate, *k*	-1.87	(1.13)
Year	-.0004	(.0095)
Number of observations	1361	
Number of stations	245	

Notes: Weighted by number of measurements.
Standard errors (in parentheses) robust to clustering at the station level.
Country effects included.

insignificant at 5%. Thus, the implications of the analysis are little changed by the use of instrumental variables.

4. Conclusion

The empirical analysis presented here supports the view that trade promotes environmental cooperation. At rivers downstream of international borders, pollution is lower if the countries on either side of the border have more extensive trade. Estimated elasticities of BOD to trade intensity (exports plus imports between the two countries as a share of upstream GDP) are in the range of $-.1$ to $-.3$. The coefficients on trade intensity are statistically significant at downstream stations, but not at the much smaller number of upstream stations. The results are consistent with countries free riding, but overcoming it with sufficient trade. Neither varying the trade intensity measure nor addressing the possible endogeneity of trade using gravity instruments substantially alters the implications.

In addition, the study examines conditions in which trade is effective. Trade may be more important across borders that are not within the EU, suggesting that trade linkages are unnecessary to resolve EU environmental disputes. However, the results did not suggest that trade plays a larger role when the upstream country is more open politically.

A negative association between trade and pollution was not observed under two conditions. First, stations on the border between countries may have increases in pollution with increased trade. This effect seems likely to be the result of a scale effect of trade: economic activity and hence pollution may be higher near the border river. Second, when monitoring station (rather than country) fixed effects are included in the analysis, the trade intensity variables are not significant. However, the fluctuations over time in trade that identify this equation may not influence the ability of countries to coordinate environmental policies.

The results have positive implications for international environmental policy. First, they suggest that, despite incentives to free ride, countries are able to exercise influence over each other's pollution control levels. Thus, they provide some reason for optimism about negotiated solutions to international spillovers. Second, the results suggest that improvements in coordination from expanded trade may yield environmental benefits to weigh against environmental costs from the pollution havens effect.

References

Antweiler, W., B.R. Copeland, and M.S. Taylor, Is free trade good or bad for the environment? *American Economic Review* **99** (2001), 877–908.

Barrett, S., International cooperation for sale, *European Economic Review* **45** (2001), 1835–5 .

______ and K. Graddy, Freedom, growth, and the environment, *Environment and Development Economics* **5** (2000), 433–56.

Bernauer, T., Protecting the Rhine River against chloride pollution, in R.O. Keohane and M.A. Levy, eds., *Institutions for Environmental Aid: Pitfalls and Promise* Cambridge, MA: MIT Press, 1996, pp. 201–232.

Bowie, G.L. et al., *Rates, Constants, and Kinetics Formulations in Surface Water Quality Modeling* (Second Edition), Washington, DC: U.S.EPA, 1985.

Carraro, C. and D. Siniscalco, Strategies for the international protection of the environment, *Journal of Public Economics* **52** (1993), 309–28.

Chander, P. and H. Tulkens, A core-theoretic solution for the design of cooperative agreements on transfrontier pollution, *International Tax and Public Finance* **2** (1994), 279–293.

Congleton, R. D., Political institutions and pollution control, *Review of Economics and Statistics* **74** (1992), 412–21.

______, Toward a transactions cost theory of environmental treaties: Substantive and symbolic environmental agreements, *Economia Delle Scelte Pubbliche* **13** (1995), 119–139.

Conte Grand, Mariana, Do regional environmental agreements have any effects on water quality? Working paper, Universidad del CEMA, 1999.

Copeland, B.R., Pollution content tariffs, environmental rent shifting, and the control of cross border pollution, *Journal of International Economics* **40** (1996), 459–76.

______ and M.S. Taylor, *Trade and the Environment: Theory and Evidence* (Princeton, NJ: Princeton University Press, 2003).

Damania, R., P.G. Fredriksson, and J.A. List, Trade liberalization, corruption, and environmental policy formation: Theory and evidence, *Journal of Environmental Economics and Management* **46** (2003), 490–512.

Dean, J.M., "Testing the impact of trade liberalization on the environment: Theory and evidence," in P.G. Fredriksson, ed., *Trade, Global Policy, and the Environment* (Washington, DC: The World Bank, 1999).

Frankel, J.A. and D. Romer, Does trade cause growth? *American Economic Review* **89**, (1999), 379–99.

Frankel, J.A. and A.K. Rose, Is free trade good or bad for the environment? Sorting out the causality, Working paper, Kennedy School of Government, 2002.

Freedom House, *Freedom in the World,* http://www.freedomhouse.org/ratings (Last viewed March 18, 2004).

Grossman, G.M. and A.B. Krueger, Economic growth and the environment, *Quarterly Journal of Economics* **110** (1995), 353–77.

Hamner, J. and A.T. Wolf, Patterns in international water resource treaties: The Transboundary Freshwater Dispute Database, *Colorado Journal of International Environmental Law and Policy 1997 Yearbook*, 1998.

Heston, A., R. Summers and B. Aten, *Penn World Table Version 6.0*, Center for International Comparisons at the University of Pennsylvania (CICUP), December 2001.

Hoel, M., International environment conventions: The case of uniform reductions of emissions, *Environmental and Resource Economics* **2** (1992), 141–159.

Limão, N., Trade policy, cross-border externalities and lobbies: Do linked agreements enforce more cooperative outcomes? Working paper, University of Maryland, 2002.

Mäler, K.-G., International environmental problems, *Oxford Review of Economic Policy* **6** (1990), 80–108.

Murdoch, J.C., T. Sandler and K. Sargent, A tale of two collectives: Sulphur versus nitrogen oxides emission reduction in Europe, *Economica* **64** (1997), 281–301.

Neumayer, E., Does trade openness promote multilateral environmental cooperation? *The World Economy* **25** (2002), 815–32.

Reppelin-Hill, V., Trade and environment: An empirical analysis of the technology effect in the steel industry, *Journal of Environmental Economics and Management* **38** (1999), 283-301.

Selden, T.M. and D. Song, Environmental quality and development: Is there a Kuznets curve for air pollution? *Journal of Environmental Economics and Management* **27** (1994), 147–62.

Sigman, H., International spillovers and water quality in rivers: Do countries free ride? *American Economic Review,* **92** (2002), 1152–1159.

———, Transboundary spillovers and decentralization of environmental policies, Rutgers University working paper, 2003.

Statistics Canada, *World Trade Analyzer 1980-96* (Ottawa: Statistics Canada, 1998) (CD-ROM).

Tobler, W., U. Deichmann, J. Gottsegen, and K. Maloy, The Global Demography Project, National Center for Geographic Information and Analysis, UC Santa Barbara, Technical Report TR–95–6 (1995).

Vaughan, W.J., "The RFF Water Quality Ladder," Appendix B in R.C. Mitchell and R. Carson, *The Use of Contingent Valuation Data for Benefit/Cost Analyses in Water Pollution Control* (Washington, DC: Resources for the Future, 1986).

Wolf, A., *International Freshwater Treaties Database,* http://www.transboundarywaters.-orst.edu (Last viewed June 14, 2004).

[16]

Cross-Country Policy Harmonization with Rent-Seeking*

Patrik T. Hultberg and Edward B. Barbier

Abstract

In a reciprocal market model with imperfectly competitive firms, domestic policies will differ across countries that are economically and politically diverse. We explore the implications of this standard result with regard to harmonization of environmental policies between corrupt and non-corrupt countries. Imposing a more stringent policy on a non-corrupt government will be welfare reducing to the 'receiving' country, but may be welfare enhancing for the 'imposing' country. However, where environmental standards are under the control of a corrupt government, it is possible that harmonization is welfare enhancing to both countries.

KEYWORDS: corruption, environmental policy, rent-seeking, reciprocal market model, harmonization

*Acknowledgements: We are grateful to, in particular, Amitrajeet Batabyal, Don Fullerton, Robert Godby, Linwood Pendleton, Sherrill Shaffer, John Tschirhart and three anonymous referees for their valuable comments. The paper also benefited from discussions with Scott Barrett. We thank Lee Bailiff and Margie Reis for their assistance.

As the volume of international trade increases, more concern is expressed regarding cross-country policy harmonization. In the trade literature this is generally defined as making the regulatory requirements or government policies of different jurisdictions identical or at least more similar (Leebron, 1996). Although there are numerous rationalizations for harmonization calls, the two most prominent are charges of "unfair trade" and fears of a "race to the bottom" (see Bhagwati, 1996). Allegations of unfair trade advantages, traditionally limited to subsidies and dumping, are now being extended to include international differences in environmental and labor standards. The "race to the bottom" fear takes this argument a step further by predicting that countries will "harmonize" downward to the lowest standards as they compete for trade and investment (Esty, 1999). The primary motivation for rich countries to join in such an environmental "race to the bottom" would be to counteract any "pollution haven" effect. As outlined by Copeland and Taylor (2003, p. 15), "according to the pollution haven hypothesis ... high-income countries have relatively stringent pollution policy and this shifts dirty good production to poor countries via international trade."

In contrast, environmentalists contend that the pollution haven effect and any race to the bottom could be avoided through deliberate policies for the "greening of trade" and the "upward harmonization" of environmental regulation across countries at the stringent policy levels set by richer countries (Arden-Clarke, 1991). Today the linking of trade negotiations to environmental policy harmonization objectives is increasingly influencing actual trade negotiations between countries. One such example is the North American Free Trade Agreement (NAFTA) with its side agreement, the North American Agreement on Environmental Cooperation (NAAEC). Although the NAAEC stops short of advocating complete harmonization of environmental policy between Canada, Mexico and the United States, the agreement does promote upward convergence of environmental regulations in Mexico to US and Canadian levels.[1]

Such efforts to combine environmental policy harmonization with international trade agreements suggest two interesting issues that have not been fully explored by the economics literature. First, what is the likely outcome of one trading partner imposing higher policy standards on the other as a condition for greater trade cooperation? And, second, what are the circumstances under which this sub-optimal global strategy can nevertheless lead to potential welfare gains for both countries?

The purpose of our paper is to address these issues by exploring the effects of harmonization of domestic policies on two imperfectly competitive trading nations following economic integration. We do this through a two-country, two-firm reciprocal market model incorporating strategic environmental policy that is solved through a three-stage game. First, we employ the model to demonstrate explicitly why domestic policies should differ across countries. Thus, if one country is characterized by low

[1] The NAAEC was clearly established in response to special interest groups (environmentalists, labor groups and industrialists) and reflect the fears of "unfair trade" and "eco-dumping" arising from the NAFTA agreement. For example, the Canadian NAAEC office states that the NAAEC establishes "a level playing field with a view to avoiding trade distortions and promoting environmental cooperation" (NAAEC).

absorption capabilities or social preferences that favor a clean environment, it will be better off with stringent environmental standards (high emission taxes), whereas its trading partner should choose less stringent policies, and perhaps even subsidies, if the country has a higher absorption capacity or greater tolerance for environmental damage. This result implies that when an open economy raises its pollution taxes to "harmonize" with the regulations imposed by its trading partner, the economy will be worse off, while its foreign rival could be better off.[2] This is particularly the case if the two trading nations differ in their level of development. If a less-developed nation is able and willing to accept relatively high levels of pollution but is forced to impose stringent environmental policies approximating their more wealthy trading partners, perhaps due to a relatively weak bargaining position, then the poorer country is locked into an inefficient outcome while the richer country may actually gain at the expense of the poorer nation. Despite this, it may be beneficial for such a poor nation to sign a bilateral trade agreement that requires environmental policy harmonization, if access to export markets in the wealthy country compensate for the inefficiency of harmonization. However, in this case the gains from trade to the less-developed nation are less than under its optimal policy choice.

When we include rent-seeking behavior in our two-country model, we observe that environmental regulations are likely to be less strict in more corrupt societies as predicted in the literature (Damania *et al.*, 2000, López and Mitra, 2000). We also show that if one country's adopted environmental policy is inefficient due to rent-seeking activities, then it is possible that a higher environmental standard imposed by its trading partner could raise efficiency (in a second-best sense). Under such circumstances, externally imposed pollution policy harmonization may actually enhance efficiency.

Finally, it is important to note from the outset that our approach to the environmental policy harmonization problem differs from the existing economics literature in two important respects.

First, the focus of most studies is on whether or not environmental policy harmonization is an optimal strategy (Kanbur *et al.* 1995; Ulph 1997 and 2000), including under the presence of special interest lobbying (Damania and Fredriksson 2003; Ulph 1998; Johal and Ulph (2002)). In contrast, our objective is not to derive "optimal" harmonization rules between trading partners, but rather to illustrate a different yet realistic scenario, in which one trading partner that has adopted unilaterally a more stringent policy seeks to impose this policy on the other trading partner as a condition for entering into a trading agreement with the latter country.[3] We believe that such a scenario is likely to occur more frequently with the advent of additional trading

---

[2] These welfare effects are important despite the fact that a first-best solution would involve a supranational organization that mitigates the inherent prisoner's dilemma problem.

[3] A realistic proposition since the domestic firm's profit is increasing in the stringency of the rival country's environmental policy choice (see e.g. Ulph, 2000). In addition, Johal and Ulph (2002) show that if environmentalists are allowed to lobby either the domestic or the foreign government to achieve higher environmental standards, they will in fact lobby the other country more intensively. The reason is, again, that domestic profits are directly related to foreign policy stringency. Our paper will explicitly model the profit-shifting effect, but not the foreign lobbying.

and economic integration agreements globally that require stricter environmental and other policy harmonization as a condition for extending the agreements to additional regional participants. This trend has been part of international economic policy since the US administration made a formal commitment to discuss environmental issues together with trade negotiations under NAFTA, which established that trade and environmental goals must be pursued in tandem (Esty 1999). Since the relationship between trade and harmonization of domestic policies is both most manifest and compelling in the area of environmental policy, we frame our discussion of policy harmonization in terms of the choice of domestic regulation of environmental quality.

Second, although our findings concur with the existing literature that allowing for special interest group politics does not, in general, justify calls for complete harmonization (Ulph, 1998, Johal and Ulph, 2002), our novel insight is that harmonization (though not first-best) may be beneficial when one trading partner is susceptible to domestic lobbying pressure. For example, Johal and Ulph (2002) report that for a variety of potential asymmetries in both information and lobbying, international cooperation is always preferable over non-cooperation, but such international cooperation does not generally involve complete policy harmonization.[4] However, our study differs from that of Johal and Ulph in some important respects. In addition to the fact that Johal and Ulph focus on the optimal policy of a supranational agency and we do not, their paper incorporates implicitly a prisoner's dilemma framework and interprets the "democratic deficit" problem to mean that the lack of lobbying is detrimental to the democratic process. In contrast, we avoid explicitly solving for the well-known prisoner's dilemma problem that faces countries in a model of "environmental dumping". Also, we view lobbying and rent-seeking activity to be hurtful to the democratic process, as it distorts the government's net benefits function. For example, whereas we argue that industrialists benefit from rent-seeking activity, Johal and Ulph find that industrialists are worse off from having more political influence.[5] Damania and Fredriksson (2003) argue, as we do, that industries have an incentive to lobby for laxer environmental policy. Damania and Fredriksson endogenize this domestic lobbying activity and show that under some circumstances trade liberalization may actually raise pollution taxes in the home country.[6] Our result is in

[4] Although Johal and Ulph (2002) is related to our study, the main focus is the lobbying asymmetries that may exist when environmental policy is set at either the national or the supranational level.

[5] The intuition is that when two countries face incentives to lower their environmental policy stringency, the effect is that the environment suffers and output levels are greater than the monopoly outcome. In the third market model, this reduces welfare, since both environmental damage increases and firm profits decrease. Since firm lobbying exaggerates this problem, such lobbying is harmful even for the firm. In our model countries and firms cannot avoid this prisoner's dilemma problem without international cooperation (which we do not allow). Also, in our reciprocal market model, an increase in consumer surplus would mitigate these adverse effects.

[6] Damania and Fredriksson (2003) have a domestic duopoly where each firm wants to free ride on the other firm's lobbying, which therefore requires some (tacit) collusion to make lobbying sustainable. They then show that trade liberalization, by increasing competition, reduces the degree of collusion and therefore may raise pollution taxes. Their policy conclusion is that when foreign and domestic goods are close substitutes trade liberalization is likely to increase the stringency of environmental policy. Although

essence the flip side of their outcome: imposing harmonization on a corrupt country with lax environmental policies, especially when accompanied by tariff reduction, may in fact be beneficial.

The next section introduces the two-country, two-firm model and discusses the determination of unilateral strategic environmental policy in the two countries. Section 2 examines the effects of environmental policy harmonization, when the two trading nations are not influenced by corruption. Section 3 extends this analysis to the case where rent-seeking behavior by a firm located in one country is allowed. The final section discusses the policy implications of our analysis and the emerging conclusions.

1. The Reciprocal Market Model

We consider a two-country (*j=A,B*), two-firm (*i=1,2*) model comprising of a three-stage game in which a firm chooses the level of contribution (bribe) in the initial stage in order to affect its government's choice of domestic (environmental) policy in the second stage. Given this choice of policy, the two rival firms compete by setting their optimal quantities in the third stage. Firm 1 is located in market A and firm 2 is located in market B, but both firms sell a homogeneous good in both markets. We choose this reciprocal market model in order to include the effects on consumers in our welfare discussions. For the same reason similar models are used by Batabyal (1996, 1998) and Motta and Thisse (1994).[7]

This three-stage reciprocal market model is solved through backward induction. In the third stage of the game, each firm chooses the quantities sold in countries A and B in order to maximize its respective profit function. The two firms' behavior is modeled as a Cournot duopoly where the two firms choose aggregate output levels q^1 and q^2, respectively. We assume that the firm's total cost function, $C^i(q^i,t)$, takes the form $(c^i+t^{kj})\cdot q^i$, where c^i and q^i are the per-unit production cost and total output for firm *i*, respectively, and t^{kj} is any tariff rate in country *j* associated with international shipping from *k* to *j* (which satisfies $t^{kj}=0$ if $k=j$). We assume that the transportation costs associated with shipping a good are either negligible or constant, and are therefore subsumed in the costs of producing q^i. In addition, the pollution abatement cost facing firm *i*, $A(q^i,e^j)$, is assumed to take the form $e^j\cdot(q^{iA}+q^{iB})$, where e^j is the emissions tax/subsidy in country *j* and where q^{iA} and q^{iB} denote firm *i*'s sales in countries A and B, respectively. Given these assumptions, firm 1 and firm 2, respectively, face the following profit functions:

Damania and Fredriksson's focus on the political process in the domestic country is quite different from our approach, it is worth noting that they also employ a three-stage game where lobbying occurs in the first stage, government sets environmental tax in the second and firms set quantities in the third stage (Cournot competition).

[7] Most current models of strategic environmental policy choice employs a third market model, but as noted in the text, the disadvantage of a third market model is that it cannot include any effects on domestic consumer surplus.

(1) $\pi^1 = p^A q^{1A} - c^1 q^{1A} - e^A q^{1A} + p^B q^{1B} - (c^1 + t^{AB}) q^{1B} - e^A q^{1B}$

(2) $\pi^2 = p^A q^{2A} - (c^2 + t^{BA}) q^{2A} - e^B q^{2A} + p^B q^{2B} - c^2 q^{2B} - e^B q^{2B}$,

where p^j is the inverse demand function in country j, which we assume is given by

(3) $p^j = a - b \cdot Q^j = a - b \cdot (q^{1j} + q^{2j})$.

This choice of a linear model allows us to explore our questions in a tractable model that follows the relevant literature's accepted framework. Note, in particular, that we consider a fixed amount of pollution per unit of output, which implies that the emissions tax is equivalent to an output tax. For firm 1, the optimization results in the following first-order conditions: $d\pi^1 / dq^{1A} = 0$ and $d\pi^1 / dq^{1B} = 0$, and similarly for firm 2. The resulting reaction functions are solved simultaneously for each market to obtain the following optimal quantities for both firms in the two markets:[8]

(4) $$q^{1A*} = \left(\frac{1}{3b}\right)\left(a - 2c^1 - 2e^A + c^2 + t^{BA} + e^B\right) \qquad q^{2A*} = \left(\frac{1}{3b}\right)\left(a - 2c^2 - 2e^B - 2t^{BA} + c^1 + e^A\right)$$

$$q^{1B*} = \left(\frac{1}{3b}\right)\left(a - 2c^1 - 2e^A - 2t^{AB} + c^2 + e^B\right) \qquad q^{2B*} = \left(\frac{1}{3b}\right)\left(a - 2c^2 - 2e^B + c^1 + t^{AB} + e^A\right).$$

These are the optimal quantities given the choice of environmental policy by each country, and assuming constant tariff rates t^{kj}. Of course, any changes in environmental or tariff policy in either country will affect the optimal quantities chosen by each firm. For example, firm 1's output in its domestic market, q^{1A*}, increases as t^{BA} and e^B increase, since both these terms makes firm 2 in country B less competitive. Similarly, an increase in e^A would make firm 1 less competitive, thus reducing its optimal quantity supplied to both the domestic and foreign markets. The partial derivatives of the above optimal quantities with respect to e^A and e^B are:

(5) $$\frac{\partial q^{1j*}}{\partial e^A} = -\frac{2}{3b} < 0 \qquad \frac{\partial q^{2j*}}{\partial e^A} = \frac{1}{3b} > 0$$

$$\frac{\partial q^{1j*}}{\partial e^B} = \frac{1}{3b} > 0 \qquad \frac{\partial q^{2j*}}{\partial e^B} = -\frac{2}{3b} < 0 \quad j = A, B.$$

[8] Thus the two outputs are strategic substitutes, so that firm i's total and marginal revenues decrease as the other firm increases its output, which means that the reaction function is downward sloping, as in Bulow *et al.* (1985).

The government then chooses an optimal policy in the second stage of the game. The government's policy choice depends on how it weighs market outcomes relative to environmental costs. The level of rent-seeking activity by the domestic firm influences the relative importance that the government assigns to the environment. In its policy choice, the government is assumed to possess full knowledge of the industrial market structure and the two firms' behavior in response to its choice of policy. Also, the choice is assumed to be credible, but myopic in the sense that when setting its own environmental policy the government does not recognize that it might influence the other country's environmental policy. Finally, the governments cannot observe the amount of corruption or how it affects the choice of domestic policy in the rival country. These assumptions allow us to solve for the two firms' optimal outputs in both markets for any given policy in those markets, and then to determine each government's optimal domestic policy. For clarity, consider country B whose government collects revenue according to the function

$$(6) \qquad GR^B\left(q^{2A}, q^{2B}, q^{1B}\right) = e^B \cdot \left(q^{2A} + q^{2B}\right) + t^{AB} \cdot q^{1B},$$

where e^B is the emissions tax imposed by country B. Further, we assume that country B sustains environmental damage given by the function

$$(7) \qquad D^B\left(q^{2A}, q^{2B}; \gamma^B\right) = \gamma^B \cdot \left(q^{2A} + q^{2B}\right)^2,$$

where γ^B is a constant. The parameter γ^j is important as it encompasses country *j*'s emission absorption capacity and/or social preferences for a clean environment.

In addition, the government may collect a bribe (*F*) offered from its rent-seeking domestic firm that influences the "weight" that the government assigns to the environmental damage function. As discussed in Section 3 below, since countries are likely to differ in their degree of corruptibility the amount of contribution offered by firms in the two countries will differ. We assume that $\alpha(F) = 1/2 + \beta \cdot F$, where β denotes the degree of corruptibility of the government. Consequently, for a non-corrupt nation $\beta=0$ and for a corrupt nation $\beta>0$. Assuming that the firm in country B chooses to offer a bribe to its government (β is sufficiently greater than zero), the government of country B solves the following optimization problem:

$$(8) \quad \underset{e^B}{Max}\, NB^B = \alpha(F)\left[\pi^2 + CS^B\left(q^{1A} + q^{2A}\right) + e^B \cdot \left(q^{2A} + q^{2B}\right) + t^{AB} \cdot q^{1B} + F\right] - \left(1 - \alpha(F)\right) \cdot \gamma^B \cdot \left(q^{2A} + q^{2B}\right)^2$$

where π^2 is firm 2's optimal profits, $CS^B(\Box)$ is the aggregate consumer surplus in country B which depends on the total amount consumed (as domestic and foreign goods are perfect substitutes). The third and fourth terms represent government revenues, generated from emissions taxes and tariffs, respectively. The third term may be positive or negative depending on whether e^B is a positive or negative emissions tax. The last term is the environmental damage function, which as noted above depends crucially on the parameter γ^B. The government's decision also includes the bribe (F) that the rent-seeking firm may offer in order to induce the domestic government to assign less weight ($1-\alpha$) to the environmental damage function in its net benefits function.[9]

After taking the first-order condition with respect to e^B, we use the partial derivatives in (5), firm 2's first-order conditions, our chosen inverse demand function and the assumption that the effect on consumers from a change in quantity consumed is the same whether the good consumed is produced domestically or abroad. This then yields country B's optimal emissions tax:

$$(9) \qquad e^{B*} = -\left(\frac{b}{4}\right)\cdot\left(q^{2A}+q^{2B}\right)-\left(\frac{1}{4}\right)\cdot CS_q^B+\left(\frac{1}{4}\right)\cdot t^{AB}+\left(\frac{(1-\alpha(F))}{\alpha(F)}\right)\cdot\left[2\gamma^B\cdot\left(q^{2A}+q^{2B}\right)\right].$$

Equation (9) indicates that the optimal environmental policy for country B is determined by the total quantities sold by the domestic firm in both markets (the profit-shifting effect), the change of consumer surplus of citizens in country B ($CS_q^B = \partial CS^B/\partial q$), the tariff on imports, and the perceived environmental damage sustained domestically. An important component of the latter term is the weight given by the government to the environment, which is influenced by the firm's rent-seeking behavior. Equation (9) thus shows that whether the optimal environmental policy will be a tax or a subsidy depends critically on the magnitude of γ^A and α. These parameters are likely to differ between the two countries.

Continuing our backward induction solution, in the first stage the rent-seeking firm chooses an optimal bribe, $F > 0$, given the optimal output levels in stage three and given its knowledge of how a bribe affects the government's choice of "weights" and therefore the environmental policy. The incentive of the firm arises because it knows that by paying the lump-sum bribe to the government, it can raise its profit, provided that the government is willing to behave in a certain way. Naturally, the firm's profit with the bribe must be greater than without the bribe. In our model, this implies that α must be sufficiently large for any given amount of bribe (β sufficiently greater than zero), and that the domestic firm can affect this weight by increasing the bribe. Consequently, the firm attempts to convince its government to adopt a policy that is even less stringent than implied by the profit-shifting incentive inherent in the imperfectly competitive market structure, leading to a shift of welfare toward producers and government (in the form of the bribe) at the expense of the consumers. If the government do adopt a less stringent policy than in the absence of rent-seeking activity, then the country will be worse off overall. As the bribe, F, is a lump sum, it influences

---

[9] We assume that all firm assets are owned by households so that the firm wants the government to care about both producer and consumer surplus in addition its revenue from taxes and bribes. An alternative approach would be to have the government make the distinction instead between producer surplus and its revenue and consumer surplus and environmental damage; that is, move consumer surplus to within the ($1-\alpha$) term.

only the firm's optimal quantities indirectly through the environmental policy. In effect, we are assuming that the bribe changes the government's "weights," but not the environmental policy directly, i.e. $F_{e^B} = \partial F/\partial e^B = 0$. Knowledge of the environmental policy relationship (9) then allows us to determine the optimal bribe for the firm. That is, firm 2 maximizes its profit according to:

$$(10) \qquad \underset{F}{Max}\, \pi^2 = p^A \cdot q^{2A} - \left(c^2 + t^{BA}\right)\cdot q^{2A} - e^B(F)\cdot q^{2A} + p^B \cdot q^{2B} - c^2 \cdot q^{2B} - e^B(F)\cdot q^{2B} - F.$$

The first order condition $\partial \pi^2/\partial F = 0$ yields:

$$(11) \qquad F^* = \frac{1}{\beta}\left[-1/2 + \sqrt{\beta 2\gamma^B} \cdot \left(q^{2A^*} + q^{2B^*}\right)\right].$$

At low levels of output no bribe will be offered, since the profit shifting effect would not outweigh the cost to the firm, but at higher output levels the firm will bribe the government if given the opportunity, *ceteris paribus*. The amount of the bribe also depends on how corrupt the government is and the level of environmental concern in the country. For example, if country B does not care about the environment or has a geography or climate that allows it not to care, i.e. $\gamma^B \approx 0$, then the weight becomes irrelevant and hence no bribe would be forthcoming. As γ^B increases, the firm would have to raise the amount of bribe in order to influence environmental policy.

2. Policy Harmonization in the Absence of Rent Seeking Behavior

Initially we consider the case of non-corrupt nations ($\beta=0$), in which case the government gives equal weight to environmental damage and market outcomes so that $\alpha=0.5$ and the term in front of marginal environmental damage in equation (9) drops out. Furthermore, here and throughout the paper, we assume without loss of generality that $\gamma^A > \gamma^B$. That is, country A is viewed as being more sensitive to environmental problems than country B, either because of the geography or climate of country A or because of the strong preferences of its population for a clean environment. Thus, country A views environmental damages as large and chooses a relatively more stringent emissions tax. It follows, again from equation (9), that countries with differing preferences or absorptive capacity for tolerating environmental damage should choose different optimal environmental policies. This in turn implies:

Proposition 1: *In the case without rent-seeking ($\beta=0$), if country A, the more environmentally concerned nation, imposes its environmental policy, e^A, on country B, then country B's net benefits will be reduced while country A's net benefits will be enhanced for a sufficiently small γ^A.*

The first part of the proposition is self-evident in that, if e^{B*} is the optimal environmental policy for country B, no other environmental policy choice can raise its net benefits. Country B will therefore be unambiguously worse off. As for the second part of the proposition, in the Appendix to this paper we derive the following condition for the effect of a change in e^B on the net benefits of country A:

$$\text{(12)}\qquad \frac{dNB^A}{de^B} = \left(\frac{1}{9b}\right)\cdot\left[\left(1-\frac{4\gamma^A}{b}\right)\cdot\left(2a-2e^A-t^{BA}\right)+6\cdot\left(e^A-t^{BA}\right)\right].$$

From the above equation (12), it is clear that country A will benefit if γ^A is sufficiently small ($\gamma^A < (b/4)$). That is, Country A will be better off as long as the profit-shifting effect is not outweighed by an increase in domestic environmental damage (see Appendix). In contrast, only if country A cares greatly about the state of its environment or is very susceptible to environmental damage, may its overall net benefits decrease as a result of any shift in production resulting from the imposition of country A's environmental policy on country B.[10]

We do not explicitly model the reasons for why country B, which is unambiguously worse off by environmental policy harmonization, may choose to accept this cross-country coordination, but the most obvious tradeoff is greater market access into country A's market.[11] Proposition 1 may help explain the recent combining of both negotiations over environmental policy harmonization and lowering trade barriers between regional partners already engaged in trade that have established reciprocal markets. In effect, if the trading partner with more stringent environmental policy (country A) wish to impose its higher emissions tax on a country (B), it may be able to achieve this by lowering its import tariffs, which can then compensate country B for adopting a more stringent environmental policy. In essence, we are establishing the case for negotiating a preferential trade agreement, such as a comprehensive regional trade agreement or an economic union, where the two countries negotiate over more than respective tariff rates. When countries engage in preferential trade agreements they tend to negotiate simultaneously about tariffs and other policies such as environmental policies. Here we simplify this negotiation by assuming that environmental policy is set first by a country with great bargaining strength, and then tariffs are negotiated in order to make sure that the trade agreement is signed. Alternatively, as discussed below, an existing trade agreement may be extended to include additional countries that are forced to accept the agreement's environmental policy if they choose to join.

---

[10] This proof assumes that after country B accepts country A's more stringent policy, country A does not change its environmental policy. That is, we do not find the "optimal" harmonization policy for country A. If country A is allowed to change its optimal environmental policy, then it would unambiguously be better off since the main effect would then be that the foreign competitor has been made less competitive.

[11] A sufficient reduction in country A's import tariffs (t^{BA}) reverses the negative output effect associated with adopting a too stringent emissions tax in country B. That is, country A could perhaps *induce* country B to accept the stringent policy by lowering its tariff barriers.

As a real world example of what we have in mind, we refer throughout the paper to the expansion of NAFTA between the United States, Canada and Mexico to a hemisphere-wide trade agreement. In this case, the NAFTA agreements will be extended to other Latin American countries, which will have the option either to comply with all the side agreements (in particular, the NAAEC) upon joining NAFTA or to decide not to join.[12] On the other hand, as we have just shown, the existing members of NAFTA cannot expect other Latin America countries to accept the NAAEC unless these countries are also offered full membership in NAFTA and the subsequent lowering of tariffs on their exports to the US, Canada and Mexico. This NAFTA example raises an additional important issue in considering the expansion of a regional trade agreement with harmonization conditions: participants in such a regional agreement that are at different stages of economic development are likely to not only differ in absorption capacity, social preferences, but also in the degree to which they control corruption in their economies. For an example of the latter, many Latin American countries appear to be more corrupt than the existing NAFTA members, especially Canada and the United States (see Table 1 in Section 3).

3. Policy Harmonization and Rent-Seeking Behavior

The previous section has established that, under certain conditions, two trading partners may forego setting environmental policy unilaterally and may instead agree to harmonize their domestic environmental policies at a stricter level as part of a wider trade-environment deal. However, there remains the possibility that a firm in one, or both, of the countries decides to lobby in order to influence the environmental policy choice of its government. We consider this effect to be important for several reasons.

First, corruption appears to be a fact of life in many countries, particularly in the poorer regions of the world. That is, just as countries appear to differ greatly in their willingness or ability to accept more pollution, they also differ in the degree to which they control corruption in their economies. For example, a recent project on governance conducted by the World Bank has assembled a measure of the control of corruption and other governance indicators for 178 developing and advanced economies (Kaufmann *et al.* (1999a,b)).[13] In Table 1, we compile the control of corruption indices for the three NAFTA countries and compare them to other Latin American countries. As the table shows, the latter countries tend to be more corrupt than the existing NAFTA members.

---

[12] We assume that the Free Trade Area of the Americas (FTAA) will contain environmental provisions similar to the NAAEC. Thus "fast-track" proposals that do not include such environmental provisions, although proposed, will in our view not be adopted.

[13] In compiling this index, Kaufmann *et al.* (1999b, p.8) use the conventional definition of corruption: "the exercise of public power for private gain." However, the authors note that "the particular aspect of corruption measured by the various sources differs somewhat, ranging from the frequency of 'additional payments to get things done' to the effects of corruption on the business environment. The presence of corruption is often a manifestation of a lack of respect of both the corrupter (typically a private citizen) and the corrupted (typically a public official) for the rules which govern their interactions ...".

Table 1. Control of Corruption Indices for NAFTA and Latin American Countries

Country	Corruption Control[a]	Country	Corruption Control
United States	1.135	Guatemala	-0.819
Canada	2.055	Guayana	-0.019
Mexico	-0.277	Honduras	-0.938
		Jamaica	-0.116
Argentina	-0.275	Nicaragua	-0.836
Bolivia	-0.438	Panama	-0.458
Brazil	0.058	Paraguay	-0.958
Chile	1.029	Peru	-0.200
Colombia	-0.490	Puerto Rico	1.118
Costa Rica	0.577	Suriname	-0.019
Dominican Republic	-0.773	Trinidad &Tobago	0.511
Ecuador	-0.819	Uruguay	0.430
El Salvador	-0.354	Venezuela	-0.725
Region			
NAFTA	0.971		
SOUTH AMERICA	-0.202		
CENTRAL AMERICA	-0.471		
CARRIBEAN	0.185		

[a] Measured in units ranging from about -2.5 to 2.5, with higher values corresponding to better control of corruption. Source: Kaufmann *et al.* (1999a and b).

Second, recent empirical evidence suggests that rent-seeking behavior and corruption may be a powerful influence on governmental policies, including environmental policies. As shown by Damania *et al.* (2000) and López and Mitra (2000), rent-seeking behavior may lead to non-optimal choices of environmental policies, as environmental regulation tends to be less strict in more corrupt societies.[14] Barrett and Graddy (2000) also show empirically the negative influence of corruption on environmental policies. As noted by Vogel (2000), environmental policies in developing countries may be particularly susceptible to the influence of rent-seeking lobbying by domestic firms and government corruption. Vogel points out that many developing countries appear to be "stuck at the bottom," which may be due in large part to the incentives for domestic firms to influence environmental policy outcomes. For many locally owned firms in poorer economies, the relative costs of environmental regulation are substantial, even if such measures might be socially optimal. Thus one reason why poorer nations do not select their "optimal" regulatory standards may be rent-seeking behavior by those firms that would in fact be made worse off by such standards. The World Economic Forum (ESI 2001) shows that a measure of countries

[14] More generally, corruption, or rent seeking behavior, has also been shown to lower the rate of overall economic growth of countries (e.g., see Scully (1988), Schleifer and Vishny (1993), and Mauro (1995)). These papers show that efficient government institutions tend to foster economic growth by raising private investment (Mauro, 1995).

overall progress toward environmental sustainability (measured by an Environmental Sustainability Index) is negatively correlated with a variable that measures corruption. This suggests that, when comparing the stringency of environmental regulations, one needs to consider that a government prone to corruption is unlikely to have strict environmental policies, or have strict policies with lax enforcement. The second objective of this paper is therefore to explore the extent to which the degree of rent-seeking behavior and government corruption affect the welfare of a country adopting stricter harmonization policies as a condition of joining an international trading agreement.

The presence of rent-seeking activity does affect a country's unilateral choice of environmental policy as shown in equation (9). Consequently, we introduce into our model the possibility that one of the countries (country B) is corrupt, in the sense that its firm can engage in rent-seeking behavior with the aim of getting the government to put less weight on the environmental damage generated by production and more weight on the market outcomes; that is, the firm convinces the government to increase α, where α represents the weight given to the market outcome and $(1-\alpha)$ the weight assigned to the environmental damage.[15] We assume only one corrupt nation, without loss of generality, since the results hinges on the two countries choosing different environmental policies, not on the relative degree of corruption. The optimal choice of environmental policy for country B is shown in equation 9, which shows that as α approaches unity, less weight is given to the marginal environmental damage. In this case, *the optimal government policy is to reduce any environmental tax or increase any subsidy*. Note that when $\alpha=1$ the results are similar to the Brander and Spencer (1985) model in a reciprocal market and with an explicit tariff rate, where the optimal policy is to use a production subsidy.[16]

Further insights into the influence of increased lobbying and corruption on both government benefits and the overall social welfare (i.e., net benefits) of country B is aided through a partial numerical simulation. The simulation was performed using values of the model parameters that yield an emissions tax of 3.1 percent for both countries (tariff rates are 8 percent and γ^B equals 0.155). These arbitrarily chosen numbers show the general result that social welfare and government benefits are in conflict (see Figure 1). Other parameter values yield very similar results.

---

[15] See Damania *et al.* (2000) and López and Mitra (2000) for further discussion of this approach to modeling the government's objective function to account for the influence of rent-seeking behavior. Following López and Mitra (2000), we could assume in our equation (9) that the there is an additional term attached to the "weight" on the environmental damage function in favor of the market outcome, where this additional term ρ is the probability of the government being re-elected, which may be linear and increasing in environmental damages, D^B. This would make our model compatible with voting models that take into account that the motive of democratically elected government is to win elections. In equation (9), to sharpen the analysis, we drop ρ but note that our model could easily be extended to include the re-election motive.

[16] The result is a production subsidy since in the reciprocal market model the firm receives the subsidy whether selling at home or abroad.

Figure 1 shows that as the degree of corruption increases, represented by a greater α (and a higher level of bribe, F, since $\partial\alpha/\partial F = \beta > 0$ for a corrupt nation), the benefits to the corrupt government increase at the expense of society at large. It is clear that the government has an incentive to be corrupt since its benefits with corruption is always greater than for all outcomes without corruption. Although not indicated explicitly in the simulation, clearly social welfare is declining as bribes and corruption increases because the government is adopting less stringent policies to control environmental damages. In essence, we have replicated in our model what is now a well-known result in the empirical literature: environmental policies tend to be less strict in more corrupt societies.

Figure 1. Social Welfare and Government Benefits under Rent-Seeking Behavior

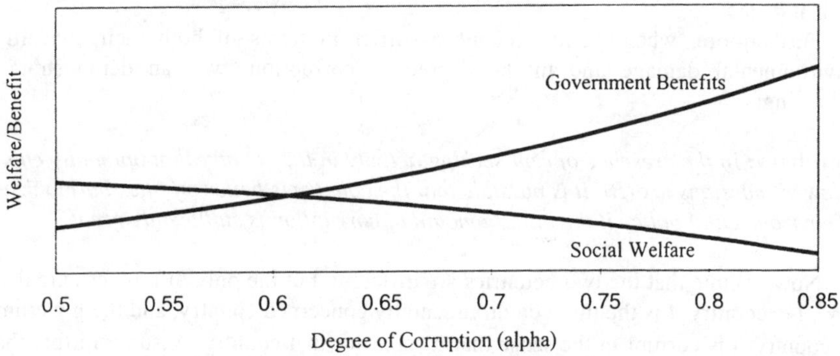

Since the net benefits to society at large decrease as a firm engages in rent-seeking behavior through bribing the government, it follows that proposition 1 no longer holds strictly for country B. That is, if country A, the more environmentally concerned nation, now imposes its more stringent environmental policy on country B, then country B's welfare may actually improve. This suggests the following:

Proposition 2: *In the presence of rent-seeking activity (β>0), it is possible that the more corrupt country (B) is better off after harmonization of environmental policy is imposed by country A.*

Consider two countries that are identical in the sense that without corruption they would choose the same (optimal) environmental policy. This assumption is made for tractability and does not affect the results.[17] Letting $\partial\alpha/\partial F = 0$ so that no corruption is present, we have that the two countries will optimally chose exactly the same emissions tax: e^A=e^B > 0. Next suppose that country *B* becomes corrupt in the sense that firm 2 pays a bribe that induces its government to care less for the environment, so that $\alpha > 0.5$. In this case, country B's government chooses a sub-optimal environmental policy $e^{B\wedge}$ that is less than e^B (=e^A). If country *A* then imposes its environmental policy on country *B*, so that country *B*'s environmental policy is no longer a choice variable to its government, then the domestic firm in country B no longer has an incentive to bribe the government. Without a bribe, the government chooses to set $\alpha = 0.5$, which means that the imposed environmental policy is again the optimal one. That is, removing the incentive for corruption by imposing environmental harmonization returns country *B* to its initial policy, i.e. e^A=e^B > 0. Country *B* is thus better off after harmonization. This result can also be seen in Figure 1 since social welfare under corruption is always lower than at $\alpha = 0.5$.[18]

Furthermore, when the two countries differ in terms of both their concern for environmental damage and in the degree of corruption, we can demonstrate the following:

Corollary: *In the presence of rent-seeking activity in the country that optimally chooses a lower emissions tax (B), it is possible that the country is better off after harmonization of environmental policy if the initial amount of corruption is sufficiently great.*

Now assume that the two countries are different, but the only differences are that $\gamma^B < \gamma^A$, i.e. country *A* is the more environmentally concerned country, and the government of country *B* is corrupt in the sense that $\alpha > 0.5$ for that country. Assume further that γ^j is sufficiently large in both countries so that their respective governments impose a positive emissions tax, i.e. $e^A, e^B > 0$. The relatively low γ^B and high α for country B imply that its government will choose a lower environmental tax than country *A* (equation 9). It follows that, if country B foregoes its optimal environmental policy and adopts the more stringent policy of country A, then the unambiguous reduction in net benefits for a non-corrupt government (proposition 1) will be mitigated by the elimination of an incentive for bribery. There is thus a trade-off present here: on the one hand, social welfare is reduced in country B by adopting a too stringent environmental policy; on the other, social welfare increases because equal weight is now in effect being given to the market outcome and environmental damages. If the benefit from setting e^B=e^A (as if $\alpha = 0.5$) is greater than the efficiency loss from setting an environmental policy that is too stringent, then the harmonization policy will actually

[17] The proof hinges on the fact that the derivative of the net benefit function with respect to the emissions tax is positive when the chosen environmental policy is less than optimal so that as the pollution tax is raised, a nation's net benefit increases with it.

[18] This discussion of course assumes that the corrupt government would in fact allow a foreign country to impose a more stringent environmental policy on its country. The reasons for this, which may be either political or economic, are not modeled in the present paper.

enhance net benefits. This will be true as long as country *B* is sufficiently corrupt, i.e. $\alpha >> 0.5$, and the only other difference between the two countries is the level of environmental concern, i.e. $\gamma^B < \gamma^A$.

However, if country B adopts country A's environmental policy, country *B* is still not operating at its first-best environmental policy choice, i.e. the level of environmental taxation it would choose if no corruption were present. Nevertheless, in the presence of rent-seeking activity it might be better for country *B*'s social welfare level if its government were to adopt the strict policy imposed by *A* if this externally imposed environmental policy removed the incentive for the government and firm to engage in rent-seeking activity, even in the absence of any import tariff reductions. However, a corrupt government would *ceteris paribus* be less likely to accept an externally imposed policy since the accompanying removal of the incentives for rent-seeking behavior would both enhance social welfare and reduce government benefits.

4. Conclusion

Since countries differ in pollution absorption capability and tolerance for environmental damage, they will optimally choose different environmental policies. Attempts at environmental policy harmonization between these countries are therefore non-optimal. Despite this well-known fact among economists, trade agreements are increasingly associated with policy harmonization efforts. We therefore explore the likely outcome on a trading partner when a high standard member imposes its policy choice on another member, e.g. as a condition for greater trade cooperation. We are able to show, within our framework with no corruption, that if a country with stringent policies manages to impose its policy on another nation, then the latter economy that values the environment less will be hurt, while the former nation may gain. This provides an explanation for why harmonization efforts are so prevalent. However, the gains from trade in this case will be lower than if both nations adopt their respective optimal environmental policies and trade according to comparative advantage or if a supranational agency was able to impose a cooperative solution on the two trading nations. Nevertheless, the analysis in the first part of the paper helps explain why two or more regional partners often combine both negotiations over environmental policy harmonization and lowering trade barriers.

In the case of an imperfect political process, there is ample empirical evidence that rent-seeking behavior can distort the environmental policy choices of nations. We model this additional dimension in terms of the government of one country being bribed to assign less weight on the marginal environmental damage, which leads to a lower pollution tax. Under this scenario, cross-country harmonization of environmental policy could be strictly welfare enhancing for both countries, if corruption is rampant in one country and the externally imposed policy removes the incentive for rent-seeking activity in that country.

Although this is a counter-intuitive outcome, and not one usually put forward by proponents of environmental policy harmonization, there is an important caveat attached to such a policy. Under current trade rules of the World Trade Organization (WTO) and General Agreement on Tariffs and Trade (GATT), it is infeasible for one trading partner to impose unilaterally its environmental policy on another unless the latter country accepted such an externally imposed environmental policy.[19] In essence, acceptance of an externally imposed environmental policy signals a change in a political regime away from rent-seeking behavior and corruption in a government with regard to the formulation of environmental policy.

We have also established an important caveat to the conventional economic argument against environmental policy harmonization: if the lower environmental policy in one trading partner is due to failures in governance, then harmonization could benefit both countries, as well as the environment. Of course, whether the latter country is willing to accept environmental policy harmonization will in itself signal its willingness to overcome inherent governance failures. Unfortunately, many developing countries appear to be persistently "stuck at the bottom" with respect to both environmental policies and corruption (Vogel 2000; ESI 2001). This suggests that the political willingness to overcome governance failures may be difficult to achieve for many developing countries, even if they are offered the incentives of a comprehensive trade-environment agreement.

The results of our paper are subject to several additional caveats. We assumed away transboundary pollution, management of a global commons, and the fact that countries with similar incentives may face a prisoner's dilemma situation where environmental policies are lower than a globally optimal policy. However, for the latter case there is no reason to believe that the world is better off by adopting the policies of the country whose geography and social preferences indicate a relatively stricter policy choice. In addition, the entire discussion hinges on global markets being imperfectly competitive. If markets are competitive, then there is no incentive for countries to distort their environmental regulations, and consequently harmonization efforts must be welfare reducing.

Appendix: Proof of Proposition 1

As noted in the text, the first part of proposition 1 is self-evident in that, if e^{B*} is the optimal environmental policy for country B, any other environmental policy must lower its net benefits. From equation (5), the partial derivatives $q^{1A*}{}_{eB} = q^{1B*}{}_{eB} = 1/3b$, so that firm 1 in country A will raise its production after country B's environmental policy is raised. This means that profits shift to country A from country B. Country *A*'s net benefits are given by

[19] This might be another reason for why modern trade agreements tend to include "harmonization" efforts. If a nation wishes to impose its environmental policy on another nation, it may be able to use a trade agreement to sidestep the WTO rules. This is indeed allowed under the General Agreement on Tariffs and Trade's Article XXIV.

$$NB^A = \pi^1 + CS^A(q^{1A} + q^{2A}) + e^A \cdot (q^{1A} + q^{1B}) + t^{BA} \cdot q^{2A} - \gamma^A \cdot (q^{1A} + q^{1B})^2,$$

which are maximized at the optimal choice e^{A*}. Intuitively, if firm 1's increase in profit outweighs the accompanying increase in environmental damage in country A, then country A will gain from country B's adoption of its stringent environmental policy. To explore this result formally, we consider how country A's net benefit function changes due to a change in environmental policy in Country B:

$$\frac{dNB^A}{de^B} = \pi^1_{q^{2A}} q^{2A}_{e^B} + \pi^1_{q^{2B}} q^{2B}_{e^B} + 2 \cdot \left(\frac{b}{2}\right) \cdot (q^{1A} + q^{1B})(q^{1A}_{e^B} + q^{2A}_{e^B}) + e^A (q^{1A}_{e^B} + q^{1B}_{e^B}) + t^{BA} \cdot q^{2A}_{e^B}$$
$$- 2\gamma^A (q^{1A} + q^{1B})(q^{1A}_{e^B} + q^{1B}_{e^B})$$

Simplifying using the partial derivatives from equation (5) and our chosen inverse demand function yields

$$\frac{dNB^A}{de^B} = \left(\frac{2}{3}\right) \cdot \left(q^{1A} + q^{1B}\right) - \left(\frac{1}{3}\right) \cdot \left(q^{1A} + q^{2A}\right) - \left(\frac{4\gamma^A}{3b}\right) \cdot \left(q^{1A} + q^{1B}\right) + \left(\frac{2}{3b}\right) \cdot e^A - \left(\frac{2}{3b}\right) \cdot t^{BA}$$

Substituting in for the optimal quantities from equation (4) and assuming that marginal costs are the same and equal to zero, $c^1 = c^2 = 0$ and also letting $e^A = e^B$ and $t^{BA} = t^{AB}$:

$$\frac{dNB^A}{de^B} = \left(\frac{1}{9b}\right) \cdot \left(2a - 2e^A - t^{BA}\right) - \left(\frac{4\gamma^A}{9b^2}\right)\left(2a - 2e^A - t^{BA}\right) + \left(\frac{2}{3b}\right) \cdot e^A - \left(\frac{2}{3b}\right) \cdot t^{BA}$$
$$= \left(\frac{1}{9b}\right) \cdot \left[\left(1 - \frac{4\gamma^A}{b}\right) \cdot \left(2a - 2e^A - t^{BA}\right) + 6 \cdot \left(e^A - t^{BA}\right)\right].$$

From the latter equation, and assuming that country *A* imposes an environmental tax and e^A-t^{BA}>0, it is clear that country A will benefit if γ^A is sufficiently small ($\gamma^A = (b/4)$). That is, as long as the profit-shifting effect is not outweighed by an increase in domestic environmental damage, country A will benefit from imposing its environmental policy on country B.

References

Anderson, K. and R. Blackhurst ed. (1992) *The Greening of World Trade Issues* Harvester Wheatsheaf, New York.

Arden-Clarke, C. (1991) *The General Agreement on Tariffs and Trade, Environmental Restriction and Sustainable Development.* WWF Discussion Paper, WWF International, Geneva.

Barrett, S. (1994) "Strategic Environmental Policy and International Trade," *Journal of Public Economics*, 54, 325-38.

________ and K. Graddy (2000) "Freedom, Growth, and the Environment," *Environment and Development Economics*, 5, 433-56.

Batabyal, A. A. (1996) "Game Models of Environmental Policy in Open Economy," *The Annals of Regional Science*, 30(2), pp. 185-200.

________ (1998) "Games Governments Play: An Analysis of National Environmental Policy in an Open Economy," *The Annals of Regional Science*, 32(2), pp. 237-51.

Bhagwati, J. (1996) "The Demands to Reduce Domestic Diversity Among Trading Nations," in J. Bhagwati and R. E. Hudec *Fair Trade and Harmonization: Prerequisites for Free Trade?* Cambridge, Mass.: MIT Press, pp. 9-40.

Brander, J. A. and B. J. Spencer (1985) "Export Subsidies and International Market Share Rivalry," *Journal of International Economics*, 18, 83-100.

Bulow, J., Geanakopolus, J. and P. Klemerer (1985) "Multimarket Oligopoly: Strategic Substitutes and Complements," *Journal of Political Economy*, 93, pp. 488-511.

Copeland, B.R. and M.S. Taylor (2003) *Trade and the Environment: Theory and Evidence.* Princeton University Press, Princeton, NJ.

Damania, R., Fredriksson, P.G. (2003) "Trade Policy Reform, Endogenous Lobby Group Formation, and Environmental Policy," *Journal of Economic Behavior and Organization*, 52, pp. 47-69

________ and J. A. List (2000) "Trade Liberalization, Corruption and environmental Policy Formation: Theory and Evidence," *CIES Discussion Paper* 0047.

Environmental Sustainability Index (ESI) (2001) Yale Center for Environmental Law and Policy, available on-line at http://www.ciesin.Columbia.edu/indicators/ESI.

Esty, D. C. (1999) "Economic Integration and the Environment," in N. J. Vig and R. Axelrod (eds.) *The Global Environment: Institutions, Law, and Policy*, Washington, D.C., CQ Press, pp. 190-209.

Johal, S. and A. Ulph (2002) "Globalization, Lobbying, and International Environmental Governance," *Review of International Economics*, 10(3), pp. 387-403.

Kanbur, R., Keen, M, and S. van Wijnbergen (1995) "Industrial Competitiveness, Environmental Regulation and Direct Foreign Investment," in I. Goldin and L. A. Winters (eds.) *The Economics of Sustainable Development*, Cambridge University Press, pp. 289-302.

Kaufmann, D., Kraay, A. and Zoido-Lobaton, P. (1999a) "Aggregating Governance Indicators". World Bank Policy Research Department Working Paper No. 2195. The World Bank, Washington DC.

_______ (1999b) "Governance Matters". World Bank Policy Research Department Working Paper No. 2196. The World Bank, Washington DC.

Leebron, D. W. (1996) "Lying Down with Procrustes: An Analysis of Harmonization Claims," in J. Bhagwati and R. E. Hudec *Fair Trade and Harmonization: Prerequisites for Free Trade?* Cambridge, Mass.: MIT Press, pp. 41-117.

López, R. and S. Mitra (2000) "Corruption, Pollution and the Kuznets Environment Curve," *Journal of Environmental Economics and Management* 40(2): 137-50.

Low, P. (2002) "International Trade and the Environment: An Overview," in P. Low (ed.) *International Trade and the Environment, World Bank Discussion Paper* (159), 1-14.

Mauro, P., (1995) "Corruption and Growth," *Quarterly Journal of Economics*, 107, pp. 407-37.

Motta, M. and J-F. Thisse. (1994) "Does Environmental Dumping Lead to Delocation?" *European Economic Review*. 38, pp. 563-76.

North American Agreement on Environmental Cooperation – Canadian Office (NAAEC), "Agreement," http://www.naaec.gc.ca/eng/index_e.htm.

Scully, G.W. (1988) "The Institutional Framework and Economic Development," *Journal of Political Economy*, 92(3), pp. 652-62.

Shleifer, A. and R. W. Vishny (1993) "Corruption," *Quarterly Journal of Economics*, 108(3), pp. 599-617.

Ulph, A. (1997) "International Environmental Regulation when National Governments Act Strategically," in J. Braden and S. Proost (eds), *The Economic Theory of Environmental Policy in a Federal System*, Edward Elgar, pp. 66-98.

________ (1998) "Political Institutions and the Design of Environmental Policy in a Federal System with Asymmetric Information," *European Economic Review*, 42, pp. 583-92.

________ (2000) "Harmonization and Optimal Environmental Policy in a Federal System with Asymmetric Information," *Journal of Environmental Economics and Management*, 39, pp. 224-41.

Vogel, D. (2000) "Environmental Regulation and Economic Integration," *Journal of International Economic Law*, 3(2), pp. 265-79.

Name index